THE SCIENCE OF ANIMAL AGRICULTURE

Second Edition

THE SCIENCE OF ANIMAL AGRICULTURE

Second Edition

Ray V. Herren

Delmar Publishers

an International Thomson Publishing Company I(T)P®

Albany • Bonn • Boston • Cincinnati • Detroit • London • Madrid
Melbourne • Mexico City • New York • Pacific Grove • Paris • San Francisco
Singapore • Tokyo • Toronto • Washington

NOTICE TO THE READER

Cover Image: Courtesy of PhotoDisc
Cover Design: Carolyn Miller

Delmar Staff:
Publisher: Susan Simpfenderfer
Acquisitions Editor: Jeff Burnham
Developmental Editor: Andrea Edwards Myers
Production Manager: Wendy Troeger
Production Editor: Carolyn Miller
Marketing Manager: Kathryn Hans

Printed in the United States of America

For more information contact:

Delmar Publishers
3 Columbia Circle, Box 15015
Albany, New York 12212-5015

International Thomson Publishing Europe
Berkshire House
168-173 High Holborn
London, WC1V7AA
United Kingdom

Nelson ITP, Australia
102 Dodds Street
South Melbourne,
Victoria, 3205 Australia

Nelson Canada
1120 Birchmont Road
Scarborough, Ontario
M1K 5G4, Canada

International Thomas Publishing France
Tour Maine-Montparnasse
33 Avenue du Maine
75755 Paris Cedex 15, France

International Thomson Editores
Seneca 53
Colonia Polanco
11560 Mexico D. F. Mexico

International Thomson Publishing GmbH
Königswinterer Straße 418
53227 Bonn
Germany

International Thomson Publishing Asia
60 Albert Road
#15-01 Albert Complex
Singapore 189969

International Thomson Publishing Japan
Hirakawa-cho Kyowa Building, 3F
2-2-1 Hirakawa-cho, Chiyoda-ku,
Tokyo 102, Japan

ITE Spain/Paraninfo
Calle Magallanes, 25
28015-Madrid, Espana

1 2 3 4 5 6 7 8 9 10 XXX 03 02 01 00 99 98

Library of Congress Cataloging-in-Publication Data

Herren, Ray V.
The science of animal agriculture / Ray V. Herren.—2nd ed.
p. cm.
Includes bibliographical references and index.
ISBN 0-8273-8612-5
1. Livestock. 2. Animal culture. 3. Domestic animals.
I. Title
SF51.H47 1998 98-27189
636—dc21 CIP

This book is dedicated to my father, the late Banks Herren who passed away before it was completed. He taught me in the best way a father can teach his son—by example.

Table of Contents

Preface

The second edition of *The Science of Animal Agriculture* is directed toward teaching the basic science concepts involved in the production of agricultural animals. It contains the latest information regarding the scientific aspect of the agricultural industry.

All facets of modern agriculture are based on science. From the most rudimentary of cultural practices to the most complicated biotechnology techniques, scientific research has produced the phenomenon known as American agriculture. The science of agriculture has brought humans from the stage of wandering and gathering food to modern civilization. Much of what we know about how living organisms reproduce and grow has come about through our quest to be more efficient in the production of food and fiber.

The Science of Animal Agriculture contains chapters dealing with the latest concepts in animal biotechnology. Topics include animal behavior, classification, consumer concerns, animal welfare, genetics, scientific selection, reproduction, growth and development, nutrition, meat science, parasites, and disease. In addition, the scientific basis for the production of the different types of agricultural animals is presented.

Since *The Science of Animal Agriculture* emphasizes the scientific principles involved in animal production, the text is appropriate for agriculture courses that receive science credit.

Each chapter begins with a listing of the student objectives in both basic science and agricultural science, along with the key terms used in the chapter. Although the two are difficult to separate, the student objectives in basic science are considered to be those concepts a student might be expected to learn before graduation. These concepts are presented in a contextual setting with photos, charts, diagrams, and figures to illustrate the concepts. The chapters are concluded with a variety of questions that may be used for reviewing the chapter. In addition, practical activities are described that will enhance the students' grasp of the concepts presented in the chapter.

The text is accompanied with an instructor's guide and Classmaster CD-ROM which includes a teacher's resource kit containing complete lesson plans for each chapter. Each lesson begins with an interest approach and is followed with the unit (chapter) objectives. Next is a complete outline in lesson plan format of all the material in the chapter. All the tables, charts, and drawings in the text are in the guide in the form of transparency masters that can be used by the teacher as he or she teaches the unit. Answers to all the questions at the end of the chapter are included in the lesson plan. Many lesson plans also contain ideas for long-term projects built around the unit (chapter) material. A computerized test item bank is also available on the CD-ROM for the generation of various level exams for each unit or chapter.

To make the instructional package complete, also included in the Classmaster CD-ROM is the student laboratory manual for the text, which is filled with practical exercises. Several options are available for activities dealing with each chapter. When used as a whole, the text, resource guide, test bank, and laboratory manual should provide an instructor with most of the material needed to teach a high-quality course in the science of animal agriculture.

About the Author

Ray V. Herren, the author of this text, has been actively involved in the animal industry for most of his life. He grew up on a diversified farm where he played a major role in the production of livestock. He obtained a B.S. degree in Agricultural Education from Auburn University, a Master's degree in Agribusiness Education from Alabama A&M University, and a Doctorate in Vocational Education (with an emphasis in Agricultural Education) from Virginia Polytechnic Institute and State University.

Dr. Herren has taught agriculture at the high school level and also has taught animal science courses at the university level. As a faculty member in Agricultural Education at Oregon State University, he served as the coach of the University Livestock Judging team and taught several animal practicum courses. He currently serves on the faculty in the Department of Agricultural Leadership Education and Communication at the University of Georgia.

Acknowledgments

The author gratefully acknowledges the following for their assistance in creating the second edition of this text:

Dr. Steve Bolden of Seaboard farms for his technical advice on poultry; Dr. David Swayne of the United States Department of Agriculture for his technical advice on Veterinary Medicine; Brandi Mullins for her contributions and advice on horses; Michelle Long for her help with the Instructors Guide; and Valerie Mosley for her assistance with photographs.

The author and Delmar Publishers would like to thank those individuals who reviewed the manuscript and offered suggestions, feedback, and assistance.

Phillip B. Allen
Seeger Memorial High School
West Lebanon, Indiana

John Busekit
Catteraugus Central High
Catteraugus, New York

Randall Cale
Gackle High School
Gackle, North Dakota

Jeffrey B. Miller
Sullivan High School
Sullivan, Indiana

LeRoy Uhlenkamp, Jr.
Long Prairie-Grey Eagle High School
Long Prairie, Minnesota

Harvey Wallace
Cheney High School
Cheney, Washington

Chapter 1

Animal Agriculture as Science

Student Objectives in Basic Science

As a result of studying this chapter, you should be able to

1. Define science.
2. Explain how science has made the lives of people better.
3. Tell how humans first began to use science.
4. Explain the concept of the scientific method.
5. Distinguish between basic and applied science.
6. Cite scientific discoveries that have made food better and less expensive for the consumer.
7. List the pharmaceuticals that are derived from animals.

Student Objectives in Agricultural Science

As a result of studying this chapter, you should be able to

1. Explain why agriculture is a science.
2. Explain what is meant by the land grant concept.
3. Explain how the land grant institutions began.
4. Discuss the advances made by American agriculture during this century.
5. Analyze how agricultural research has benefited the consumer.
6. List the developments that have revolutionized animal agriculture.

Key Terms

agriculture
domesticated
hypothesis
experiment
control group
experimental group
lactation
agricultural animals
hormones
omnivorous
poultry
broiler
beef
veal
vaccinating
immunity
serum
vaccines
environment
genes
offspring
artificial insemination
sires
semen
dams
gestation
progeny

Science, as defined by *Webster's Ninth New Collegiate Dictionary*, is the study or theoretical explanation of natural phenomena. Nature has given us a wonderfully complex world in which to live. Our environment is governed by natural laws that control everything from gravity to the weather. These laws also control the way plants and animals live and grow on our planet. Through science, these laws are investigated and new ways are found to use them to make our lives easier and better.

Americans enjoy one of the highest living standards in the world. Our per capita income ranks near the very top among all the nations of the world. Most of the comforts and enjoyment of our lives have come about as a direct result of science.

Agriculture is the oldest and most important of all sciences. Without a sound basis of agricultural knowledge, all humans would be struggling to find enough food, shelter, and clothing to survive. Because we have discovered so much about the world around us, our food, shelter, and clothing are produced so efficiently that a very small percentage of the population is required to produce these necessities. That leaves the rest of us to develop new knowledge in different areas that allows us to have better transportation, communications, and recreation.

From the very earliest time, humans have studied the environment in which they lived. Their first thoughts were to survive, and this meant that first they had to find food. Early people ate the fruits, seeds, and animals they found in their environment. As these supplies of food became exhausted, the people were forced to leave the area and find a new place where plants and animals were more abundant. They undoubtedly observed the behavior patterns of animals and from these observations devised means to locate and kill them. As the patterns of animal movement and responses were studied, people reasoned that if they could domesticate animals, the need for moving with the herds and hunting as a group could be eliminated.

A close observation of the animals determined which would be best to tame. As animals were **domesticated**, people reasoned that there were certain methods they could use to grow the animals more efficiently. Through the use of trial and error, the best ways of caring for the animals were discovered and passed along from parents to children. Only relatively recently have people gone about the study of animals in a systematic way.

Progressive scientific research began in the United States about the middle of the 1800s. At that time, the universities in the United States taught a curriculum known as the Classics where the main subjects studied were Latin, Greek, History, Philosophy, and Mathematics. People began to realize that there was a need for institutions of higher learning where students could study areas that had practical applications. The nation was emerging as an industrial and agricultural based economy. To make progress in these areas, young people needed to be taught how to produce food and manufactured goods in a more efficient manner.

In the late 1850s, a senator from Vermont, Justin Morrill, introduced a bill that would provide public land and funds for establishing universities to teach practical methods of manufacturing and producing food and fiber, Figure 1–1. The bill passed in 1862 and became known as the Land Grant Act or the Morrill Act. During that same year, President Lincoln signed into law a bill that established the United States Department of Agriculture (USDA). Soon, almost all of the states in the country established land grant colleges.

As the students enrolled and classes began, a severe problem was recognized—no

Figure 1-1. Senator Justin Morrill of Vermont introduced a bill to provide land grant colleges.

one really knew anything about agriculture! Most of the knowledge about growing plants and animals had been passed from generation to generation and represented people's beliefs rather than proven knowledge. To solve this problem, Congress passed the Hatch Act in 1872. The Hatch Act authorized the establishment of experiment stations in different parts of the states that had land grant colleges. The purpose was to create new knowledge through a systematic process of scientific investigation. These experiment stations put to use what has come to be known as the scientific method of investigation.

Later, in 1914, Congress passed the Smith-Lever Act, setting up the Cooperative Extension Service, which serves to disseminate information learned from new research to the population. This completed a system known as the *land grant concept*, which held that the purpose of a land grant university was to teach, conduct research, and carry the new information to the people in the state through the extension service.

In 1917, the Smith-Hughes Act was passed, establishing Vocational Agriculture as a program in the public high schools as a means of teaching new methods of agriculture.

The Scientific Method

The scientific method is a systematic process of gaining knowledge through experimentation. The method is used to make sure that the results of an experimental study did not occur just by chance and that something caused the change. This process involves formulating a hypothesis, designing a study, collecting data, and drawing conclusions based on an analysis of the data, Figure 1–2.

A scientist begins by identifying a problem that needs to be solved. He or she may have an idea or suspicion of what causes the problem or what might solve the problem. This suspicion is called a **hypothesis** and serves as the basis for investigating a problem. The hypothesis is then subjected to a test called an **experiment** that attempts to isolate the problem in question and determine the solution.

For instance, if a scientist wanted to know whether milking cows twice a day would obtain more milk than milking only once a day, he or she might milk a group of cows twice a day and record the results. However, just milking twice a day and recording the results would not prove much. The real question is whether or not milking twice a day produces more milk than only milking once a day. If two groups of cows were used—one group that is milked twice a day and one group that is milked only once a day—the results would be more meaningful. The cows milked once a day are called the **control group** and the cows milked twice a day are called the **experimental group**.

Other conditions must also be met to ensure that the data collected in the experiment

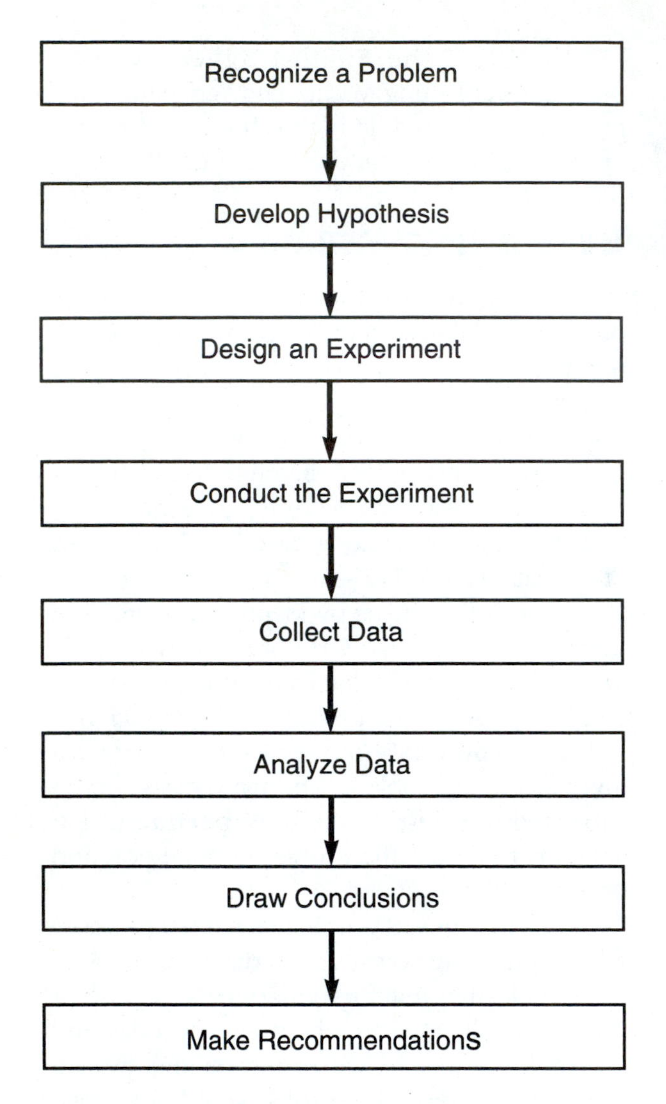

Figure 1-2. The scientific method requires several steps.

are valid. The cows must be of the same breed, the same age, the same stage of their **lactation** cycle, and the same size. Also there should be enough cows in each group that the differences in the cows would average out. When the cows are selected and the scientist feels they are enough alike and of sufficient number, the treatment given each group must be the same.

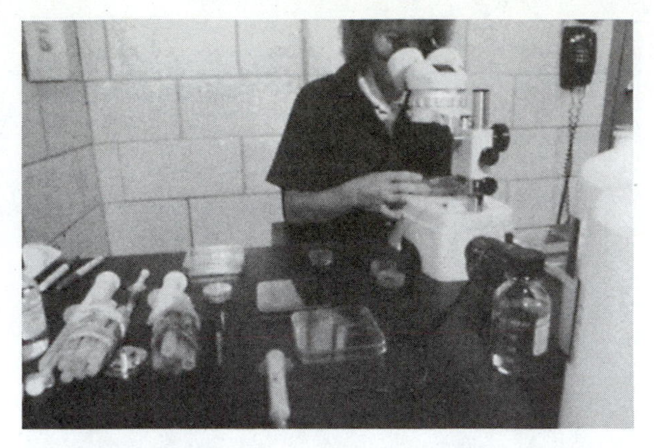

Figure 1-3. Basic research investigates why or how processes occur. *James Strawser, Cooperative Extension Service, The University of Georgia.*

This means that the cows will stay in the same type of environment, eat the same ration, consume the same amount of water, and be treated in the same manner. Then if the cows that are milked twice a day give more milk, the scientist may conclude that milking cows twice a day results in more milk.

The scientific method has been used thousands of times to develop the methods that are used to produce **agricultural animals**. As more knowledge was gained, the experiments became more involved and complicated. Today scientific research is classified into two broad areas: basic research and applied research.

Basic research deals with the investigation of why or how processes occur in the bodies of the animals, Figure 1–3. For instance, basic research is used to discover the specific **hormones** that control growth in animals. Applied research deals with using the discoveries made in basic research to help in a practical manner, Figure 1–4. If basic research discovers the growth-regulating hormones, applied research uses this knowledge to develop ways of using natural or artificial hormones to increase the growth efficiency of agricultural animals.

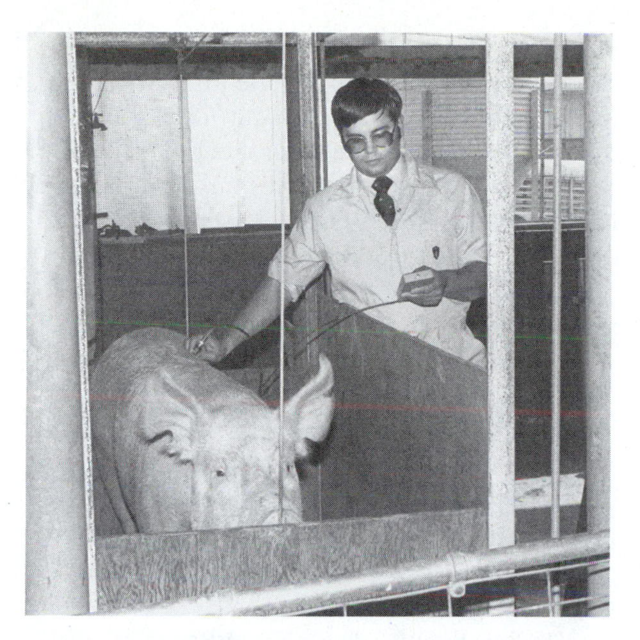

Figure 1-4. Applied research investigates how basic research can be put to use. *Rick Jones, Cooperative Extension Service, The University of Georgia.*

In a very real sense, agriculture is the application of almost all the research and knowledge associated with plants and animals. Aside from the field of medicine, about the only application of basic research in the life sciences is that of agriculture.

Basic research carried through to application in animal science has benefited humans in many ways—the most obvious of which is food. Humans are **omnivorous** animals. This means that we eat both plants and animals. The production of animals provides people with a reliable, abundant source of high-quality food. Advancements made through scientific research have resulted in not only an abundance of animals for food but also relatively low prices for food. Better production methods mean better efficiency in the raising of agricultural animals. This greater efficiency means a constant supply of affordable food.

Figure 1-5. The weight of weaned calves has more than doubled since 1925. *Cooperative Extension Service, The University of Georgia.*

As a result of scientific research and the application of the findings of research, gigantic strides have been made in the efficient production of food from animals. The Council for Agricultural Science and Technology has compiled a list of some of the advancements that have been made since 1925. The following paragraphs outline some of the progress.

Beef cattle liveweight marketed per breeding female increased from 220 pounds to 482 pounds. This means that for every cow raised, we are now selling more than twice the beef we did in 1925, Figure 1–5. These increases have come about as a result of the scientific selection of breeding animals, a better understanding of beef cattle nutrition, and better control of parasites and diseases.

A better understanding of all phases of the lives of the animals has led to higher-quality and less-expensive beef for the consumer. The lower cost no doubt accounts to a large degree for the tremendous increase in the annual per capita consumption of beef. Since 1925, consumption has doubled from 60 pounds of carcass weight

Figure 1-6. Research has brought about a 100 percent increase in the weight of market lambs. *National Livestock and Meat Board.*

equivalent to 120 pounds of carcass weight equivalent. Americans now have a more nutritious diet at less cost.

Sheep liveweight marketed per breeding female increased from 60 to 130 pounds. This represents an increase of 100 percent over the production in 1925. Around the turn of the century, sheep were raised primarily for wool. Then through research in selection, sheep began to be produced that were raised mainly for meat, Figure 1–6. Research efforts have concentrated on raising better meat-type animals.

Milk marketed per dairy cow increased from 4,189 to 10,500 pounds—an increase of 2 1/2 times the amount of milk marketed in 1925. In the period from 1950 to 1975, the number of dairy cows was reduced by about half and the amount of milk produced remained about the same, Figure 1–7. This greater efficiency has resulted in quite a bargain for the consumer. A gallon of milk now costs 36–84 cents less than it would using 1950s technology.

Swine liveweight marketed per breeding female increased from 1,600 to 2,850 pounds. Since 1950 the amount of feed required to pro-

Figure 1-7. The same amount of milk is now produced by half the number of cows. *James Strawser, Cooperative Extension Service, The University of Georgia.*

Figure 1-8. Since 1950 the average time required to produce a 220-pound hog has been reduced from 170 to 157 days. *Rick Jones, Cooperative Extension Service, The University of Georgia.*

Figure 1-9. Modern broilers are more than a pound heavier and are raised in half the time on half the amount of feed than in 1925. *Dr. Frank Flanders, Agricultural Education, The University of Georgia.*

duce a 200-pound market hog has been reduced by about 50 pounds. Also during this time, the average time to produce a 220-pound hog has been reduced from 170 to 157 days, Figure 1–8. Because pigs can be raised in a shorter time on less feed, the cost of production is less. This savings is passed along to the consumer in less-expensive pork.

No other segment of agricultural industry has made more dramatic advancements than that of the **poultry** industry. Since 1925 the time for **broiler** chickens to reach market weight was cut in half from 15 weeks to 7 1/2 weeks. The amount of feed required per pound of liveweight gain for broiler chickens was cut in half from 4 pounds to around 2 pounds. The weight of broilers at marketing increased from 2.8 pounds to 3.38 pounds. This means that we can now produce a heavier broiler in half the time on half the feed than we could in 1925, Figure 1–9. Not only do consumers get a better broiler at much less cost but they also get a bird that is completely dressed and ready to cook.

The annual production per laying hen doubled from 112 to 232 eggs, and the feed required to produce a dozen eggs decreased from 8.0 to 4.2 pounds. Advances made through scientific research have made the production of eggs in an enclosed building possible. This allows the hens to be much better managed since the producer can control the environment in which the eggs are produced.

The weight of marketed turkeys increased from 13.0 to 18.4 pounds. This increase in gain was achieved on less feed (from 5.5 to 3.1 pounds) and in nearly half the time (from 34 to 19 weeks).

The countries in North and Central America, Europe, and Oceania (Australia, New Zealand) have only 29.9 percent of all the world's cattle, yet these countries produce 68 percent of the world's **beef** and **veal**. Similar statistics are true for the other areas of animal agriculture. These countries are where almost all of the scientific knowledge about the growing of agricultural animals has been discovered. The people in these areas are the best fed and enjoy the highest living standard of any people in the world as a result of the new knowledge acquired through basic and applied research. Before countries can develop and prosper, a sound basis for producing food must first be achieved. Many of the poorer countries are attempting to imitate the agricultural systems of the more advanced countries.

Through the years there have been many discoveries and developments that have aided the advancement of animal agriculture. Some of the progress has been the result of many small discoveries that go together to provide greater efficiency in the production of animals. Other advancements have come about as the result of great strides made through scientific breakthroughs. The following are some of the milestones that served to revolutionize animal agriculture.

Animal Immunization

Up until the last half of the 1800s, diseases devastated herds of all types of agricultural animals all over the world. Once disease started in an area, all the animals in the surrounding countryside contracted the disease because there was no method of preventing the spread.

During the 1870s and 1880s a French scientist named Louis Pasteur developed a means of **vaccinating** animals to make them immune to disease. Using the scientific method, Pasteur hypothesized that animals that had contracted a disease and survived must have built up an **immunity** to the disease. Using the blood from sheep that had contracted and survived the deadly disease of anthrax, he developed a **serum**.

His experiment consisted of using two groups of healthy sheep. One group was injected with the serum and was later injected with anthrax organisms; the other group received only an injection of the anthrax organisms. The group that had received the serum remained healthy; the group that had not received the serum died.

Using the discovery of Pasteur, other scientists began to conduct research on other diseases, and during the next century numerous new **vaccines** were developed to control most of the diseases that are contracted by agricultural animals. Animals that can be raised in a disease-free **environment** can be raised at a much lower cost and at quite a bit less risk to the producer, Figure 1–10.

Refrigeration

A problem that plagued the producers of animals since the time humans first began raising them was how to preserve the meat and other products. When a large animal was slaughtered, not all of the meat could be eaten at once. Particularly in the summer months the meat spoiled very quickly, and if it wasn't consumed quickly, most would go to waste. In colder climates the animals would be killed in the winter, and the freezing temperature would

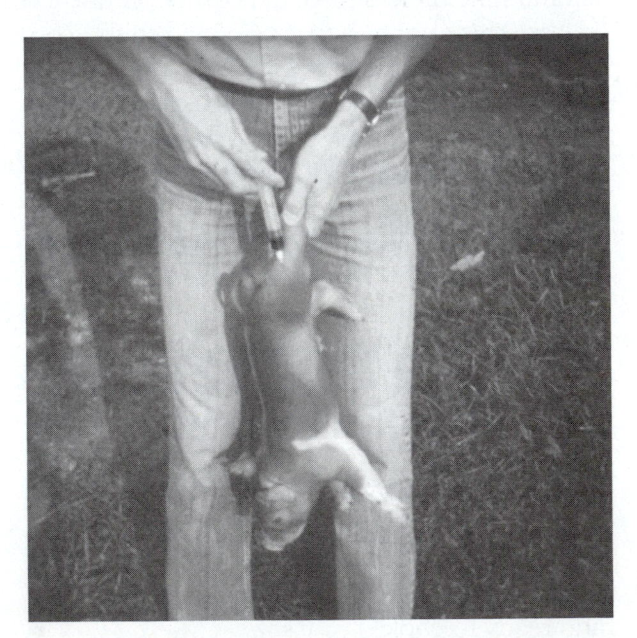

Figure 1-10. Through the use of vaccinations animals can now be raised disease-free.

preserve the meat until the spring thaws. About the only other way of preserving the meat was to salt or dry the meat. Both of these methods were time-consuming and did not produce a very palatable product. Another problem that occurred later was that of getting the meat to market. Up until around the turn of the century, live animals had to be delivered to the population centers. This meant driving the animals to market or driving them to a railhead where they could be transported live to the market, slaughtered, and sold as fresh meat. If the meat didn't sell quickly, it was lost to spoilage.

The first attempt at cooling meat was the use of ice that had been cut from frozen lakes during the winter and stored in icehouses. The ice blocks were suspended from the ceiling in the meat storage rooms in an effort to keep the meat cool. This effort was not very successful. During the 1880s mechanical refrigeration was developed and used in slaughterhouses to store meat.

A few years later the refrigerated boxcar was invented. This innovation allowed the transportation of meat anywhere in the country any time during the year, Figure 1–11. Now not only could animals be slaughtered any time of the year, but the meat could be stored for a long period of time. This meant that meat could be distributed to everyone in the country and a larger supply of meat was needed.

Artificial Insemination

Advancements in the type of animal produced are brought about through the transfer of superior **genes** from the parents to an **offspring**. With the advent of **artificial insemination** in the 1930s, the transfer of genes from

Figure 1-11. The development of refrigeration has allowed the preservation of meat in warm weather. *USDA photo.*

superior **sires** was greatly multiplied. Through modern techniques of **semen** collection, storage, and distribution almost any producer can have access to the very best genes in the industry. This innovation is one of the reasons for the phenomenal advancements of the dairy industry. Most of the dairy animals born in the United States are the results of artificial insemination.

EMBRYO TRANSFER

Although artificial insemination increased access to superior sires, advances through the use of superior **dams** were slow because of the **gestation** period of the female. With the development of the embryo transfer process, one superior dam can produce many offspring in one year, Figure 1–12. This process combined with artificial insemination allows producers to make extremely rapid gains in the quality of their herds at a relatively low cost.

THE USE OF COMPUTERS

Computers were developed during the 1940s, but it was not until the 1980s that the use of the computer became such an integral part of our lives. In the animal industry the use of the computer has had a profound effect on many different aspects. For example, the computer has made all areas of research move more rapidly. Data that once took days and even weeks to analyze can now be computed in a matter of seconds. Computer-simulated experiments and models have helped decrease the cost and time involved in scientific research.

The selection of superior dams and sires is made more convenient and accurate through computerized production records of **progeny**. Sires for artificial insemination can be matched with dams for embryo transfer through the use of computers. Many breed associations keep detailed records of all the animals in their registry on computer file.

Feed formulation is now done by computer. Not only is the balancing of feed rations done through the use of the computer, but computers control the mixing and regulation of ingredients in the feed, Figure 1–13. This allows a much more accurate blending of the nutrients needed by animals.

Perhaps the greatest impact computers have made is in the area of information

Figure 1-12. With the development of the embryo transplant process, one superior dam can produce many offspring in one year. *James Strawser, Cooperative Extension Service, The University of Georgia.*

Figure 1-13. Modern feed mills such as this use computers to balance and mix feed rations. *James Strawser, Cooperative Extension Service, The University of Georgia.*

retrieval. Through the internet, producers can access an almost unbelievable amount of information without leaving home. This system connects the producer to information from all over the world. Producers obtain summations of the latest scientific research, find sources for feed, supplies, and breeding animals; locate markets; and even buy and sell animals online. New ideas on management practices and production can be gathered from other producers all over the world. Decisions on buying and selling are now easier because the producer can get immediate price quotes and predictions over the internet. In fact, all banking and financial transactions can be done electronically.

In addition, most major breed associations place information on the internet so producers can have up-to-the-minute information on events within the associations. Production and progeny records on the most valuable sires and dams are available to aid in decision making on breeding programs. Information on problems, such as disease outbreaks, can be instantly passed on to the producer over the internet. Notices of sales, show schedules, and upcoming events can be placed on the worldwide web for the use of the producers. Computers have brought the information of the world into the homes of the producers.

Efficiency in the production of agricultural animals is not the only benefit derived from scientific research, Figure 1–14. The lives of humans have been greatly enhanced by research dealing with animals through the development of pharmaceuticals from animal by-products. A pharmaceutical is a substance that is used as a drug to make the life of a person better. Many of the drugs routinely prescribed by doctors are derived from animals, Figure 1–15. For instance, cortisone, which relieves the suffering of people with arthritis, is made from the gallbladder of cattle. Until recently, when a synthetic form of cortisone was developed, the only source for this drug was cattle.

Figure 1-14. The development of artificial insemination techniques has allowed a superior sire to sire hundreds of offspring. *James Strawser, Cooperative Extension Service, The University of Georgia.*

Similarly, before a synthetic form of insulin was developed, it was made from the pancreas of animals. A periodic intake of insulin is necessary for people who have diabetes. Insulin from hogs is of particular value because this form most closely resembles human insulin. Some diabetics are allergic to synthetically produced insulin and can only take insulin derived from hogs.

Many of the hormones used in the treatment of human disorders come from animals that have been slaughtered for food. When for some reason the human body does not produce enough hormones to control or stimulate a body function, the hormone must be supplied from an outside source. Without the ready supply of drugs derived from animals, many people's lives would be shortened and the quality of their lives would be lessened.

Another result of scientific research is that animal parts can be used as replacements for human parts. Heart valves from pigs have been used for about 20 years as replacements for faulty human heart valves. Pig heart valves are highly superior to mechanical valves because mechani-

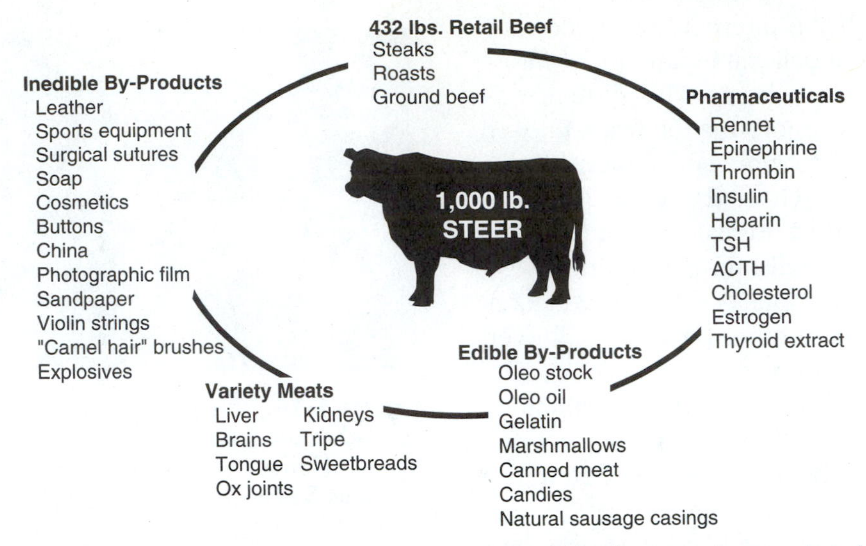

Figure 1-15. Animals provide much more to humans than just beef. *National Livestock and Meat Board.*

cal valves tend to accumulate residue from the flow of blood. As a result, the mechanical valve sticks and the result can be fatal. In contrast, a pig heart valve does not accumulate as much residue because it came from a living, functioning animal. Although the valve is treated to make it into an inanimate object before it is put in a human heart, the traits that allow its use are retained. Skins from pigs are now being used to treat humans who have received severe burns. The pig tissues are used as skin covers until the new human skin has had a chance to grow.

SUMMARY

No other branch of science truly touches our lives more than that of the science of agriculture. It has revolutionized every aspect of our lives. Other countries are fierce competitors in electronics, automotives, and manufacturing, but no one in the world even comes close to competing with American agriculture. Our research, combined with our free enterprise system, has made us the envy of the world in agricultural production.

◆ Review Exercises

TRUE OR FALSE

1. Science is the study or theoretical explanation of natural phenomena. ______.
2. The Land Grant Act or Morrill Act provided public land and funds for establishing universities that would teach practical methods of manufacturing and producing food and fiber. ______.
3. The Cooperative Extension Service came into being with the Smith-Lever Act in 1914. ______.
4. A hypothesis is a scientifically proven cause of a particular problem. ______.
5. Basic research is used in every aspect of society; agriculture is only a small part. ______.
6. Research and applied science have made it possible to raise more dairy cows on the same

amount of land although the milk produced per cow has remained the same. ______.

7. Before countries can develop and prosper, a sound basis for producing food must first be achieved. ______.
8. Almost all of the progress in research has come about through milestone breakthroughs. ____.
9. Animals that can be raised in a disease-free environment can be raised at a much lower cost and at quite a bit less risk to the producer. ______.
10. One of the old ways of preserving meat was salting it, but this method was time-consuming and did not produce a very palatable product. ______.
11. With the advent of artificial insemination in the 1930s the transfer of genes from superior sires was greatly multiplied. ______.
12. Efficiency in the production of agricultural animals is the only benefit derived from scientific research. ______.
13. A pharmaceutical is a substance that is used as a cosmetic to make the life of a person better. ______.
14. Animal parts can be used as replacements for human body parts. ______.
15. While the United States is a leader in electronics, automotives, and manufacturing, it is behind many countries in the agriculture field. ______.

FILL IN THE BLANKS

1. As the patterns of animal __________ and _________ were studied, people reasoned that if they could _________ animals, the need for moving with the _________ and hunting as a group could be _________.
2. To make progress in industry and agriculture, young people needed to be taught how to __________ food and __________ goods in a more _________ manner.
3. The Hatch Act authorized the establishment of _________ _________ in different parts of all the states with _________ _________ _________.
4. The scientific method has been used _________ of times to _________ the methods that are used to produce _________ _________.
5. Basic research deals with the _________ of why or _________ events occur, while applied research deals with using the _________ made in basic research to help in a __________ manner.
6. Advancements made through scientific research have resulted in not only an _________ of animals for _________ but also relatively low _________ to pay for _________.
7. The increase of beef cattle liveweight has come about as a result of the scientific _________ of __________ animals, a better understanding of beef cattle _________, and better control of _________ and _________.
8. Producers can now grow heavier broilers in _________ the time on _________ the feed than they could in _________.
9. During the 1870s and 1880s, a French scientist named __________ _________ developed a means of _________ animals to make them _________ to _________.
10. Refrigerated boxcars meant that not only could animals be _________ any time of the year, but the meat could be _________ for a _________ period of _________.
11. Embryo transfer combined with artificial _________ allows producers to make extremely rapid gains in the _________ of their _________ at a relatively low _________.
12. Computer-simulated experiments and ___________ have helped decrease the ___________ and ____________ involved in scientific _________.

13. Cortisone relieves the suffering of people with __________ and is made from the ________ bladder of ________.

14. Heart valves from _______ have been used for about ________ years as replacements for faulty ________ heart valves.

15. A pig heart valve does not accumulate as much ________ as a mechanical valve does because it came from a living ________, ________ animal.

DISCUSSION QUESTIONS

1. Why is agriculture considered to be the oldest of the sciences?

2. How has a strong knowledge about agriculture helped advances in other areas?

3. What five acts passed by Congress have helped make advances in agriculture?

4. What is the three-part purpose of the land grant institution?

5. What does the scientific method involve?

6. What advances have been made in the beef industry since 1925? in the sheep industry? in the dairy industry? in the swine industry?

8. List five developments that have had a great impact on animal agriculture.

9. What drugs are derived from animals?

10. Why are pig heart valves superior to mechanical valves in replacing faulty human heart valves?

STUDENT LEARNING ACTIVITIES

1. Interview your parents (and if possible, your grandparents). Make a list of the advancements they perceive as having made their lives better since they were your age. Determine how many of these advancements came about as a direct or indirect result of agriculture.

2. Interview a livestock producer. Ask him/her what new improvements he/she has seen during the past five years. Also ask him/her what problems he/she has that could be solved through scientific research.

3. Design a scientific experiment to solve a problem. Be sure to include your hypothesis, how you would conduct the experiment, and how you would use the results.

4. Go to your school computer lab and access the internet. Locate a research study on animal science. Also, locate four breed associations and print the information available.

Chapter 2

The Classification of Agricultural Animals

Student Objectives In Basic Science

As a result of studying this chapter, you should be able to

1. Explain the importance of scientifically classifying animals.
2. Define and explain the use of the binomial system of classification.
3. List all of the five kingdoms that are used to classify all living organisms.
4. Explain the different categories used in the scientific classification of animals.
5. List characteristics of animals that place them in different classifications.
6. Describe methods of classifying animals by means other than scientific classification.

Student Objectives in Agricultural Science

As a result of studying this chapter, you should be able to

1. Explain how agricultural animals are classified scientifically.
2. Explain how breeds of livestock were developed.
3. Explain the purposes of breed associations.
4. Describe the classification of agricultural animals according to use.

Key Terms

organisms
binomial nomenclature
genus
species
kingdoms
animalia
plantae
monera
protista
fungi
phyla
phulon
notochord
classes
orders
cud
families
scientific name
polled
breeding true
underline
selective breeding
breed associations
blood typing
dual-purpose animals
wether

There are millions of different types of animals and other living things in the world. Most of these **organisms** have been identified, grouped, and classified in an attempt to more effectively study them and communicate about them. Plants, animals, and other organisms are classified or grouped together by characteristics they have in common. They may be characterized by their physical characteristics, by the uses people make of them, or by other categories used to put similar animals together.

There are several ways of classifying agricultural animals. First of all, these animals are part of the overall population of the animals of the world. Therefore they are classified scientifically just as all other animals are. One big difference is that agricultural animals have been domesticated for some type of human use. These domesticated animals have been developed into breeds having distinctive characteristics and distinctive uses. Breeds and the uses of the various types of animals are other ways of classifying and identifying animals.

SCIENTIFIC CLASSIFICATION

Animals are given names according to a scientific classification system. This system is known as **binomial nomenclature**. Binomial means two names; nomenclature refers to the act of giving a name. This system was developed by a Swedish botanist named Carolus Linnaeus who grouped organisms according to similar characteristics. Linnaeus used two Latin names for identifying each individual organism. The first of the two names is the **genus**; the second part of the name is the **species**.

Latin was used because at the time of Linnaeus, Latin was the international language of scholars and so many of the languages of the world are based on Latin. People from many different countries who speak many different languages can recognize the scientific names. A *Sus scrofa* will be recognized as a domesticated pig by people who speak German as well as those who speak English, Figure 2–1. The two-name system also helps people from different areas of the same country accurately identify animals or other living things.

Common names are often confusing and can be misleading. For instance, in Oregon a gopher is a burrowing animal having long claws and long teeth that eats insects and roots. In Alabama, a gopher is a large rat that inhabits barns and other outbuildings. To people who live in Florida, a gopher is identified as a type of burrowing terrapin. Without a standard naming system, people from outside one of these areas would be confused about which of the three animals was being referred to. However, if the gopher in Oregon is referred to as *Thomomys species*, if the gopher in Florida is referred to as *Gopherus polyphemus*, and if the gopher in Alabama is referred to as *Rattus norevegicus*, there can be little doubt as to the identity of the animal. A scientific system of classification allows the exact identity of an animal to be recognized anywhere in the world.

Figure 2-1. Scientists all over the world recognize this animal as *Sus scrofa. Poland China Association.*

The scientific classification of all living things is an orderly and systematic approach to identification. Broad groups of animals are classified together in categories of common characteristics. Each of these broad groups is then broken down into categories of animals having similar characteristics. Then each of these groups is further broken down into smaller categories. The process is repeated until the groups cannot be categorized into smaller groups. Following is an explanation of the system and how agricultural animals fit into the classification system.

Kingdoms

All living things are first classified into five broad categories called **kingdoms**. For many years, scientists recognized only two kingdoms: the plant kingdom and the animal kingdom. As new discoveries were made and more organisms were classified, scientists realized that some animals could not be classified as either plant or animal. They didn't fit into either of the two existing kingdoms. In order to revise the system, three new kingdoms were added. Scientists now recognize five kingdoms of living organisms. These five kingdoms are

- **Animalia**—all multicelled animals
- **Plantae**—multicellular plants that produce chlorophyll through photosynthesis
- **Monera**—bacteria and blue-green algae
- **Protista**—paramecia and amoebae
- **Fungi**—mushrooms and other fungi

Obviously, all agricultural animals are large multicellular animals and belong to the kingdom Animalia. The kingdom Animalia includes all animals ranging from a tiny gnat to the huge whales that inhabit our oceans. Because of this great diversity it is necessary to group the different animals in the kingdom Animalia into smaller groups known as **phyla**.

Phyla

The kingdom Animalia is divided into twenty-seven different phyla (*singular* phylum) according to the animals' characteristics. The word phylum comes from the Greek word **phulon,** meaning race or kind. Several of the phyla are divided into subphylas. Animals in phyla or subphylas are grouped by broad characteristics shared by the animals.

For instance, the phylum Arthropoda consists of animals that have a hard external skeleton called an exoskeleton. This phylum includes insects, spiders, crayfish, crabs, centipedes, etc. Another phylum, the Mollusca, include the animals that have soft bodies that are protected by a hard shell, e.g. starfish, snails, and clams. The segmented worms, such as the earthworm, belong to the phylum Annelida. All agricultural animals (with the exception of certain specialty animals such as earthworms and oysters) belong to the phylum Chordata. Animals in this phylum have a stringy rodlike structure called a **notochord** that is made of tough elastic tissue that is present in the embryo.

This phylum is divided into subphylas, one of which is the subphylum Vertebrata. This subphylum includes animals with backbones. The jointed backbone that supports the animal is developed from the notochord of the embryo. The animals with backbones belong to the subphylum Vertebrata and include the animals generally found on farms and ranches. However, the subphylum Vertebrata is comprised of animals as diverse as sharks and monkeys.

Classes

The phyla and subphyla are further divided into **classes**. Examples of the classes in the subphylum Vertebrata are

- Amphibia—includes frogs, toads, and salamanders
- Reptilia—includes turtles, snakes, and lizards
- Aves—includes the birds
- Mammalia—includes animals that have hair, nurse their young, and give live birth

Agricultural animals such as horses, cattle, goats, sheep, pigs, and dogs belong to the class Mammalia. The class Mammalia is also a very expansive classification group. It includes mice, elephants, tigers, whales, and humans.

Orders

Classes are divided into smaller groups that categorize animals within a class that possess certain characteristics. These groups are called **orders**.

The class Mammalia contains eighteen different orders including Primates, which encompasses humans. Cattle, goats, sheep, and pigs belong to the order Artiodactyla. Animals are placed in this order because they have an even number of toes. Although sometimes referred to as hooves, the feet on these animals all have an even number of divisions (usually two) to their hooves or feet. Within this order there are three suborders:

- Suiformes—includes pigs and hippopotami
- Tylopoda—includes camels and llamas
- Ruminantia—includes deer, cattle, sheep, and goats.

The common characteristics of animals in the suborder Ruminantia are that they chew a **cud** and have several compartments to their digestive system, which allows them to eat grass, hay, and other roughages.

Horses and donkeys have only one toe (hoof) and belong to the order Perissodactyla. Also in this order are zebras, tapirs, and rhinoceroses.

Families

At this point in the classification system, the characteristics of the animals that are grouped together begin to narrow and the animals have a lot more in common. However, there is still a considerable amount of difference between a cow and a deer (both of which belong to the suborder Ruminantia) or between a rhino and a horse (both of which belong to the order Perissodactyla). Orders and suborders are further broken down into **families**. There are many families within each order and suborder. The suborder Ruminantia is divided into five families:

- Cervidae—includes deer, elk, and moose
- Antilopinae—includes the antelopes
- Tragulidae—includes certain types of goats
- Giraffidae—includes the giraffe
- Bovidae—includes cattle, buffalo, sheep, and domestic goats.

However, sheep and goats are put in the subfamily Caprinus, Figure 2–2.

Genus and Species

The final categories of the scientific classification system are genus and species. The genus and species compose what is known as the **scientific name**. All identified animals have been given this two-part classification. Families are broken down into genera and each genus is further divided into species. For instance, sheep are placed in the genus *Ovis;* goats are classified in the genus *Capra*. Domestic sheep are separated from the various types of wild sheep by species. The species of domestic sheep is *aries*. Within the family Bovidae, cattle are classified in the genus *Bos*.

Figure 2-2. Goats belong to the same family (Bovidae) as cattle but are classified in a different subfamily.

They are further separated by species. Cattle of European origin are classified as *Bos taurus;* cattle that originated in India are classified as *Bos indicus*, Figure 2–3.

Classification by Breeds

Within species of animals there are a lot of differences. For instance, a Great Dane and a Poodle are both classified as *Canis familiaris.* Just think of all the differences between the two breeds! Breeds of agricultural animals can show almost as much difference within the species. Color patterns, size, horned or **polled**, and country of origin can all be characteristics used to distinguish different breeds of cattle.

A breed of animal is defined as a group of animals with a common ancestry and common characteristics that breed true. **Breeding true** means that the offspring will almost always look like the parents. For instance, the Hereford breed of cattle is characterized by being brownish red with a white face and white **underline**, Figure 2–4. If a male and a female Hereford are mated, the offspring are expected to be red and have the characteristic white face and white underline.

Figure 2-4. The Hereford is unique in that it is brownish red in color with a white face and white underline. *American Hereford Association.*

Common Name	Pigs	Cattle	Horses	Sheep	Chickens	Turkeys	Rabbits	Honeybees	Catfish
Kingdom	Animalia	Animalia	Animalia	Animalia	Animalia	Animalia	Animalia	Animalia	Animalia
Phylum	Chordata	Chordata	Chordata	Chordata	Chordata	Chordata	Chordata	Arthropoda	Chordata
Class	Mammalia	Mammalia	Mammalia	Mammalia	Aves	Aves	Mammalia	Insecta	Osteichthyes
Order	Artiodactyla	Artiodactyla	Perissodactyla	Artiodactyla	Galliformes	Galliformes	Lagomorpha	Hymenoptera	Siluriformes
Family	Suidae	Bovidae	Equidae	Bovidae	Phasianidae	Meleagrididae	Leporidae	Apidae	Ictaluridae
Genus	*Sus*	*Bos*	*Equus*	*Ovis*	*Gallus*	*Meleagris*	*Oryctolalgus*	*Apis*	*Ictalurus*
Species	*scrofa*	*taurus, or indicus*	*caballus*	*aries*	*domesticus*	*gallopavo*	*cuniculus*	*mellifera*	*furcatus*

Figure 2-3. Scientific classification of agricultural animals.

Selective Breeding

All breeds of hogs probably came from a common ancestor; all breeds of sheep probably came from a common ancestor; all breeds of cattle probably came from a common ancestor. After the animals were domesticated, breeds were developed by the people who took care of the animals. The characteristics of the different breeds were probably developed because the people who raised them wanted those particular characteristics. Those animals showing the desired traits were kept for breeding and the others were slaughtered and eaten.

A group of producers may have liked the black color of some of their beef animals and only bred those that were black. After a few generations of this **selective breeding**, only black calves were produced. Sometime during this process someone may have noticed that some of the calves did not develop horns. Subsequently, if only black cattle with no horns were used for breeding, after several generations a group of cattle developed that were always black and had no horns. From this group came the modern Angus breed of cattle.

Purebreds

Those animals whose ancestors are of only one breed are referred to as a purebred. **Breed associations** have been developed to promote certain breeds of animals. These associations usually set the standards for animals that are allowed to be registered as a purebred animal of that particular breed. If breed associations did not set standards for their animals, the breed might disappear in a few years. For example, the American Duroc Association says that in order for an animal to be registered as a Duroc, the animal must be red in color with no white on the body. By only allowing a certain type of animal to be registered as a purebred, the breed association is assured that the characteristics that make the animals a certain breed will continue.

Blood Typing

Not only physical characteristics are used in breed identification. A process known as **blood typing** is also used to determine the ancestry of animals. Individual animals and humans have different types of blood known as blood groups. The blood of different types or groups will have different characteristics that are passed on genetically from the parent to the offspring. As the blood is analyzed, these differences show up. This process is useful in determining the parentage of a particular animal.

Since the black color of Angus cattle is dominant, the sire of a black calf from an Angus cow would be difficult to determine just by looking at the calf. However, the blood type of a calf sired by an Angus bull would be different from the blood type of a calf sired by a bull of a different breed. In fact the blood type of two Angus bulls might even be different. Determining the parentage of animals using blood typing is usually considered to be about 90 percent accurate.

New breeds of animals are constantly being developed from combining animals of different breeds. The Murray Grey breed was developed by systematically breeding Shorthorn cattle and Angus cattle until the offspring had the desirable characteristics and bred true, Figure 2–5. Just as in earlier history, a breed was developed that had the characteristics that certain people wanted.

Crossbreeding

Sometimes species can be successfully crossed to produce new breeds. For example, the breeds of cattle developed in Europe are scientifically classified as *Bos taurus*. These cat-

PLATE 1 Angus Bull
Courtesy of the American Angus Association

PLATE 2 Charolais Bull
Courtesy of the American-International Charolais Association

PLATE 3 American Salers Bull
Courtesy of the American Salers Association

PLATE 4 Limousin Bull
Courtesy of the North American Limousin Foundation

PLATE 5 American Polled Hereford Bull
Courtesy of the American Polled Hereford Association

PLATE 6 American Simmental Bull
Courtesy of the American Simmental Association

PLATE 7 Ayrshire Cow
Courtesy of the Ayrshire Breeders' Association

PLATE 8 Brown Swiss Cow
Courtesy of the Brown Swiss Cattle Breeders' Association

PLATE 9 Guernsey Cow
Courtesy of the American Guernsey Association

PLATE 10 Holstein-Friesian Cow
Courtesy of the Holstein Association USA, Inc.

PLATE 11 Jersey Cow
Courtesy of the American Jersey Cattle Club

PLATE 12 Milking Shorthorn Cow
Courtesy of the American Milking Shorthorn Society

PLATE 13 Poland China Swine
Courtesy of the Poland China Record Association

PLATE 14 Duroc Swine
Courtesy of the United Duroc Swine Registry

PLATE 15 Chester White Swine
Courtesy of Swine Genetics

PLATE 16 Landrace Swine
Courtesy of the American Landrace Association

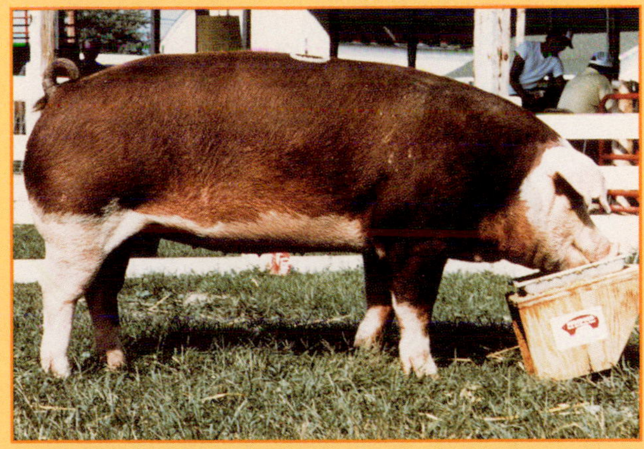

PLATE 17 Hereford Swine
Courtesy of the National Hereford Hog Record Association

PLATE 18 Spotted Swine
Courtesy of Swine Genetics

PLATE 19
Columbia Ram
Courtesy of the Columbia Sheep Breeders'

PLATE 20
Dorset Ram
Courtesy of the Sheep Breeder Magazine

PLATE 21
Hampshire Ram
Courtesy of the Sheep Breeder Magazine

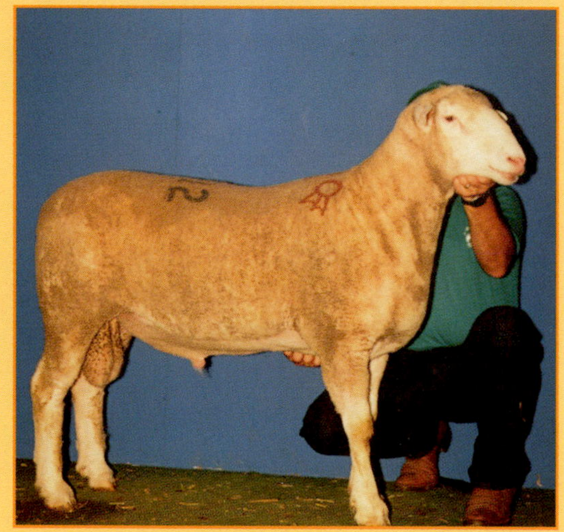

PLATE 22
Polypay Ram
Courtesy of the Sheep Breeder Magazine

PLATE 23
Rambouillet Ram
Courtesy of the Sheep Breeder Magazine

PLATE 24
Suffolk Ram
Courtesy of the Sheep Breeder Magazine

PLATE 25 Alpine Goat
Courtesy of the American Dairy Goat Association

PLATE 26 Nubian Goat
Courtesy of the American Dairy Goat Association

PLATE 27 Saanen Goat
Courtesy of the American Dairy Goat Association

PLATE 28 Lamancha Goat
Courtesy of the American Dairy Goat Association

PLATE 29 Toggenburg Goat
Courtesy of the American Dairy Goat Association

PLATE 30 Oberhasli Goat
Courtesy of the American Dairy Goat Association

PLATE 31 Appaloosa
Courtesy of the Appaloosa Horse Club/photo by Tom Poulsen

PLATE 32 Quarter Horse
Courtesy of the American Quarter Horse Association

PLATE 33
Arabian
Courtesy of Johnny Johnston

PLATE 34
American Paint
Courtesy of Don Shugart

PLATE 35 Standardbred
Courtesy of the U.S. Trotting Association

PLATE 36 Thoroughbred
Courtesy of the Illinois Racing News

PLATE 37
White Leghorn Rooster
Photo by Dr. Charles Wabeck

PLATE 38
White Plymouth Rock Hen
Photo by Dr. Charles Wabeck

PLATE 39
Broad Breasted Large White Turkey
Courtesy of the Minnesota Turkey Research and Promotion Council

PLATE 40
Broad Breasted Bronze Turkey
Courtesy of Watt Publishing Co.

PLATE 41
White Pekin Duck
Courtesy of Jurgielewicz Duck Farm

PLATE 42
Toulouse Goose
Photo by Dr. Charles Wabeck

PLATE 43
English Spot
Courtesy of the American Rabbit Breeders Association

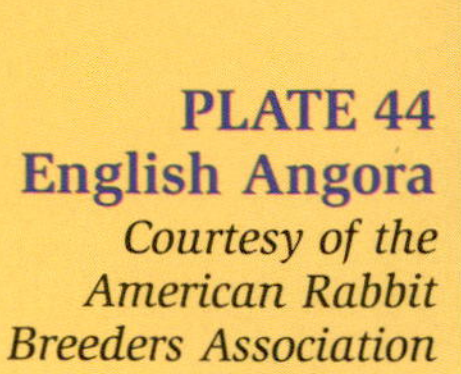

PLATE 44
English Angora
Courtesy of the American Rabbit Breeders Association

PLATE 45 Dutch
Courtesy of the American Rabbit Breeders Association

PLATE 46 Checkered Giant
Courtesy of the American Rabbit Breeders Association

PLATE 47 English Lop
Courtesy of the American Rabbit Breeders Association

PLATE 48 New Zealand White
Courtesy of the American Rabbit Breeders Association

Figure 2-5. The Murray Grey breed was developed by systematically breeding shorthorn and Angus cattle. *American Murray Grey Association.*

Figure 2-6. The Brahman is a different species of cattle (*Bos indicus*). *American Brahman Breeders Association.*

tle have characteristics that make them desirable as beef animals that do well in the climates of Europe. In the tropical climate of India, another species of cattle (*Bos indicus*) developed that had characteristics that made the animal useful in that part of the world, Figure 2–6. Cattle breeders in the subtropical regions of the United States (the Southwest and Southeast) recognized the need for an animal that had characteristics of both the *Bos taurus* and the *Bos indicus*.

One of the first successful breeds of this type was the Santa Gertrudis, which was developed by systematically crossing the Shorthorn breed of cattle (*Bos taurus*) with the Brahman breed of cattle (*Bos indicus*), Figure 2–7. This new breed of cattle combined the growth and carcass quality of the *Bos taurus* with the hardiness of the *Bos indicus*. Since that time, many other breeds have been developed using these two species.

Another example is the mule, which was developed by breeding a mare (female horse, *Equus caballus*) with a jack (a male donkey, *Equus asinus*). The resultant animal—the mule—combined the size and strength of the horse with the toughness and surefootedness of the donkey, Figure 2–8.

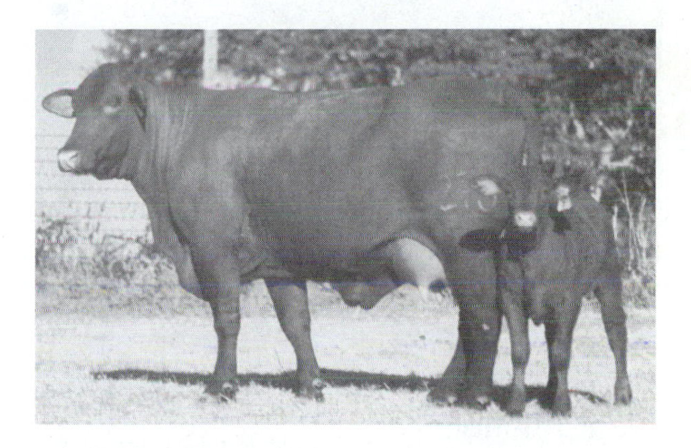

Figure 2-7. The Santa Gertrudis was developed from the Shorthorn and Brahman breeds. *Santa Gertrudis Breeders International.*

Classification According to Use

Breeds of domesticated animals are sometimes grouped together because of the use humans make of them. Agricultural animals are raised for several different reasons and are therefore classified by the uses made of them.

Figure 2-8. The mule was developed by breeding a mare with a jack. *Russ Reagan, Missouri Department of Conservation.*

Meat Animals

These animals are raised primarily for slaughter and human consumption. For instance, with the exception of those raised as laboratory animals, almost all pigs are raised for pork and have little use otherwise. Sheep on the other hand may be raised for various purposes. Breeds such as Rambouillet and Merinos are grown primarily for their wool, Figure 2–9. Suffolk and Hampshire breeds are grown primarily as meat animals. Other breeds of sheep are produced for their milk, which is used in such products as Roquefort cheese.

Another example can be found in cattle. Hereford cattle are raised for beef because they produce a lot of muscle and only enough milk to feed their calves. On the other hand, Ayrshire cattle have considerably less muscle than Herefords, but produce a tremendous amount of milk. Because of this, Ayrshires are raised for their milk and not for beef, Figure 2–10. Likewise some breeds of chickens produce a lot of eggs but little meat; these breeds are used as layers. Other breeds produce a lot of meat and are raised for slaughter.

Figure 2-9. Some sheep are raised for their wool, some for their meat, and some for their milk. *U.S. Targhee Sheep Association.*

Figure 2-10. Ayrshires were developed for milk production instead of meat production. *Ayrshire Breeders Association.*

Figure 2-11. In some areas of the world, animals such as the water buffalo are still a valuable source of power.

Work Animals

Another classification according to use is the work animal. In the past, work animals have been an essential part of agriculture. Even today in some parts of the world, animals are the primary means of transportation and tillage of the soil. Donkeys provide power to pull carts and are ridden. Camels provide means of bearing heavy loads. Oxen, camels, water buffaloes, and donkeys all are used to pull wagons and plows, Figure 2–11.

Horses

In this country animals are still used to assist humans in work. On farms and ranches in the United States horses are still a valuable means of working cattle. In some instances, they are still used as draft animals. They are also used for recreational purposes. Given all of these uses for horses, it is easy to see that they are classified according to the type of work they do. Cutting horses, such as the American Quarter Horse, are used to herd and work cattle, Figure 2–12. Larger breeds, such as Belgians and Clydesdales, are used to pull wagons and heavy loads. They are classified as draft horses. Some breeds, such as the Morgan, Tennessee walker, and the American Saddlebred, are used for riding and are classified as saddle horses. Others, such as the Hackney and the Standardbred, are used for

Figure 2-12. In the United States, horses are still used to work cattle.

pulling sulkies or light carriages and are known as harness horses, Figure 2–13.

Dogs

Dogs are also used to herd cattle, hogs, and sheep. On a sheep ranch, a good sheepdog is quite a valuable asset. Not only are they used to round up and sort sheep but they also protect them from predators, Figure 2–14.

Dual-Purpose Animals

As animals were domesticated, many were developed to be **dual-purpose animals**. Not only could cows provide milk; they could also serve to pull plows, carts, or other implements. Also the surplus young could be slaughtered and eaten. On most modern farms and ranches, agricultural animals are specialized, i.e., they serve only one purpose. For instance, cattle are raised either for milking or for beef, but seldom for both. Exceptions do exist, however. Most sheep that are raised for meat are also shorn for their

Figure 2-13. Many horses work to provide people with recreation. *Kentucky Horse Park.*

Figure 2-14. Dogs are classified as work animals when they are used to herd other animals. *The Working Border Collie Magazine.*

wool. While the wool is not as high-quality as that of a sheep raised primarily for wool, the producer does obtain some income from the sale of the wool. Likewise calves from dairy cattle often are slaughtered for veal or beef.

In many parts of the world, dual-purpose animals still play a major part in the agricultural economy. In the harsh deserts of the Middle East, camels are raised and used by the people. Camels provide a source of power for carrying or pulling loads and a source for milk and meat, Figure 2–15.

Summary

There are many ways of classifying animals, and agricultural animals are no exception. Whether they are classified according to size, use, color, breed, or scientific classification, the system is an organized method of helping to identify types of animals. Without such a system, there would be much confusion surrounding the names and identification of animals.

Figure 2-15. In the Middle East, camels are used for work, milk, and meat.

◆ Review Exercises

TRUE OR FALSE

1. Plants, animals, and other organisms are classified or grouped together according to characteristics they have in common. ______.
2. Latin was used originally for scientific names, but the modern system uses English. ______.
3. The five kingdoms now recognized by scientists are Animalia, Plantae, Monera, Protista, and Fungi. ______.
4. The kingdom Animalia is divided into twenty-seven different phyla according to the animals' characteristics. ______.
5. Animals with backbones belong to the subphylum Vertebrata and include the animals generally found on farms and ranches. ______.
6. The class Mammalia contains eighteen different orders including Primates, which encompasses humans. ______.
7. The common characteristics of animals in the suborder Ruminantia are that they have one toe, they have only one compartment in their digestive system, and they have no hair. ______.
8. Families are broken down into genera; each genus is further divided into species. ______.
9. Within species of animals there are very little distinguishing differences. ______.
10. Animals whose ancestors are of only four or less breeds are referred to as purebred. ______.
11. Determining the parentage of animals using blood typing is usually considered to be 40–50 percent accurate. ______.
12. Breeds of domesticated animals are sometimes grouped together because of the uses humans make of them. ______.
13. Holstein cattle are raised primarily for their meat. ______.
14. Machinery has replaced the use of animals such as horses and buffaloes for transportation and tillage of the soil. ______.
15. Cattle are raised either for milking or for beef, but seldom for both. ______.

FILL IN THE BLANKS

1. Binomial means two ____________, and nomenclature means the ______________ of giving a ______________.
2. A scientific system of classification allows the exact ___________ of an ____________ to be recognized anywhere in the _____________.
3. As new discoveries were made and more ____________ classified, scientists realized that some animals could not be classified as either ___________ or ___________ and did not fit into either of the two existing _____________.
4. All agricultural animals are large _______________ animals and belong to the kingdom ___________________.
5. Animals in the phylum Chordata have in common a stringy ___________ like structure called a ________________ that is made of tough _____________ tissue that is present in the _______________.
6. The class Mammalia includes animals that have ________________, _______________ their young, and give ______________________________.
7. The suborder Ruminantia is broken into _______________ families, two of which are the Cervidae, to which _______________, elk, and _________________ belong, and the Bovidae, which includes _______________, buffalo, __________________________, and _____________________ goats.
8. If male and female Hereford cattle are mated, their offspring would be expected to be _______________ and have the characteristic _______________ face and_______________ underline.

9. The characteristics of the different breeds were probably ______________ because the ___________ who raised them wanted those particular __________________.
10. Not only physical characteristics are used in breed identification. A process known as ___________ _____________ is also used to determine the ____________________ of __________________.
11. The Santa Gertrudis was developed by systematically crossing the ______________ __________ breed of cattle (_____________ ___________) with the _______________ breed of cattle (____________ ___________).
12. Hereford cattle are raised for ___________ because they produce a lot of ____________ and only enough ____________________ to feed their ______________________.
13. Sheepdogs are used to round up and _______________________ sheep and to ____________ them from ______________.
14. As animals were domesticated, many served dual purposes; cows provided __________, but they could also serve to pull ______________, _____________, or other ________________.
15. In the Middle East, camels provide a source of ____________________ for carrying or _________________ loads, and a source of ________________ and ______________.

DISCUSSION QUESTIONS

1. What is the binomial nomenclature system of classification?
2. Why is the scientific classification of animals essential in studying and communicating about them?
3. List the five kingdoms and describe the type of organisms included in each.
4. What are the scientific names for the following agricultural animals: cattle, pigs, horses, sheep, dogs?
5. Explain how and why breeds of animals were developed.
6. What are the purposes of breed associations?
7. How is blood grouping used to classify animals?
8. Give two examples of two different species of animals that have been bred to produce a new breed or type of animal.
9. What are three classifications of animals according to their use?
10. What are three classifications of horses according to their use?
11. Explain the term "dual-purpose animal" and give some examples.

STUDENT LEARNING ACTIVITIES

1. Choose a particular animal (a Hereford cow, a Duroc boar, a Suffolk **wether**). List all of the different ways this animal could be grouped or classified. Discuss your classification methods with others in your class.
2. Choose two types of agricultural animals (such as a sheep and a pig) and list all their common characteristics you can think of. Also make a list of all the ways in which the animals are different. Compare your lists with others in the class.
3. Write to three breed associations and ask for information on the disqualifications of animals for those breeds. Compare the requirements of the different associations. As a class project, try to determine which breed associations are the most restrictive about qualifications.
4. Talk to several purebred livestock producers to determine the characteristics they like best about the breeds they raise. Compare your findings with those of others in your class.

Chapter 3

The Beef Industry

Student Objectives in Basic Science

As a result of studying this chapter, you should be able to

1. Explain the importance of beef in the human diet.
2. Explain how the environment helps determine where animals are produced.
3. Define ecological balance.
4. Describe how cattle make use of feed stuff that cannot be consumed by humans.

Student Objectives in Agricultural Science

As a result of studying this chapter, you should be able to

1. List the per capita consumption of products from beef animals grown in the United States.
2. Explain the importance of the beef industry to the economy of the United States.
3. Justify the use of agricultural land to produce beef.
4. Describe the various segments of the beef industry.

Key Terms

veal
sire breeds
dam breeds
exotics
purebred operations
cow-calf operations
stocker operations
feedlot operations
stocker

BEEF IN THE AMERICAN DIET

Americans are a nation of beef eaters. Each year the average person in this country consumes 63.5 pounds of beef and **veal**. Each year we buy more beef than fresh poultry, pork, and seafood combined. In fact, beef accounts for about six percent of all supermarket sales. Over the past few years, the consumption of beef has been increasing both in supermarket sales and as meals in commercial restaurants. In addition to the large amount of beef consumed in this country, almost a million metric tons of beef are exported each year. This represents a value of nearly $2.5 billion.

Lean beef is very dense in nutrients. A pound of beef may equal or surpass the nutritive content of the feed consumed to produce the meat. Meat is among the most nutritionally-complete foods that humans consume. Foods from animals supply about 88 percent of vitamin B_{12} in our diets because this nutrient is very difficult to obtain from plant sources. In addition, meats and animal products provide 67 percent of the riboflavin, 65 percent of the protein and phosphorus, 57 percent of the vitamin B_6, 48 percent of the fat, 43 percent of the niacin, 42 percent of the vitamin A, 37 percent of the iron, 36 percent of the thiamin, and 35 percent of the magnesium in our diets.

Very few nations in the world even come close to United States in the per capita consumption of beef and other meats. To a large degree, this is an indication of the prosperity of the American people. In the past, livestock ownership has been a sign of prosperity and in many cultures even today, a person's wealth is measured by the number of cattle owned.

In the United States' history, the beef industry has played a prominent role in the development of its economy. Cattle have been in the new world almost as long as the European settlers. The animals were brought across the oceans to feed the settlers in their new homes. Until around the time of the Civil War, most beef was raised on family farms for the purpose of feeding the family. As the population became more urbanized, it became more difficult for people to raise their own meat. Also, they became more affluent and could afford to buy their food rather than raise it. The large, grassy areas of the west were being settled, and cattle were a natural product to raise on the vast plains of native grasses.

Currently, there are over 100,000,000 head of beef on over 1,000,000 farms and ranches in the United States, Figure 3–1. The number of operations far exceeds any other segment of animal agriculture, with the cattle industry accounting for the largest segment of all the agricultural industry in the United States. Most cattle are raised on family-owned farms and ranches. In fact, about 80 percent of all cattle businesses have been in the same family for the past twenty-five years. Annually, the United States produces nearly 25 percent of the world's beef supply with less than 10 percent of the world's cattle.

LOCATION OF THE BEEF INDUSTRY

The United States is well suited for the production of animals that supply beef. In the West, vast areas of land are used to graze cattle. Throughout the Midwest, millions of acres of corn are grown on some of the most productive farmland in the world. In the southern portion of the country, beef producers take advantage of the mild climate to produce grass and hay to help feed the millions of head of cattle raised there.

In addition, when compared to the rest of the world, Americans spend a small percentage of their annual income for food. This means that they can afford to buy the type of food they

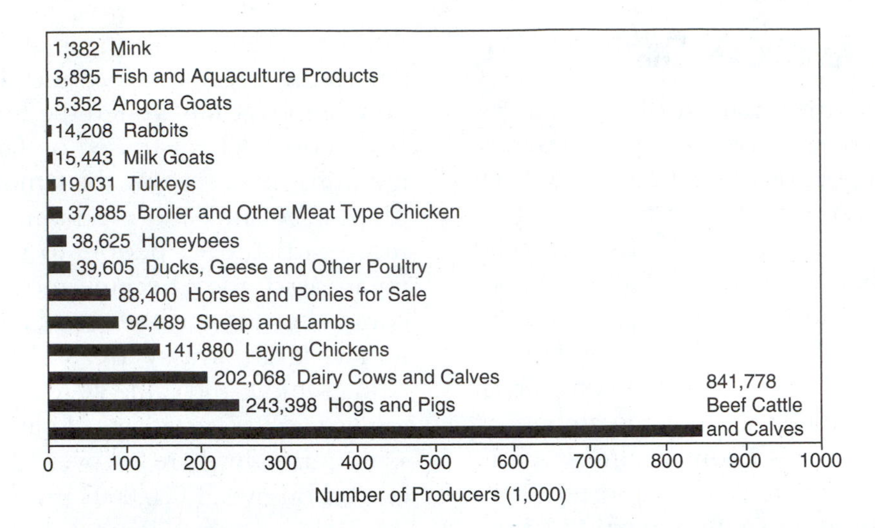

Figure 3-1. There are many producers of livestock in the United States. Cattle producers are by far the most numerous.

prefer. Obviously they prefer meat. Critics of the beef industry contend that we are being wasteful in feeding several pounds of feed to animals in return for a pound of meat. They say that the grains fed to animals could be better used to feed people. A pound of beef requires from six to nine pounds of feed to produce. Beef producers counter the argument by saying that land used to graze agricultural animals would be of little use for other agricultural purposes. Almost half the land in the United States is classified as land that is not practical for growing cultivated crops. Without the production of grazing animals, this land would be wasted instead of being used as a food-producing resource, Figure 3–2. Furthermore, they point out, livestock are finished (fattened) using grains that are not considered good for

Figure 3-2. Cattle can make use of land that is not suited for growing row crops.

human consumption. The better grades and types of grains are used for human consumption, but the lower grades of corn and grains such as grain sorghum are fed to livestock.

Beef animals also make use of by-products such as meal that is a result of the cooking oil market. The harvested crops such as soybeans and cottonseed are pressed until most of the oil is removed, and the resulting cake is ground into feed for livestock. Also, such by-products as beet pulp from the sugar industry and citrus pulp from the orange and grapefruit juice industry are fed to cattle. If not fed to livestock, these valuable by-products might go to waste.

As outlined in chapter 1, other products are obtained as well from animals. Most of the hides from the animals are used in making pharmaceuticals, leather for belts, shoes, and other articles of clothing, and upholstery materials.

Beef Cattle Breeds

Over three-quarters of the cash receipts for marketing meat animals comes from the sale of beef. These cattle are produced on almost a million farms and ranches across the United States. Contrary to popular belief, most of the beef animals do not originate from large ranches raising vast herds of thousands of cattle. The average size of the beef herds in this country is around 100 head. These producers represent a wide variety of different breeds and types of beef animals. In the United States, there are over 40 different breeds grown, besides all of the different combinations of crosses of these breeds. Livestock producers choose the breed they grow based on the type of market where the animals will be sold, the type of environmental conditions in which the animals will be produced, and the personal likes and dislikes of the individual producer.

Some breeds are large and produce a large carcass, some mature at a smaller size and produce a smaller carcass. Both have a place in the market and both are produced. Some breeds are more adapted to hot, humid climates and some breeds tolerate the cold and snow better than others. A producer may like the color pattern or the docile nature of a particular breed and prefer to produce that breed.

Some breeds make excellent mothers while other breeds grow rapidly and produce high-quality, meaty carcasses. Because of this, some breeds are referred to as **sire breeds** and some are referred to as the **dam breeds**. Obviously, a cross-breeding program helps producers take advantage of the good points of both types of animals. There are three broad classifications of beef breeds grown in the United States: the British breeds, the continental European breeds, and the zebu breeds.

The British breeds include the Angus, Hereford, and the Shorthorn, Figure 3–3. These animals are generally of a docile nature and produce high-quality carcasses at a medium size. These were the first breeds brought to this country and there are more of this class than any other.

Figure 3-3. The Shorthorn is classified as a British breed. *American Shorthorn Association.*

Figure 3-4. The Limousin is a good example of the large, meaty, continental breeds. *North American Limousin Association.*

The continental European breeds include the Limousin (Figure 3–4), the Simmental, Charolais, and the Chianina.

These breeds, once known as the **exotics**, were brought to this country because of their size and ability to grow. At maturity, most of the breeds in this class become quite large. The largest of the breeds, the Chianina, may reach the weight of 4000 pounds for the bulls and 2400 pounds for the cows, Figure 3–5. They are generally crossed with the British breeds.

Figure 3-5. The Chianina are the largest of all the breeds of cattle. *American Chianina Association.*

Figure 3-6. The Brahman is a type of zebu cattle that is a different species from the British and continental breeds. *James Strawser, Cooperative Extension Service, The University of Georgia.*

The zebu breeds are those that are scientifically classified as *Bos indicus*, a separate species from the traditional *Bos taurus* of the other breeds. The most common zebu type of cattle in the United States is the Brahman, Figure 3–6. These cattle are characterized by a large fleshy hump behind the shoulder and loose folds of skin. These cattle tolerate heat and humidity quite well and are resistant to insects. These characteristics make them well suited to the hot, humid climate of the southeastern part of the United States and the hot, dry climate of the Southwest. Brahmans have been used as the basis for-the development of several breeds such as the Santa Gertrudis, Brangus (Figure 3–7), Simbrah, and Beefmaster. These developed breeds combine the ruggedness of the *Bos indicus* with the carcass quality and docile nature of the *Bos taurus*

Segments of the Beef Industry

The Beef industry has four major segments: **purebred operations**, **cow-calf operations**, **stocker operations**, and **feed-**

Figure 3-7. The Brangus is an example of a breed developed from Brahmans. *International Brangus Breeders Association.*

Figure 3-9. Cattle shows serve to educate breeders and to make improvements in the cattle industry. *Calvin Alford, Cooperative Extension Service, The University of Georgia.*

lot operations. Purebred cattle are produced in the first phase of the industry, Figure 3–8. The purpose is to produce what is known as the seed stock cattle. These represent the cattle that are to be used as the dams and sires of calves that will be grown out for market. As mentioned earlier, different breeds have different advantages, and the growing of purebred stock allows breeders to concentrate on improving and accentuating the advantages of a particular breed. Each year, at numerous shows across the nation, purebred cattle breeders compete with each other by displaying their animals in the show ring. Expert judges select the animals they consider to be the best type for that breed. Shows serve both as a means of education and as a way of implementing change in the industry as economic conditions change and new research reveals new insights into the type of animals that should be selected, Figure 3–9.

Figure 3-8. The purebred breeders produce animals that will be used as dams and sires. *American Gelbvieh Association.*

The second phase is the cow-calf operations where the calves that will eventually be grown out and sent to market are produced, Figure 3–10. Most of these calves are crossbred animals from purebred parents of different breeds. A large part of this industry is centered in the southern and western states. The mild winters of the South are ideal for calving in the winter. In most areas of the South, calves are born in January and February to take the advantage of weather too cold for flies and parasites, but not too cold for the calves to thrive. In addition, the calves will be old enough to begin grazing in the spring when the grass begins to grow again.

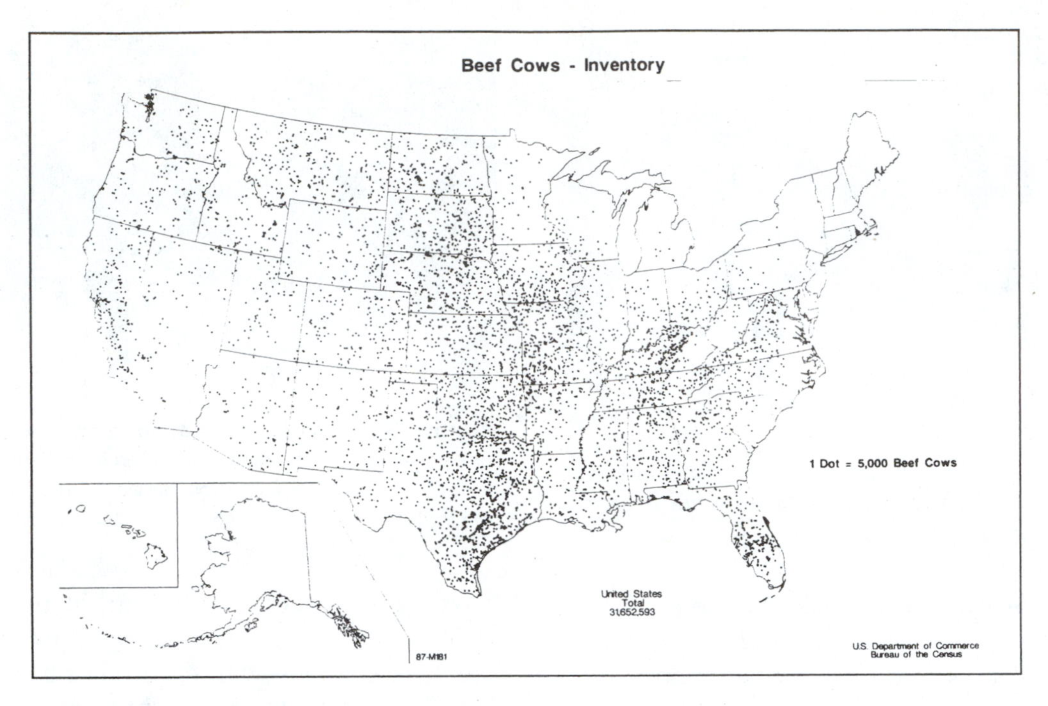

Figure 3-10. Beef cattle are grown all across the country. *Bureau of the Census.*

Cows are fed primarily roughage in the form of grass or hay. The ample rains and mild temperatures of the South provide ideal conditions to produce large amounts of green forage, Figure 3–11. Some grasses, such as fescue and rye, grow very well in the winter months and supply a good source of feed for cows as they gestate or produce milk for their young. Much of the cropland of the hill country of the south has been converted to pasture and forest. These lands were so susceptible to erosion that it was no longer practical to produce row crops and the growth of woodlands and pastures offered a way to make the land useful and productive.

Although the largest numbers of cow-calf operations are found in the South, cow-calf operations are found all across the country. In the West, producers can take advantage of the vast amounts of government lands that are open to grazing for a small fee. Often, cows are left on free range (not fenced in) to have their calves, which are then rounded up, weaned, and sold.

Figure 3-11. Climatic conditions in the South are ideal for growing grass for cows and calves. *James Strawser, Cooperative Extension Service, The University of Georgia.*

Calves are usually sold upon weaning. They are weaned in the weight range of 300 to 500 pounds. Buyers prefer calves that have been castrated and vaccinated and are in good enough condition to move to a new environment.

The next phase of the industry is that of the **stocker**. Stocker operations provide a step between the weaning of the calves and their finishing (or fattening) prior to slaughter. In order for an animal to start depositing fat in the right places, the animal must be mature enough to have stopped growing. Weaned calves that weigh between 300 and 500 pounds are placed on pasture land and fed a ration designed to allow for skeletal and muscular growth. The stocker purchases the animals from the cow-calf producer and sells them to the feedlot operator. The job of the stocker is that of providing a transition period for the calves between the time they are weaned from their mothers and before they are put in the feedlot. During this time, the animals are fed on a relatively high roughage diet and supplied with the proper balance of protein, carbohydrates, vitamins, and minerals that will ensure that the animals make sufficient gains to be placed in the feedlot where they will be finished.

It is not uncommon for feedlot owners to also be the operators of stocker operations. Such arrangements are economical because fewer transportation costs are incurred if the two types of operations are close together. The trend in the industry has been away from the stocker industry since recent research has developed production methods that allow cows to wean heavier calves. A calf that is weaned weighing 700 pounds may very well go directly into the feedlot without going through a stocker operation.

The feedlot operation is the final phase before the animals are sent to slaughter, Figure 3–12. Here the animals are fed on a high concentrate ration designed to put on the proper amount of fat cover. The producers usually want their animals to be marketed when the cattle reach a sufficient fat cover to allow the animals to grade low choice. Many feedlots in this country are situated in the Midwest, Figure 3–13. The reason is that this is the section of the country that produces the most grain and it is usually more economical to feed the animals there rather than ship the grain across country. An exception is the state of Texas; it has more feedlots than any other state.

Figure 3-12. Cattle are finished for market in feedlots. *Farmer Stockman Magazine.*

Some feedlots may be located at other parts of the country in order to take advantage of by-product feeds. For example, feedlots in Idaho take advantage of the potato industry and feed by-products from the processing of potatoes. Likewise, in Florida cattle are fed citrus pulp that is left over from the processing of orange juice.

Feedlots range in size from a hundred or fewer head to feedlots that feed thousands of cattle every year. Long bunker feeders are automatically filled by automated systems or from trucks, Figure 3–14. The animals are supplied with all the high-quality feed they will ingest. They are also given medicines to pre-

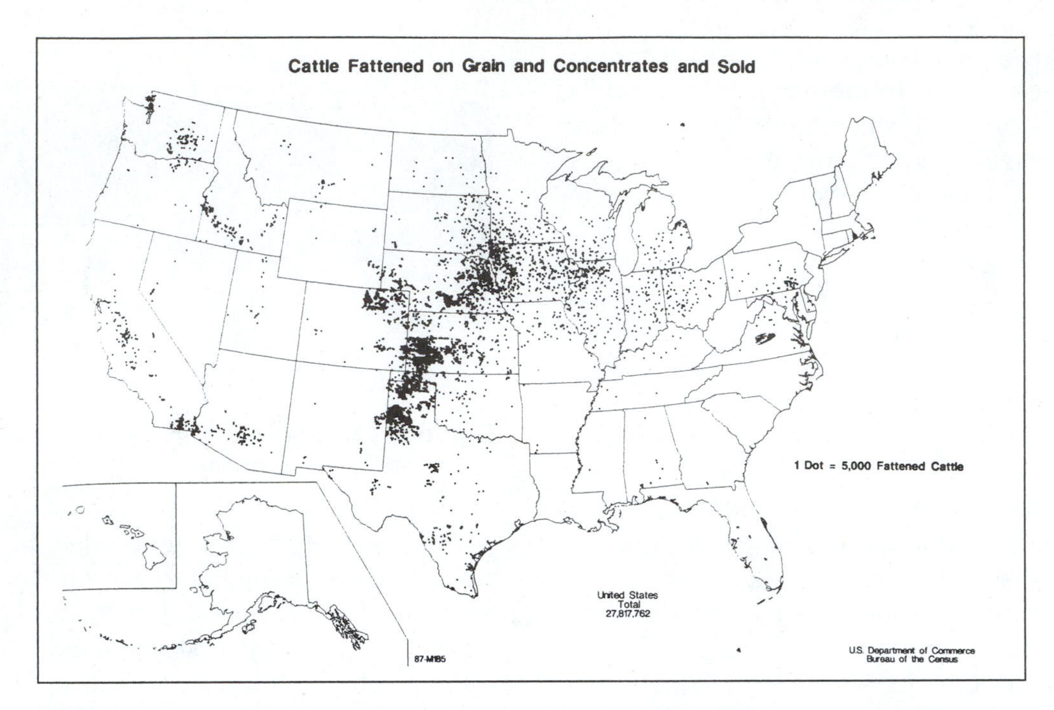

Figure 3-13. Feedlots are concentrated in the central part of the United States. *Bureau of the Census.*

vent disease and to ward off both internal and external parasites.

When the animals have reached the proper degree of finish, they are quickly moved from the feedlot to the slaughterhouse. When the animals are slaughtered, they are generally around 18 to 24 months in age and can weigh from 800 to 1500 pounds, Figure 3–15.

This age and size offers consumers the type beef they prefer.

Figure 3-14. Feedlot cattle are fed from a long trough filled by a truck.

Figure 3-15. A properly finished steer that is ready for market. *North American Limousin Association.*

SUMMARY

The beef industry represents a large part of our diet and our economy. The nutritious food that comes from cattle provides nutrients that are difficult to obtain from other food. Our vast continent provides an ideal environment for the production of beef cattle. Often these cattle make use of feed stuff that would otherwise go to waste. The many phases of the industry provides jobs for millions of people all over the country. The beef industry is a dynamic, growing sector of our country and should remain so for many years to come.

Review Exercises

TRUE OR FALSE

1. Each year, Americans buy more beef than fresh poultry, pork, and seafood combined. ________.
2. One disadvantage of beef is that it is low in Vitamin B_{12}. ________.
3. The average ranch in the United States runs about 1000 head of cattle. ________.
4. The three classifications of beef breeds are British, Continental, and Zebu. ________.
5. The Chianina is an example of a Zebu breed. ________.
6. In most areas of the south, calves are born in the late fall. ________.
7. The largest number of cow calf operations are located in the Midwest. ________.
8. In a cow calf operation, calves are usually sold at weaning. ________.
9. Cattle buyers usually prefer calves that have been castrated and vaccinated and that are in good enough condition to move to a new environment. ________.
10. In order for an animal to start depositing fat in the right places, the animal must have matured or stopped growing. ________.

FILL IN THE BLANKS

1. Food from animals supplies about 88 percent of the ________ in our diets because this nutrient is very difficult to obtain from plant sources.
2. The British breeds include ________, ________, and the ________.
3. The Charolais and the Limousin are examples of ________ breeds.
4. All zebu breeds are scientifically classified as ________, a separate species from the traditional *Bos taurus* of the other breeds.
5. Brahman cattle have been used as a basis for the development of several breeds such as ________.
6. The beef industry is divided into four major segments: ________, ________, ________, ________.
7. Cows in a cow-calf operation are fed primarily ________.
8. Calves are usually weaned at a weight range of about ________ pounds.
9. When cattle are slaughtered, they are generally around ________ months of age and weigh about ________ pounds.
10. The largest of the beef breeds, the________, can reach weights of up to 4000 pounds.

DISCUSSION QUESTIONS

1. Why is the United States an excellent location for raising livestock?
2. What reasons are there for using agricultural land to produce beef?
3. Discuss the nutritional value of beef in the diet.
4. List and describe the four major segments of the beef industry.
5. Why are most feedlots in the United States located in the Midwest?
6. Explain how the environment determines where animals are produced.

STUDENT LEARNING ACTIVITIES

1. For the period of one week, keep a list of the amounts of all the different meat eaten by your family. Which type of meat does your family eat the most? Ask your parents to tell you the reasons why they buy the type meat they do.
2. Determine which breed of cattle would be most appropriate for your area. Give several reasons for your choice.
3. If you became a cattle producer what type of operation (feeder calves, stockers, feedlot, purebred, etc.) would you prefer? Give the reasons for your choice.
4. Conduct an internet search for information on a particular breed of cattle. Report to the class.

Chapter 4

The Dairy Industry

Student Objectives in Basic Science

As a result of studying this chapter, you should be able to

1. Describe the process by which milk is produced.
2. Identify the hormones that control lactation.
3. Describe the composition of milk.
4. Explain the process of pasteurization.
5. Describe the biological processes used to produce cheese.

Student Objectives in Agricultural Science

As a result of studying this chapter, you should be able to

1. Identify the major areas of dairy production in the United States.
2. Explain how the producer uses the reproductive process to maintain milk production.
3. Trace the steps used to milk cows in the modern dairy.
4. List the uses made of milk.
5. Tell how milk is processed and marketed.
6. Explain how cheese is made.

Key Terms

yogurt
balanced ration
silage
linear evaluation
heifers
embryo transplant
colostrum
antibodies
alveoli
prolactin
lumen
lobule
tertiary ducts
gland cistern
sphincter muscle
teat
oxytocin
pituitary gland
letdown process
epinephrine
milking parlors
stanchion
mastitis
specific gravity
homogenization
homogenized milk
pasteurization
starter culture
fermentation
enzyme
rennet
curd
whey

The dairy industry is a large component of American agriculture. The sales of dairy products account for about 13 percent of all receipts for farm commodities. The dairy industry is different from other segments of animal agriculture in that the product harvested is intended by nature for no other purpose than to be used as food. Cows are raised and cared for in order to obtain milk that is produced as food for young calves, Figure 4–1. As indicated in an earlier chapter, scientific research has advanced dairy cows to the point where they can produce many times more milk than is needed for calves.

Milk is often described as nature's most perfect food because of its nutritive value. Although milk is 87 percent water, the other 13 percent consists of solids containing proteins, carbohydrates, and water-soluble vitamins and minerals. Because of the nutritive value and rich flavors, Americans consume large quantities of dairy products. Each year on the average we each consume 28.8 gallons of milk, 23.7 pounds of cheese, 16 pounds of ice cream, 4.3 pounds of butter, and 4.3 pounds of **yogurt**. This adds up to a lot of milk production. In addition, milk comes from the cow as a processed food and needs very little additional processing.

Figure 4-1. Cows produce milk as food for their young. *Cooperative Extension Service, The University of Georgia.*

Milk is produced and processed in every state in this country. The five leading milk-producing states are Wisconsin, California, New York, Minnesota, and Pennsylvania, Figure 4–2. These five states produce more milk each year than all of the other states combined. Unlike the meat industry, the dairy industry relies more on forage than grain to produce a product. These states produce a lot of forage. They also have high concentrations of population in large cities located in the states.

About 85–90 percent of dairy cattle in the United States are Holstein, Figure 4–3. These large, docile animals with the familiar black and white markings give a larger amount of milk with a smaller amount of milk fat than other breeds. The lower milk fat was once considered to be a disadvantage, but is now considered to be an advantage because of modern consumer demand for low fat and skim milk.

FEEDING

In the past, dairy cows were generally kept on pastures where they could make use of grass, which is converted into milk, Figure 4–4. However, the modern trend is for large dairies to keep cows in lots or barns and feed the animals a **balanced ration**. One of the main feeds of dairy cattle is **silage**, Figure 4–5. Silage is corn, grain sorghum, or other forage that is chopped—stalk and all—while the plant is green and growing. The chopped silage is then placed in a silo or ground bunker where it undergoes a fermenting process. This means that while the green chopped silage is stored, a chemical process takes place in which complex compounds in the forage are broken down into simpler compounds. This helps preserve the feed and maintains the palatability or eating quality of the feed. The feeding of silage is timed so that the milk from the cows will not

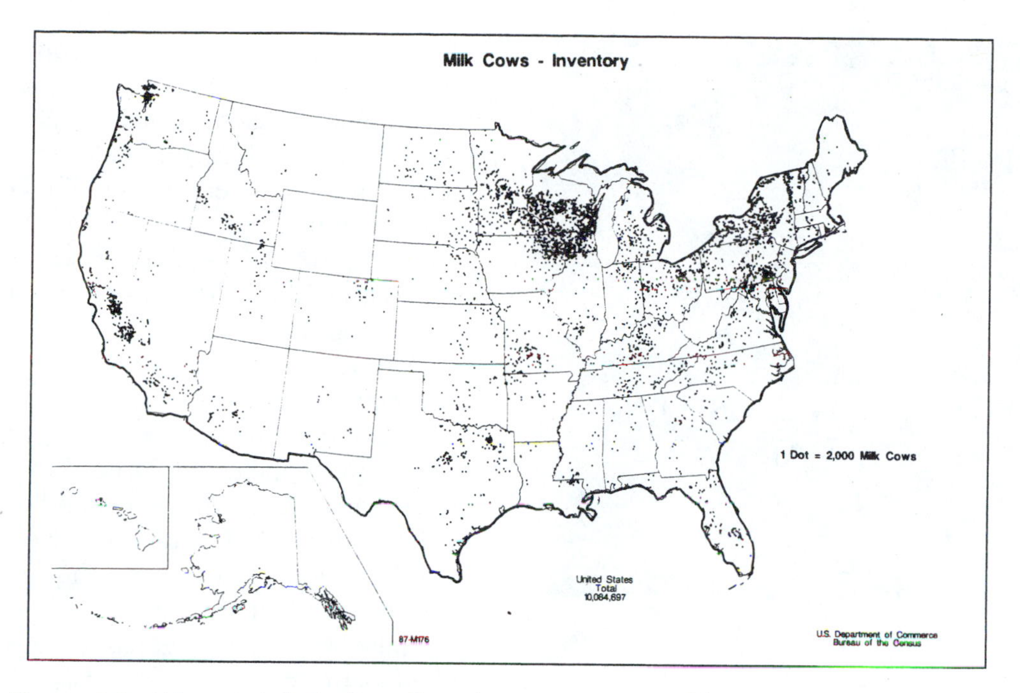

Figure 4-2. Wisconsin is the leading milk-producing state. *Bureau of the Census.*

have an off flavor. Off flavors can occur if the silage is fed to the animals too close to the time they are milked.

Gestation

Milk is the food produced for the feeding of the young. In order to maintain the production of milk, the cows must go through the gestation process and give birth each year. Artificial insemination is widely used to breed dairy cows. Superior sires can be selected at a minimal cost, and the producer does not have to maintain bulls for breeding.

Figure 4-3. Most of the dairy cattle in the United States are Holstein. *James Strawser, Cooperative Extension Service, The University of Georgia.*

Figure 4-4. Dairy cattle are generally kept on pastures. *James Strawser, Cooperative Extension Service, The University of Georgia.*

Figure 4-5. The main feed of dairy cattle is silage. *Gehl Co.*

The Holstein Association conducts a program called **linear evaluation**. An association representative visits the operation and visually evaluates the cows. The representatives are highly trained and competent individuals who give a thorough evaluation of the animals. Certain traits of each animal are given a score based on the ideal. A computerized system can then tell the producer which bull is best to use in breeding the cows, Figure 4–6. Through this system, a producer can make rapid gains in the production of the herd by using the offspring as replacement **heifers**. If a producer wishes to make even greater advances, the use of **embryo transplant** is an option.

Once the calves are born, they are allowed to remain with the cow for one to two days and are then taken from the mother and raised separately, Figure 4–7. The female calves are often raised as replacements and the male calves raised and sold for slaughter.

Milk from a cow that has just given birth is called **colostrum**. Colostrum is a milk that contains a concentration of **antibodies** that are passed to the young from the mother. Since the young calf can only absorb these antibodies during the first twenty-four hours of life, it is

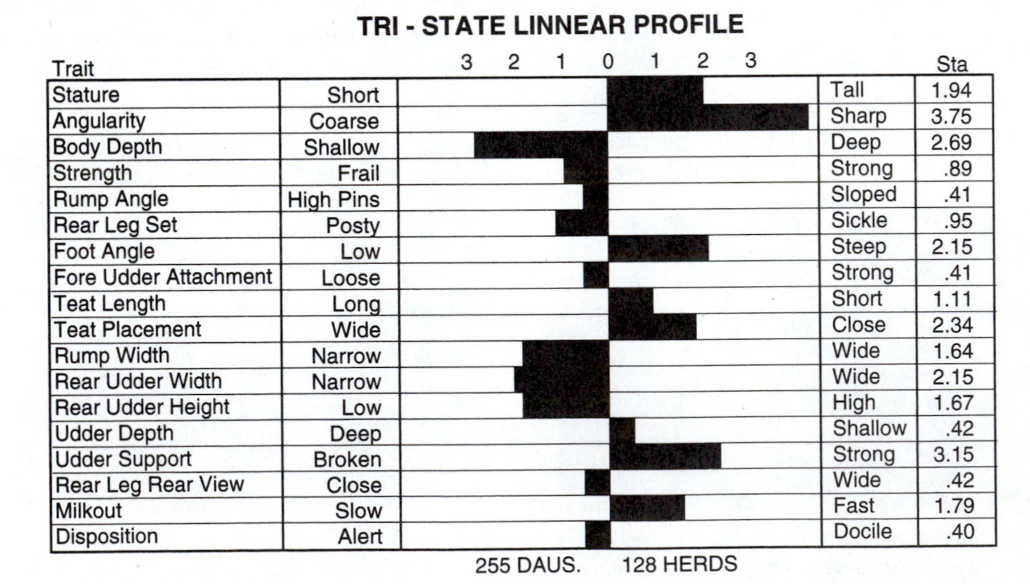

TRI - STATE LINNEAR PROFILE

Trait		3 2 1 0 1 2 3		Sta
Stature	Short		Tall	1.94
Angularity	Coarse		Sharp	3.75
Body Depth	Shallow		Deep	2.69
Strength	Frail		Strong	.89
Rump Angle	High Pins		Sloped	.41
Rear Leg Set	Posty		Sickle	.95
Foot Angle	Low		Steep	2.15
Fore Udder Attachment	Loose		Strong	.41
Teat Length	Long		Short	1.11
Teat Placement	Wide		Close	2.34
Rump Width	Narrow		Wide	1.64
Rear Udder Width	Narrow		Wide	2.15
Rear Udder Height	Low		High	1.67
Udder Depth	Deep		Shallow	.42
Udder Support	Broken		Strong	3.15
Rear Leg Rear View	Close		Wide	.42
Milkout	Slow		Fast	1.79
Disposition	Alert		Docile	.40

255 DAUS. 128 HERDS

Figure 4-6. A linear evaluation profile of a Holstein bull. Note that this bull has sired 255 daughters. *Tri-State Breeders.*

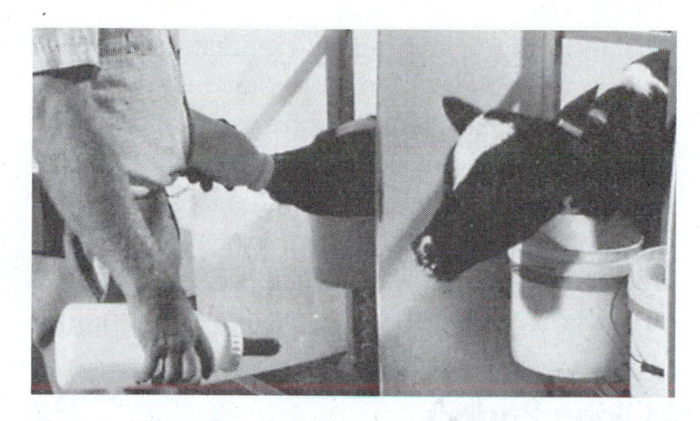

Figure 4-7 When the calves are born, they are weaned and raised on a bottle. *James Strawser, Cooperative Extension Service, The University of Georgia.*

important that the calf be allowed to suckle often during that period. Also, the milk is not generally considered fit for human consumption, so it is not allowed to enter the milk designated for market.

Milk Production

Milk is produced in the udder of the cow in small clusters of grapelike structures called **alveoli**. Blood from the cow circulates through the udder. The alveoli take raw materials from the bloodstream and synthesize these materials into milk, Figure 4–8. Three hundred to five hundred pounds of blood are circulated through the udder for every pound of milk produced.

A hormone called **prolactin** causes the aveoli to begin to secrete milk as a cow nears the time of giving birth. As long as the cow is milked or the calf nurses, prolactin stimulates the aveoli to produce milk. The longer the period from birth, the less prolactin is produced. Over a period of time milk production is decreased, so the cow is bred again to restart the process.

As milk is secreted by the alveoli, it is drained into the **lumen**—or hollow cavity—in the alveoli, Figure 4-8.

The lumens (or *lumina*) are connected to the stem that connects the cluster of alveoli together. This cluster is called the **lobule**. The lobule contains ducts—called the **tertiary ducts**—that drain into larger ducts which carry the milk to an area called the **gland cistern** where the milk is stored. A circular muscle known as a **sphincter muscle** prevents the milk from leaking into the **teat**, Figure 4–9.

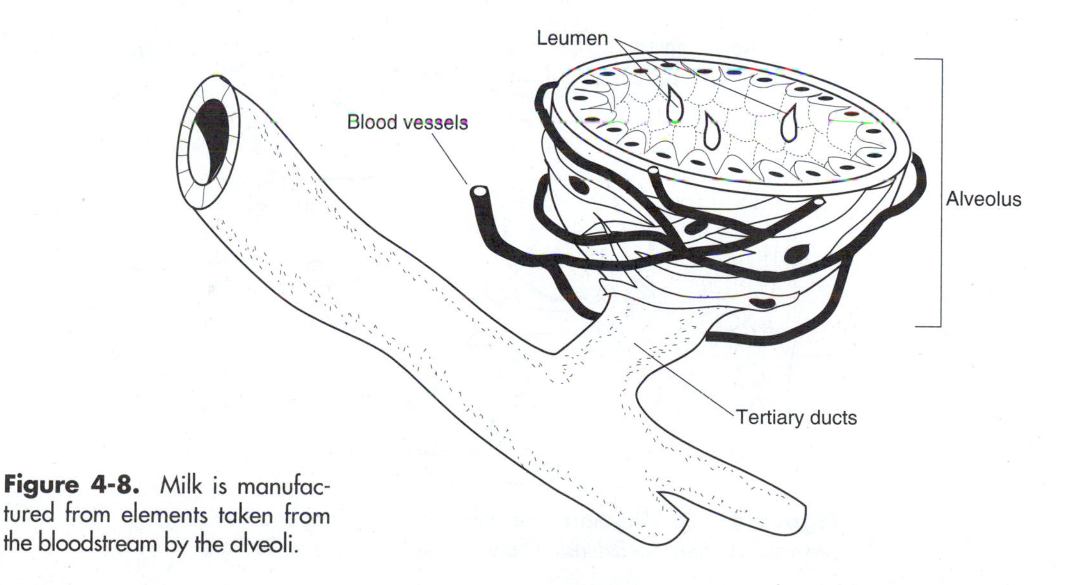

Figure 4-8. Milk is manufactured from elements taken from the bloodstream by the alveoli.

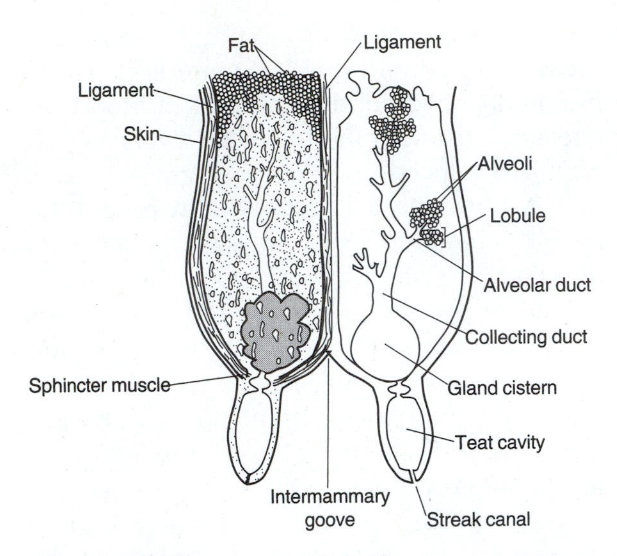

Figure 4-9. Cross section of a udder.

The Letdown Process

As the mother prepares to be milked or to nurse, a hormone called **oxytocin** is released by the **pituitary gland** into the bloodstream. Oxytocin causes the alveoli to release milk into the ducts and cisterns, and it causes the sphincter muscle in the teat to relax. The teat is relatively hollow and allows the milk to pass out as the calf sucks or the milking machine pulsates. The release of oxytocin is caused by stimuli such as a calf rubbing the cow, the washing of the udder prior to milking, or other pleasant stimuli associated with milking. This is called the **letdown process,** Figure 4–10.

Milking Parlors

If the animal becomes frightened or upset, a hormone called **epinephrine** is released that inhibits milk from being let down. For this reason, it is essential that the milking area be clean and comfortable for the cows. Milkers must handle the cows as gently as possible to prevent them from becoming upset. Most milking areas (called **milking parlors**) are designed for easy handling of the cows and for the cows' comfort, Figure 4–11. Milking parlors are designed so that the cow can enter a **stanchion** where the cows stand while being milked. In some modern dairies, a computer chip in a tag

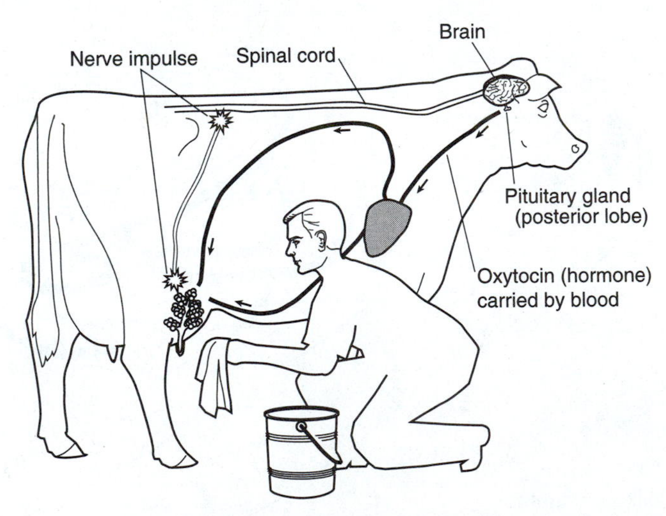

Figure 4-10. The hormone oxytocin stimulates the letdown process. *Cooperative Extension Service, The University of Georgia.*

Figure 4-11. Milking parlors are designed for efficiency and for the cows' comfort. *James Strawser, Cooperative Extension Service, The University of Georgia.*

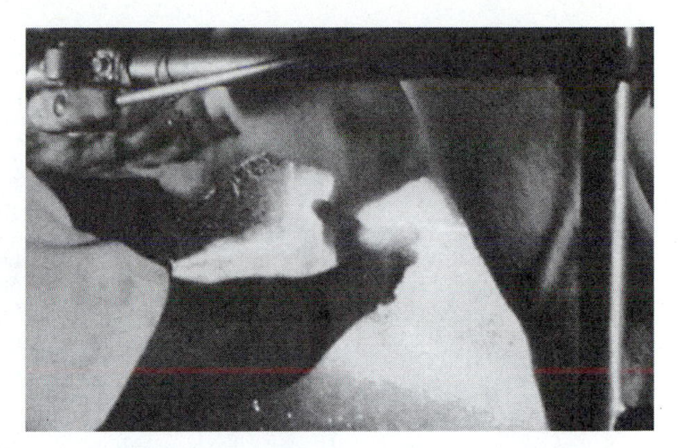

Figure 4-13. The udder is washed and dried prior to milking. *James Strawser, Cooperative Extension Service, The University of Georgia.*

around the cow's neck activates the dumping of the cow's ration into her trough as she enters a stall. The computer is programmed to recognize each individual cow by the chip around her neck, and it gives her the specific amount of ration designed for her.

A common type of arrangement in the milking parlor is the herringbone design. In this design, the cattle stanchions are arranged side-by-side at an angle resembling the pattern of the rib bones on the skeleton of a herring fish, Figure 4–12. The milkers work in an area below the cows so that they don't have to bend in order to place the milkers on the udders of the cows.

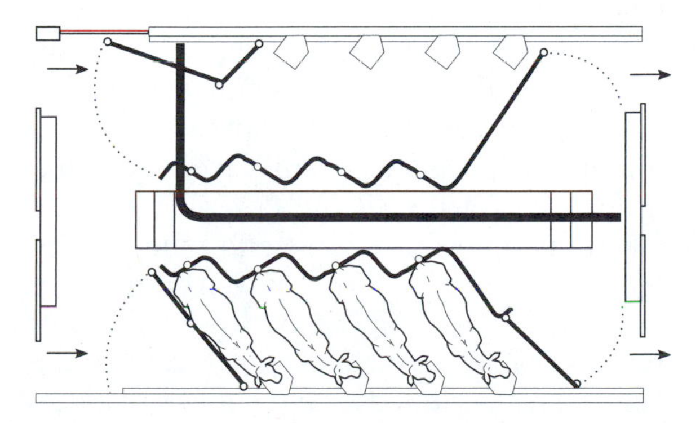

Figure 4-12. The herringbone design is a popular design for milking parlors.

Modern parlors and lots where the cows are kept are designed with the comfort and safety of the cow in mind. They contain such items as mats for the cows to lay on and/or rubber feed bins that prevent injury to the cow. As the cow comes into the parlor and the feed is dropped into a trough in front of her, the milker manually milks a small amount of milk into a cup called a strip cup. This procedure serves two purposes. First, that the milker can check for a disease called **mastitis** which is caused by injuries to the udder. Symptoms of mastitis are lumps or blood that come out in the milk. If evidence of mastitis is found, the cow is moved aside where she can be treated and her milk is not used. Second, stripping the first two to three squirts of milk removes milk that may have a high bacterial count. This milk may have a higher bacterial count because it is near the teat opening and more exposed to bacteria from the outside world.

The udder is then washed using a warm water solution, and dried thoroughly, Figure 4–13. Washing and massaging the udder helps begin the letdown process in the cow. The teat cups are then attached and the milking begins. The cups are lined with a soft material that is attached to a tube. The teat cups fit snugly on the cow and pulsate by means of a vacuum on the lining

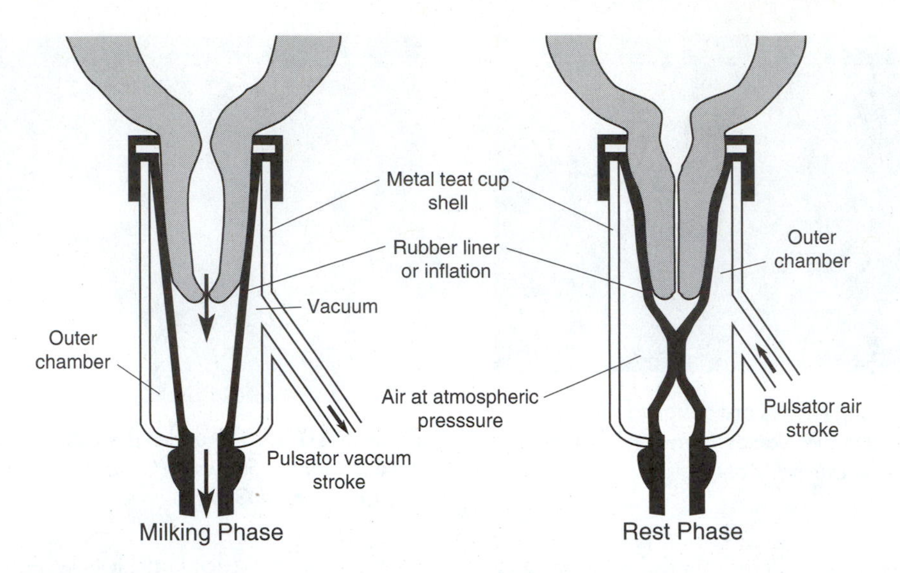

Figure 4-14. Milking is accomplished using a vacuum system that pulsates on the udder. *Cooperative Extension Service, The University of Georgia.*

of the cup to gently draw the milk from the teat, Figures 4–14 and 4–15. The milk is removed in three to six minutes, depending on the individual cow and the amount of milk she gives. Care is given that the cups are left on for the proper amount of time. If they are left on for too little time, the udder will not be milked out; if they are left on for too much time, injury to the udder can result. The teats are then treated with a disinfectant, and the cows are released. The teat cups are kept clean in order to prevent the spread of disease.

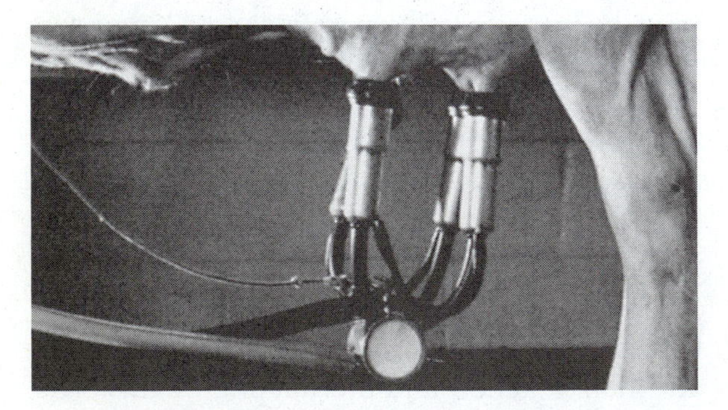

Figure 4-15. The teat cups are placed on the cow's teats. *James Strawser, Cooperative Extension Service, The University of Georgia.*

A good milker times the operation so that as the first cow in the parlor is milked out, he/she will have just attached the teat cups to the last cow to enter the parlor. The milker can then remove the teat cups from the first cow, then the second, and so forth. Some parlors are even designed to rotate so the milker can remain in one position to milk all the cows.

The milk is drawn through the lines and into a holding tank where it is rapidly cooled to about 40°F to prevent the multiplication of bacteria and to prevent the milk from souring, Figure 4–16. After all the cows have been milked, the lines, teat cups, and other equipment are cleaned thoroughly. About every other day the milk is picked up by a tanker truck and hauled to the processing plant. At the plant, the milk is tested for the number of bacteria, drug residue, and the number of somatic cells, Figure 4–17. Somatic cells are the white

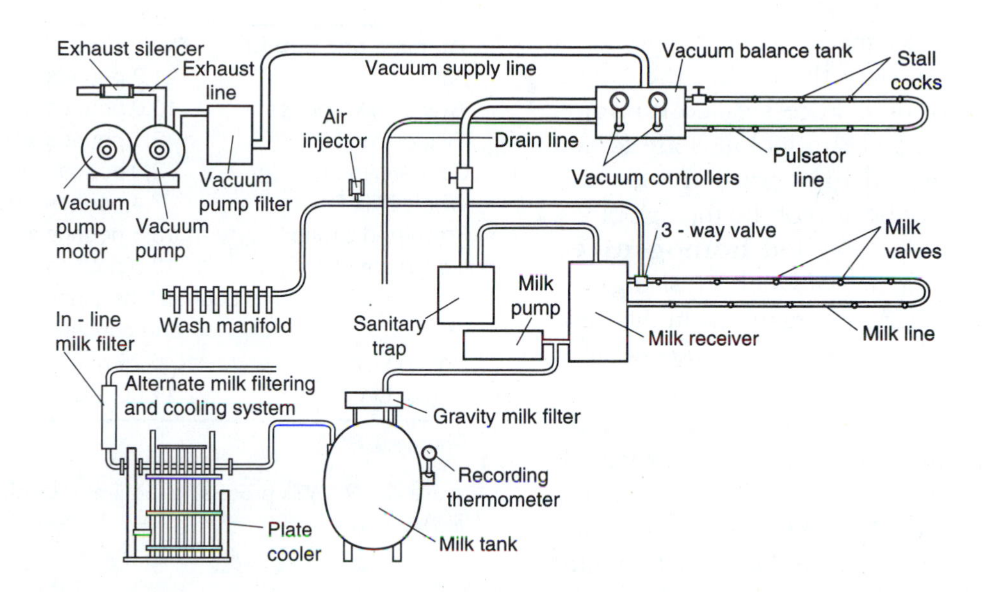

Figure 4-16. The milk is drawn through lines into a holding tank. *Cooperative Extension Service, The University of Georgia.*

blood cells produced by the cow to combat infection; their presence indicates that the cow has an infection.

When the milk arrives at the processing plant, it is thoroughly filtered to remove any foreign particles. The milk is allowed to sit so the cream may be removed from milk that is to be sold as low-fat milk. As consumers are becoming more conscious of the amount of fat in their diets, they want milk that is lower in milkfat than whole milk. In recent years, sales of low-fat and skim milk have increased sharply. Low-fat milk is milk that has had the percentage of milk fat lowered to between .5 percent and 2 percent. Skim or nonfat milk is a milk containing less than .5 percent milk fat. The milk fat that is removed from the milk is then used to make other products, such as ice cream and other cream products.

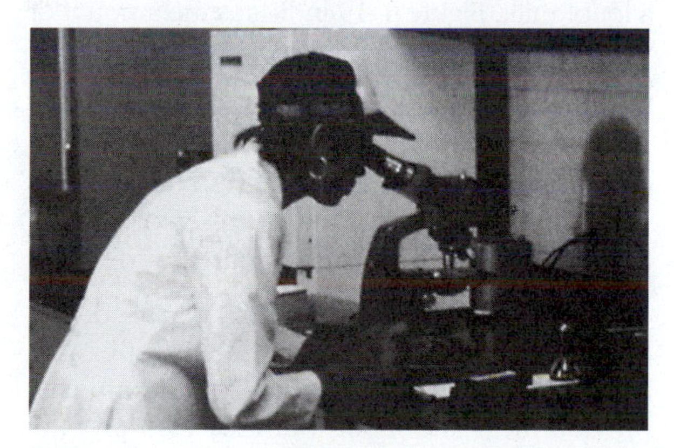

Figure 4-17. At the plant, the milk is tested for bacteria, drug residue, and the number of somatic cells. *James Strawser, Cooperative Extension Service, The University of Georgia.*

Whole milk contains about 4 percent milk fat. The globules of fat are what makes up the cream that floats to the top of raw, unprocessed milk. These globules are larger than the other molecules in the milk, and this size difference causes the cream to separate if the milk is left undisturbed for a few hours. Cream is said to have a lower **specific gravity** than the rest of the milk. Specific gravity refers to the density of a substance compared to the density of water. Substances with a lower specific gravity than water will float on water. Since cream has a lower specific gravity than milk, the cream floats to the top. In a process called **homogenization** the large cream globules are forced through a screen at high pressure and are reduced in size to the size of the milk globules. The processed milk, called **homogenized milk**, will not separate out when left sitting.

To kill any harmful organisms in the milk, the milk is heated and cooled in a process called **pasteurization**. One process of pasteurization raises the temperature of the milk to 145°F for not less than thirty minutes then promptly cools it. An alternative method raises the temperature of the milk to 161°F for fifteen seconds then rapidly cools it. The time and temperature must be precisely controlled in order to protect the nutritive value and the flavor of the milk.

Milk is graded according to the dairy from which it came. Dairies that sell Grade A milk must pass rigid standards for milk production. These involve cleanliness and other conditions under which the milk is produced. Only Grade A milk can be used for the milk that is sold as fluid or beverage milk, Figure 4–18. Milk that is graded as Grade B milk can only be used for processing manufactured dairy products. Since the production of Grade A milk far exceeds the demand for fluid milk, Grade A milk may be used in processing as well. For pricing purposes, the milk is classified as Class I, II, or III. Class I is used for beverage consumption; Class II is used for manufacturing soft products such as ice cream, yogurt, and cottage cheese; Class III is used with Grade B milk in the processing of cheese, butter, and nonfat dry milk. Processing of milk into finished products such as cheese takes a lot of milk. Figure 4–19 indicates the amount of whole milk required to produce various milk products.

Figure 4-18. Only Grade A milk can be sold for beverage milk. *James Strawser, Cooperative Extension Service, The University of Georgia.*

To make one pound	Requires
Butter	21.2 pounds whole milk
Whole Milk Cheese	10.0 pounds whole milk
Evaporated Milk	2.1 pounds whole milk
Condensed Milk	2.3 pounds whole milk
Whole Milk Powder	7.4 pounds whole milk
Powdered Cream	13.5 pounds whole milk
Ice Cream (1 gal.)	12.0 pounds whole milk (15 pounds when including butter and concentrated milks)
Cottage Cheese	6.25 pounds skim milk
Nonfat Dry Milk	11.00 pounds skim milk

Source: USDA

Figure 4-19. Milk processing requires a lot of fluid milk. *USDA.*

Dairy Goats and Sheep

All mammals produce milk for their young. Various cultures in the world use different animals as a source of milk for food. For example, desert nomads use the milk of camels for food. Not only do these versatile animals provide meat and labor, but they also provide milk for the people. In fact, a camel can thrive and produce milk in the harsh desert environment where a milk cow could not survive.

Figure 4-20. In many parts of the world, goats are an important source of milk. The Saanen is a popular breed. *American Goat Association.*

Likewise the Mongolians use horse milk as a source of food. They make yogurt and a fermented drink from the milk of the mares they keep to ride and to do work.

Other than milk cows, the animal that is most widely used to supply milk for human consumption is the milk goat, Figure 4-20. In poor or developing countries, dairy goats are a very important source of food. The animals can survive and produce milk on forage that is much lower in quality than the forage necessary to sustain dairy cows. Most of the world's goat milk is produced in Africa and Asia.

In the United States there are over 129,000 dairy goats, Figure 4–21. Some of these goats are in large herds, but most are in small herds owned by hobbyists. Most of the milk produced is for home consumption, Figure 4–22. Goat milk is very nutritious and is comparable to cow's milk. Some authorities claim that goat milk is easier to digest than cow's milk. Cheese, yogurt, and cottage cheese are made from dairy goat milk. In many parts of the world, cakes of goat cheese that were made by the producers can be bought in the local markets.

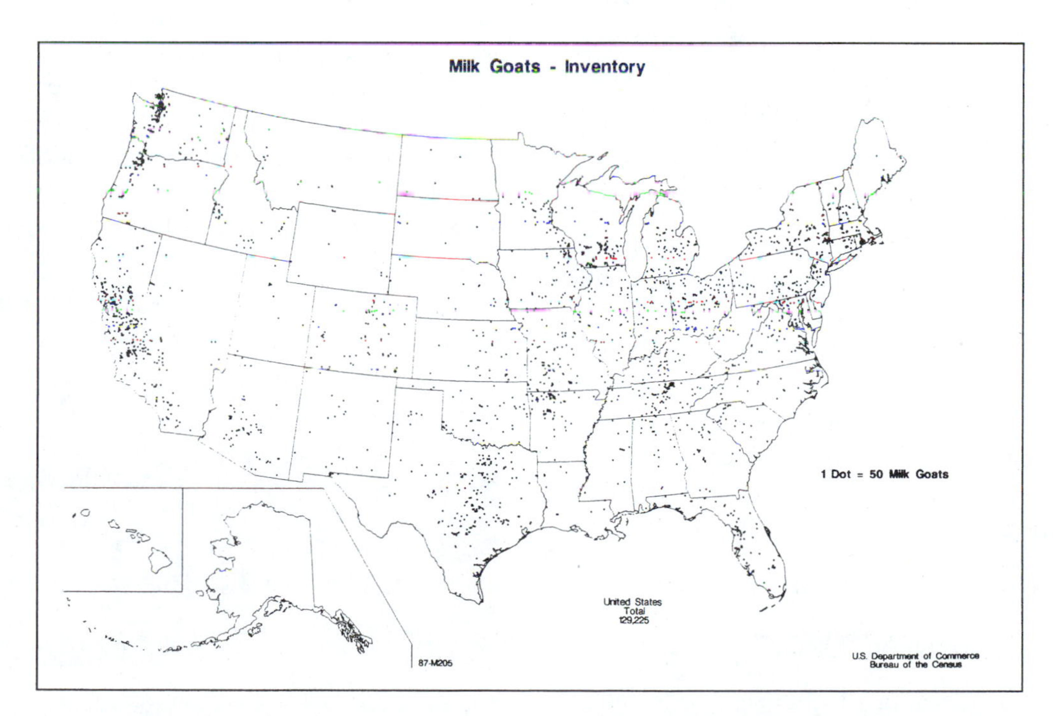

Figure 4-21. There are milk goats in almost every state of the United States. *Bureau of the Census.*

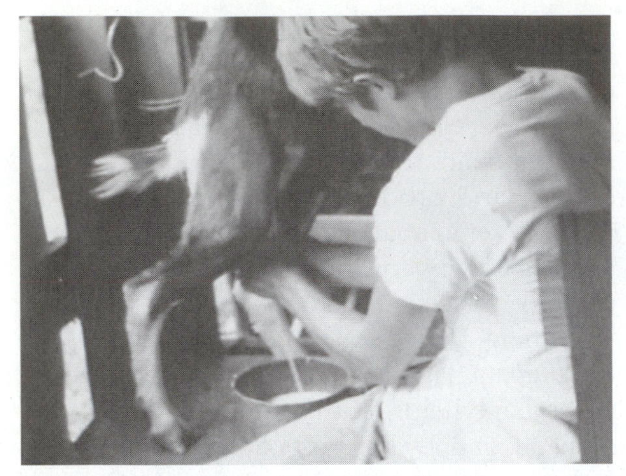

Figure 4-22. In this country, most milk goats are milked for home use. *Rick Jones, Cooperative Extension Service, The University of Georgia.*

In many parts of the world, sheep are also an important source of milk; over 100 million ewes are milked each year. Dairy sheep are milked in Europe, North Africa, the Middle East, and Asia. Although there are a few sheep dairies in the United States, the milking of sheep is not a large industry in this country. Sheep milk is used mostly in the making of cheese. Ewe's milk contains a much higher percentage of solids (18 percent) than does cow's milk (13 percent). In addition, sheep's milk has twice the fat content and 40 percent more protein than cow's milk. This makes the manufacture of cheese and other products from ewe's milk more efficient than from cow's milk. In other words more cheese can be made from a gallon of ewe's milk than from a gallon of cow's milk. Many of the world's best-tasting cheeses are made from the milk of sheep. For instance, the Roquefort cheese that is such a popular ingredient of salad dressings is made from sheep's milk.

Cheese Manufacturing

The processing of cheese is one of the oldest of all the ways humans process food. The practice goes back thousands of years and was found in many ancient cultures. Legend has it that cheese was discovered in the deserts of the Middle East when a nomad transported milk in a bag made from a calf's stomach. The bag was thrown over a camel, and as the animal walked along, the milk sloshed and churned in the bag until the solids in the milk were separated from the liquids. This was a crude means of obtaining cheese.

Today the world consumption of cheese continues to grow and to be a large part of the diet of people in many countries. In the United States the yearly per capita consumption of cheese is almost twenty-five pounds. The manufacture of cheese accounts for almost one-third of all the milk used. Cheese stores very easily and is a highly nutritious food that is high in protein content.

There are hundreds of different types of cheese. Although the differences may come about as a result of the different types of milk (cow, goat, and sheep), most differences are the result of the variations in the processing. The process begins with processed milk that has been pasteurized to prevent the multiplication of harmful bacteria. The milk is placed in a large vat where a bacteria culture is added, Figure 4–23.

Figure 4-23. Milk is placed in large vats and a starter culture is added. *Tillamook County Creamery Association.*

Figure 4-24. The solid mass that results when the liquid is drained off is called the curd. *Tillamook County Creamery Association.*

Figure 4-25. The final step is the wrapping and storage of the cheese. *Tillamook County Creamery Association.*

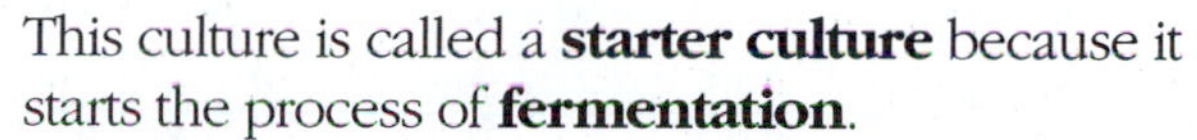

This culture is called a **starter culture** because it starts the process of **fermentation**.

Fermentation is the process that changes sugars to acids. These acids cause the proteins in the milk to coagulate, or form a solid. To further the process, an **enzyme** called **rennet** (rennin) is added. (An enzyme is a substance that speeds up or stimulates a chemical process.) Rennet is obtained from the stomachs of calves. (Remember, the discovery of cheese came about as a result of milk in a bag made from a calf's stomach.) During this step, large paddles turn the milk to ensure that the starter bacteria and the rennet are evenly distributed.

The solid resulting from this step is called **curd**, Figure 4–24. The liquid that is drained off is called **whey**. The curd is cut into small cubes by stainless steel wire knives that are passed through the mass of the curd. The purpose is to increase the surface area of the curd in order to allow the drainage of the whey. After the whey is drained off, the curd sits until it forms a solid mass again. The curd is then heated, causing it to contract and further expel the whey. The amount of heat and length of time the cheese is heated depend on the type of cheese being made. The cheese is salted and pressed into a metal form or a cloth bag.

The final step in cheese making is the curing or ripening of the cheese. The cheese is placed in an environment that is controlled for temperature and humidity, the specific conditions will vary with the type of cheese that is being made. During this time, enzymes produced from the starter bacteria bring about changes in the flavor, texture, and appearance of the cheese. The cheese is packaged in a coating of paraffin or is wrapped in cloth or plastic, Figure 4–25.

SUMMARY

The dairy industry is almost as old as civilization. Milk and milk products have always been an important part of human diets. Scientific research has brought about many changes in the production, processing, and storing of these products. Demand will almost certainly remain strong in the future for fluid milk, cheese, yogurt, ice cream, and all of the other products made from milk.

Review Exercises

TRUE OR FALSE

1. Scientific research has advanced dairy cows to the point where they can produce many times more milk than is needed to feed their calves. ______.
2. The five leading milk-producing states produce more milk each year than all the other countries in the world. ______.
3. Silage is considered to be a "safe" feed for dairy cows since it does not produce an off flavor in milk. ______.
4. Milk from a cow that has just given birth is not usually fit for human consumption. ______.
5. As long as the cow is milked or the calf nurses, prolactin stimulates the aveoli to produce milk. ______.
6. A hormone called epinephrine stimulates the release, or letdown, process. ______.
7. Washing and massaging the udder help begin the letdown process in the cow. ______.
8. Milk is drawn from the cow to a holding tank where it is rapidly cooled, then frozen. ______.
9. Pasteurization refers to the process of heating and cooling milk to kill any harmful organisms. ______.
10. Both Grade A and Grade B milk can be used for fluid or beverage milk. ______.
11. Milk cows are second only to goats for the production of milk for human consumption. ______.
12. Sheep milk is used mainly in the making of cheese in the United States. ______.
13. Cheese processing is a relatively recent development. ______.
14. Most differences in cheese flavor and texture come about through variations in the processing. ______.
15. An enzyme is a substance that slows down or stops a chemical process. ______.

FILL IN THE BLANKS

1. Milk is __________ water with 13 percent made up of __________________________, __________________________, and water-soluble __________________________ and __________________________.
2. Silage is ____________, grain ____________, or other ______________ that is chopped ______________ and all while the plant is green and ______________.
3. Artificial ____________________ is widely used to breed __________________ cows because superior ________________ can be selected at a ____________________ cost, and the producer does not have to maintain ____________ for ____________________.
4. Milk from a cow that has just given birth is called __________________________ and contains a concentration of ______________ that are passed to the ________________ from the ______________.
5. Oxytocin causes the __________ to release _______ into the ducts and ______________ and causes the __________________ muscle in the ____________________ to ______________.
6. In some modern dairies, a ______________ chip in a __________________ around the cow's __________________ activates the dumping of the cow's ration into her __________________ as she enters the __________________.

7. Mastitis symptoms include ____________ or ______________ that come out with the milk.
8. At the processing plant, milk is tested for the number of ____________________, ____________________ residue, and the number of ______________ cells (or ______________ blood cells).
9. Low-fat milk has between _________ and ___________ milk fat while skim or nonfat milk contains less than ___________ milk fat.
10. In homogenized milk the large __________ ______________ are forced through a ______________ at high ______________ and are reduced in ________________ to the size of the milk ________________.
11. Goats can survive and produce __________ on __________ that is much lower in ____________ than the forage necessary to sustain ____________ ____________.
12. Dairy sheep are milked in ____________, North ______________, the ______________ ______________, and ______________.
13. Ewe's milk contains a higher percentage of solids (__________) than cow's milk, twice the __________ content, and __________ more protein.
14. Fermentation is the process that changes ____________ to ____________ that cause the proteins in the milk to ______________ or to form into a ______________.
15. The amount of __________ and the length of ____________ the cheese is ____________ depend on the ____________ of cheese being made.

DISCUSSION QUESTIONS

1. In what way is the dairy industry different from other segments of the animal industry?
2. What are the leading states in milk production?
3. Why does a cow have to produce a calf in order to continue producing milk?
4. Why is it important that a calf receive the first milk after birth (colostrum)?
5. List the hormones that control milk production.
6. What is meant by the letdown process?
7. What is mastitis? What causes it?
8. Regarding fat content, what are three categories of milk?
9. What is meant by homogenization? pasteurization?
10. What is the difference between Grade A and Grade B milk?
11. What animals, other than cows, are used to produce milk for human consumption?
12. List the steps in cheese production.

STUDENT LEARNING ACTIVITIES

1. Obtain a cow's udder from a slaughterhouse. Using rubber gloves, dissect the udder and identify the alveoli, the lumen, the gland cistern, and the sphincter muscle.
2. Visit a large grocery store. From the dairy section make a list of all the products that are made from milk. Make a list of all the different types of cheese.
3. Prepare a list of all the processed foods in your home that contain milk. The ingredients should be listed on the food package.

Chapter 5

The Swine Industry

Student Objectives in Basic Science

As a result of studying this chapter, you should be able to

1. Explain why pork is more healthy to eat than it once was.
2. Explain why protein is important in the diet of a growing pig.
3. List the different types of amino acids.
4. Define hybrid vigor or heterosis.

Student Objectives in Agricultural Science

As a result of studying this chapter, you should be able to

1. Explain the importance of the swine industry.
2. Briefly describe the history of the swine industry in the United States.
3. Name the predominant breeds of swine.
4. Distinguish between a dam and a sire breed.
5. Describe the production methods involved with raising swine.
6. Explain the environmental impacts of a large swine operation.

Key Terms

lard
synthetic lines
mother
sire
farrowing
growing operation
finishing operations
climate-controlled houses

The pork industry represents an important and dynamic component of the animal industry in this country. In the past thirty years the number of swine operations in this country has decreased by almost two thirds. Yet the number of hogs slaughtered has actually increased. This is because producers are more efficient and because the size of the operations have increased dramatically. In 1950, there were over three million swine producers in this country. Today there are just over 155,000 producers, Figure 5–1. Over 80 percent of pigs produced are from farms that raise over 1,000 pigs per year. Each year, producers raise almost 100 million hogs that yield over 17 billion pounds of pork. According to the American Pork Producer's Council, this industry supports over 600,000 jobs and is responsible for over $64 billion in total economic activity.

Worldwide we rank second only to China in the number of hogs produced annually. In per capita consumption, we rank 13th with 63.3 pounds of pork consumed per person each year. Denmark leads the world in pork consumption with over 144 pounds consumed per person each year. Bacon, ham, and pork chops have always been popular in the American diet; however, in recent years concern has been raised over the level of fats in pork products. The National Pork Producers Council has been successful in educating consumers on the merits of pork. Although pork was once considered a fatty food, today's leaner pigs produce pork that is relatively lower in fat content and quite nutritious, Figure 5–2. In terms of meat, pork production and consumption rank second only to beef in this country. Pork consumption is distributed throughout

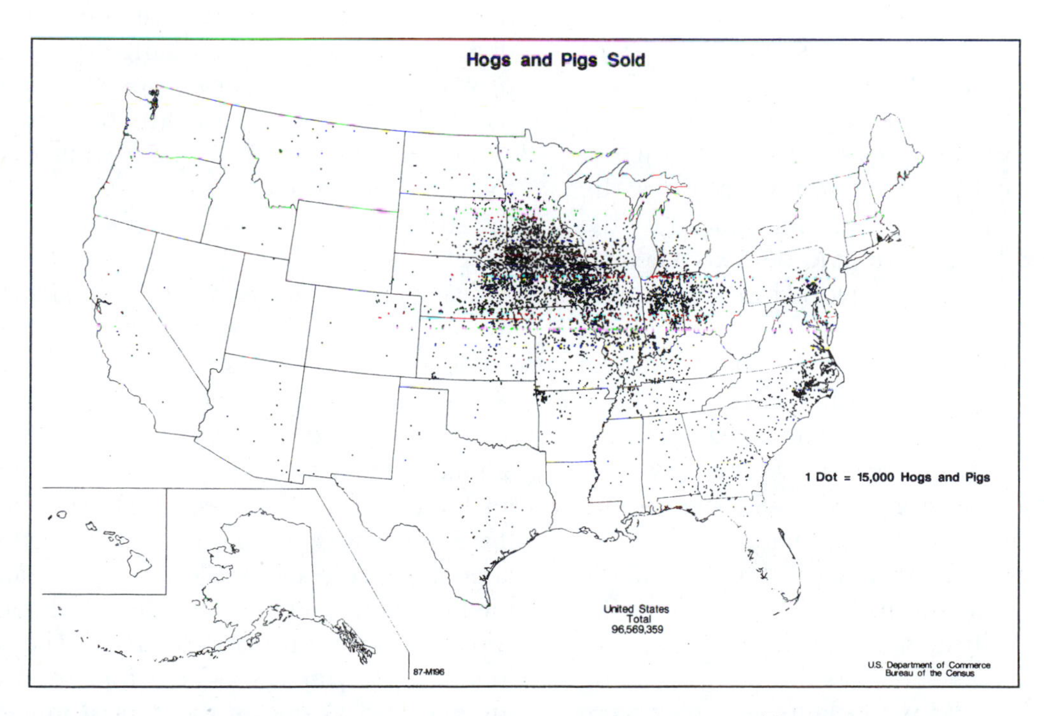

Figure 5-1. Most of the hogs are grown where corn is produced. *Bureau of the Census.*

Figure 5-2. Today's pork is much leaner than the pork produced in the past. *National Pork Producers Council.*

the country, although there are certain populations, such as Moslems and Jewish people, who do not eat pork for religious reasons.

INDUSTRY HISTORY

Pork production has been a part of American agriculture since the earliest Europeans settled in this country, Figure 5–3. Columbus brought pigs on his first voyage to the new world to be used as food for the sailors. The first pigs were introduced by the Spanish explorer, Hernando de Soto when he landed on the coast of Florida in 1539. It was reported that he brought only 13 head, but in a period of only three years, this small herd had grown to over 700 head. Native Americans developed a taste for pork and began to hunt the pigs that escaped captivity. These escaped pigs are the ancestors of the wild pigs prominent in many parts of the country today.

As settlers moved west, they inevitably took pigs with them. These animals easily adapted to differing environmental conditions. They could live off the land by eating acorns, roots, and wild plants, and they were often allowed to roam "free range" through

Figure 5-3. Pigs have been a very important part of agriculture since the beginning of our country. *Progressive Farmer Magazine.*

the woods. As the settlements became more dense, the practice of free range often caused problems with neighbors' crops. Roaming pigs caused such a problem in the colony of Manhattan in New York that a wall was built to keep the pigs out. Even after all these years, the street along this wall is still known as Wall Street.

Pigs provided food for the settlers. They were proficient breeders and each female could produce several offspring each year. When the weather turned cold, pigs were slaughtered and the meat was preserved by smoking and salting. People could eat all winter long on the preserved meat. The fat from the animals was cut into chunks, placed into a large iron kettle and rendered. This meant that the fat was heated until it melted and could be separated from the solid particles. The resulting fat, called **lard**, was kept for use in cooking and also as one of the central ingredients in the making of soap. In fact, until about

1950, the major reason for raising pigs was to obtain fat for lard. With the advent of vegetable oils, lard became less prominent in the American diet, and hogs began to be raised primarily for meat.

At one time in this country, most of the people who lived on farms raised pigs. The animals required relatively little space and fit well into most enterprises as a sideline income. It has been said that hogs have sent more farm youngsters to college than any other enterprise. Most of the feed was raised on the farm and very little had to be bought. Today, many hog producers buy their feed already mixed and delivered to their farms ready to feed.

Since the gestation period is short and there are several pigs born in each litter, the time required to build up a herd of hogs is short compared to many other agricultural animals. For this reason, an operation can be built up in a relatively short time. Also, the type of pigs produced can be changed in less time than with most agricultural animals.

For many years, most of the pork produced in the United States came from the Midwest in the states of Iowa, Illinois, Indiana, Minnesota, and Nebraska. These states produce a large amount of corn, the major grain fed to swine, and remain leaders in the production of pork. However, in recent years larger numbers of pigs are being raised in the South where mild winters help lower the cost of production. In fact, the state of North Carolina is now second only to Iowa in the number of pigs produced.

BREEDS OF SWINE

Although there are not as many breeds of hogs as there are breeds of cattle in this country, there are still several popular breeds. Modern swine are bred to be leaner and more efficient than the swine of several years ago. Efficiency means that they can grow faster, mature at an earlier age on less feed, and have more pigs per litter. Today, most pork producers raise one or more of nine major breeds. These are Yorkshire, Duroc, Hampshire, Landrace, Berkshire, Spotted, Chester White, Poland China, and Pietrain. Increasingly, producers are using what is termed as **synthetic lines** that are derived by crossing these breeds. Several commercial breeding companies develop these lines for use as breeding animals. Almost all of the pigs produced for slaughter in the United States are the result of crossbreeding of purebred or synthetic lines, Figure 5–4. Crossbreeding makes use of a biological phenomenon known as *hybrid vigor* or *heterosis* which results in offspring that are superior to what might be expected of the parents.

Breeds of swine are categorized as **mother** or **sire** breeds. Mother breeds are superior in the number of pigs in a litter, the amount of milk they produce for their young, and their docile temperament. The mother

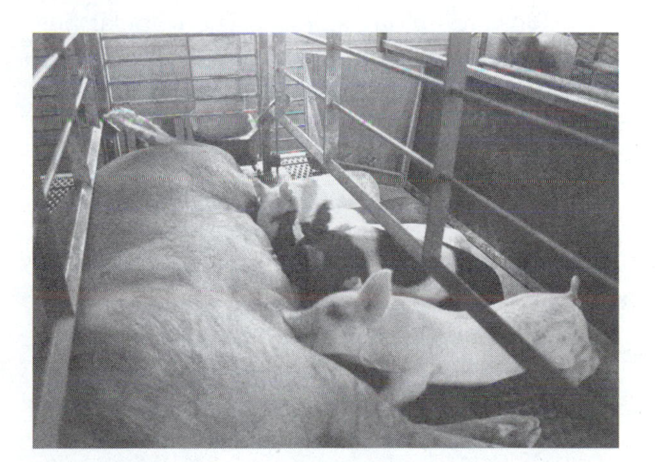

Figure 5-4. Most pigs produced in the United States are the result of cross breeding programs. *James Strawser, Cooperative Extension Service, The University of Georgia.*

Figure 5-5. Yorkshires are considered to be a mother breed because of their large litters and their high level of milk production. *American Yorkshire Club Inc..*

breeds are the white pigs that include Chester White, Landrace, and Yorkshire, Figure 5–5.

The sire breeds, such as the Duroc and the Hampshire, characteristically grow rapidly and produce well-muscled, meaty carcasses. They are also very durable and leaner.

Production Methods

Pigs are unique animals. The popular perception is that they are dirty, stupid animals that constantly overeat. These perceptions are erroneous and, in fact, are opposite the truth. Pigs got the reputation of being dirty animals because, if given the opportunity, they wallow in mud. In fact, pigs wallow because the mud helps keep them cool in hot weather and also helps keep off parasites. If given the room, pigs will use only a certain part of their pens to drop wastes and will keep the rest of their area clean. Pigs are very intelligent animals. They rank among the top of all agricultural animals in overall intelligence. In handling pigs, producers must understand how smart pigs are in order to move them and keep them in their pens.

Pigs are one of the few agricultural animals that will not overeat. Given the proper type of feed, pigs will only consume the amount of feed they need. Cattle, horses, and other animals will overeat to the point where they may become ill. This is not a problem with pigs.

Scientists have studied the unique characteristics of pigs and have successfully used the findings of these studies to design production methods to suit these animals' needs. The type of buildings, health regimens, management procedures, and diets are all designed to help pigs live healthy, comfortable, and productive lives.

Of all animals raised, pigs are the closest to humans in terms of the type of digestive, circulatory, and other systems. Tissue from pig skin is used to replace human skin that has been badly burned. Valves from the hearts of pigs are used as replacements for human heart valves that have worn out or been damaged by disease. Also, pigs are used in many areas of research for products that will eventually be used by humans.

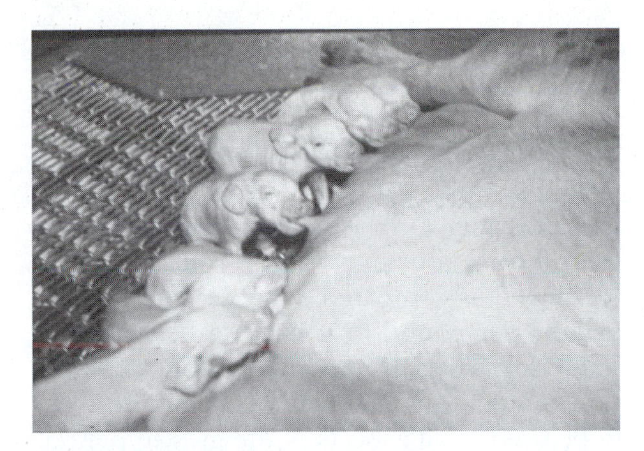

Figure 5-6. Farrowing crates prevent the mother from crushing the piglets when she lies down. *National Pork Producers Council.*

Three phases of the industry are the **farrowing**, **growing**, and **finishing operations**. The three phases can be operated separately or together. Some producers prefer to raise only feeder pigs (pigs that are weaned and sold shortly after weaning), and some prefer to buy feeder pigs and finish them as their only operation. Most pigs are farrowed in **climate-controlled houses** where the mother is kept in a crate to prevent her from injuring the piglets when she lies down, Figure 5–6. Good producers make quite an effort to provide an environment that is clean, dry, and comfortable for both the mother and the piglets.

At farrowing, the sow usually has around nine to ten piglets. At birth, the piglets are dried off and their navel cords are dipped in iodine to prevent infection. Some pigs are born with long sharp teeth called needle teeth. If left alone, these teeth may injure the sow's teats or may injure other piglets. The producer clips the teeth off to prevent these injuries. Also, the newborns are given a shot of supplemental iron that helps improve the oxygen-carrying capacity of their blood. Pigs that are born and raised on the ground usually get enough iron from the soil; however, most pigs are born on slatted, concrete, or raised deck floors and need the supplemental iron.

The pigs are generally weaned from the mother at three to four weeks of age, although some producers may wean the pigs at six weeks or as old as eight weeks. At this time, the pigs usually weigh around 10–15 pounds and are placed in nurseries. These facilities make use of a slotted floor that allows waste material to fall through. This helps keep the floor cleaner and drier. The male (boar) pigs are castrated. In this procedure, the testicles are removed to prevent aggressiveness, avoid pregnant females, and to prevent off-flavored meat when the pigs are slaughtered. All pigs have their tails removed (docked). Pigs kept in confinement operations have a tendency to bite at each other's tails. By removing the tails, the incidents of tail biting is eliminated.

In the nursery, the pigs are fed on a scientifically-balanced diet that provides the proper amount of nutrients needed by the animals at this particular stage of growth, Figure 5–7. As the animals continue to grow, their diet is changed to fit the needs of that particular stage

Figure 5-7. In the nursery, the piglets are fed on a scientifically balanced diet that provides exactly the proper amount of nutrients needed by the animals at a particular stage of growth. *National Pork Producers Council.*

of growth. By the time the pigs are moved out of the nursery at eight to ten weeks of age, they may have been fed five different diets. These diets consist of grain, protein supplements, and milk products. The protein is supplied from a mixture of plant and animal sources. The amount and type of protein required varies as the animals grow.

After weaning, the pigs weigh 40–60 pounds and are placed together with pigs of similar age, size, and sex in a *confinement operation.* This means that the hogs are kept in a pen together rather than running loose on a pasture. Sufficient space is allowed for the pigs to be comfortable and to grow at a fast rate. In this system, pigs are kept comfortable in climate-controlled houses where they are protected from heat, cold, and rain. Although these operations are quite expensive, less labor is needed to care for the pigs. The animals drink from automatic waterers that ensure a constant supply of clean water, Figure 5–8. Feed is supplied from automatic feeders where the animals obtain all the feed they want. In these houses, animals are less likely to pick up parasites or contract diseases. The pens are cleaned and disinfected periodically to help protect the health of the animals.

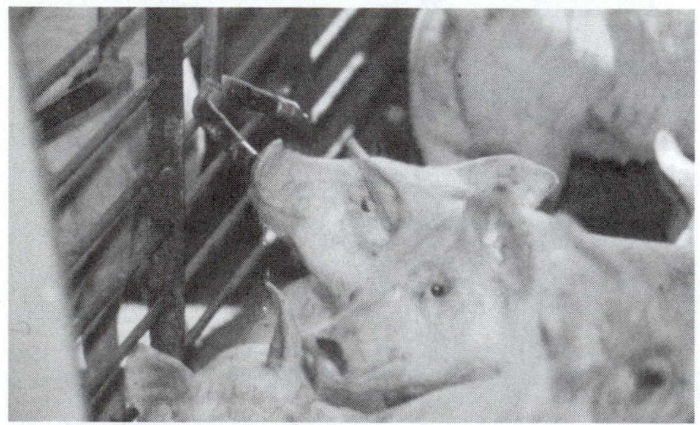

Figure 5-8. Pigs learn to drink from automatic waterers that ensure a constant supply of clean water. *National Pork Producers Council.*

Pigs are said to be more efficient than cattle. This means they put on a pound of body weight with less feed consumed. However, pigs cannot make use of large amounts of roughages like cattle and must be fed on a ration of grain. On the average, pigs will gain a pound for about every five pounds of feed consumed as compared to about nine pounds of feed per pound of gain for a beef animal. This is known as the *feed conversion ratio.*

Scientists have developed different diets to be used in all aspects of the growth period. As with the weaned pigs in the nursery, growing pigs are given different diets as their nutritional needs change in the maturation process. The pigs are sometimes segregated into *barrows* and *gilts.* Gilts are female pigs that have not had a litter of pigs, and barrows are males that have been castrated. By separating the pigs according to sex, the diets can be "fine tuned" to provide even greater efficiency, Figure 5–9.

Pigs are fed a high-protein diet to promote growth and muscle development. As the animals mature, the diet is switched to one with a lower protein content and a higher carbohydrate content. A ration rich in protein is needed in the early stages of growth to build muscle and bones. Without the proper amount of protein, neither muscles, bone, nor internal tissues and organs will develop properly. When the animal approaches maturity, less protein is needed and more carbohydrates are required because the carbohydrates help the animal develop fat as their skeletal and muscular systems mature. Some fat is required in the meat to produce the juiciness and flavor wanted by consumers.

At one time, protein was calculated in terms of the percentage of protein needed in the diet. Today, the building blocks of protein, called amino acids, are used as the basis of

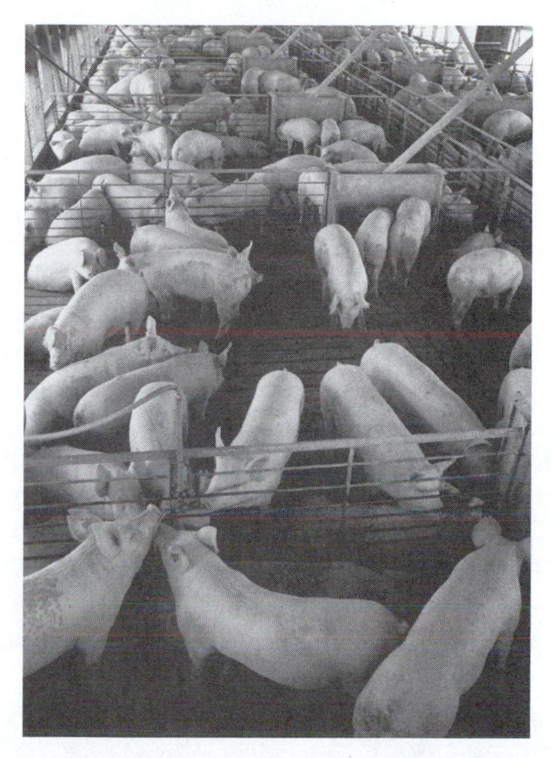

Figure 5-9. Pigs are finished for the market in confinement operations. By separating them according to sex, the diets can be fine-tuned to provide even greater efficiency. *Rick Jones, Cooperative Extension Service, The University of Georgia.*

balancing the feed diet. Amino acids are composed of carbon, hydrogen, oxygen, and nitrogen. Swine need ten types of essential amino acids from the feed they eat. Several other amino acids are synthesized by the animals' bodies from the essential amino acids.

The pigs should be *finished* (reach the proper market weight and condition) at about twenty weeks. Packers like to buy market hogs that weigh in the range of 220 to 260 pounds, Figure 5–10. Most pigs are marketed by directly selling them to the processor, although a few are still marketed through live auctions. When sold directly to the processor, hogs are often sold on *carcass merit.* This means that premium prices are paid for pigs with low amounts of fat and high amounts of muscle.

Figure 5-10. Packers like to buy pigs that weigh 220–260 pounds. *Rick Jones, Cooperative Extension Service, The University of Georgia.*

Environmental Concerns

Strict federal, state, and local laws regulate how and where pigs are raised. Hogs in close confinement can cause problems with odor and manure disposal. The larger the operation, the greater the problem. To dispose of the manure and odor, waste from the finishing pens is flushed into ponds called *lagoons*, Figure 5–11. The building and operation of the lagoons are regulated to ensure that the waste material (manure) does not pose a threat to streams and water supplies. In the lagoons, bacteria help break down the waste materials into a slurry that does not have as bad an odor as untreated manure.

The waste material is periodically pumped from the lagoons and spread on pastures or cropland as fertilizer. This not only

Figure 5-11. Waste from confinement operations is washed into lagoons. *James Strawser, Cooperative Extension Service, The University of Georgia.*

provides a means of disposing of the manure, but also supplies a high-quality, organic fertilizer to crops. This form of waste disposal is a type of recycling of nutrients that helps protect the environment.

SUMMARY

Pigs have been a part of American agriculture from the beginning and still hold a large portion of the agricultural industry. They are relatively efficient, highly intelligent animals that are often wrongly depicted. Modern pork production systems provide comfortable, clean facilities for all phases of the industry. Diets are scientifically balanced to give the animals the nutrients they need. The future of this industry is bright, and pigs will continue to play an important role in agriculture and the diet of Americans.

◆ Review Exercises

TRUE OR FALSE

1. In the past thirty years, the number of hogs slaughtered has decreased. _____.
2. The United States ranks second only to China worldwide in the number of hogs produced annually. _____.
3. In terms of meat, pork consumption ranks second only to beef in this country. _____.
4. The first pigs were probably introduced to the United States by the Vikings. _____.
5. Until about 1950, hogs were raised primarily for bacon. _____.
6. An example of a *mother breed* of swine is the Yorkshire. _____.
7. At farrowing, the sow usually delivers three to four piglets. _____.
8. Packers like to buy market hogs that weigh in the range of 220–260 pounds. _____.
9. A major problem with raising hogs is that they frequently overeat. _____.
10. A *gilt* is a castrated male swine. _____.

FILL IN THE BLANKS

1. The country of __________ ranks first in pork consumption.
2. The biological phenomenon known as ___________ results in offspring that are superior to the parents.
3. Three major mother breeds are the ________, ________, and the ________.
4. The three phases of the swine industry are _________, __________, and _________.
5. Pigs that are weaned and sold shortly after weaning are called _________.
6. Some baby pigs are born with long sharp teeth called ________ _________. These teeth

are clipped, and the newborn pigs are given a shot of supplemental _______.

7. Pigs are said to be more _________ than cattle, meaning that they put on a pound of body weight with less feed consumed.
8. A female that has not yet had a litter of pigs is referred to as a __________.
9. Today, the building blocks of protein, called _______ _______, are used as a basis for balancing the swine diet.
10. To help alleviate environmental problems, waste from finishing pens is flushed into ponds called ________.

DISCUSSION QUESTIONS

1. What are the different characteristics of a sire breed and a mother breed? How can these characteristics be used in a crossbreeding program?
2. List some of the popular misconceptions about hogs. Be sure to tell why these perceptions are untrue.
3. Why is pork considered to be healthier now than in the past?
4. Why are pigs often used in medical research for products that will eventually be used for humans?
5. Describe the three phases of the pork industry.
6. List the advantages and disadvantages of a confinement operation as opposed to a free-range operation.

STUDENT LEARNING ACTIVITIES

1. List all the pork products your family consumes in a month. Be sure to list products such as sausage, bacon, bologna, and other processed meats.
2. Create a list of all the hog operations in your area. Define the type of operation (feeder pig, finishing, purebred) and determine which type is the most popular. Explain why this type is popular in your area.
3. Interview a purebred producer and determine why he/she grows that particular breed.
4. Do an internet search and locate information on a particular breed of swine. Report to the class.

Chapter 6

The Poultry Industry

Student Objectives in Basic Science

As a result of studying this chapter, you should be able to

1. Compare the process of egg development in birds and mammals.
2. Describe the biological processes involved in the production of eggs in birds.
3. Describe how the chick embryo develops in the egg.
4. Relate how nature protects eggs from the environment.
5. Describe the ideal conditions for the production of bacteria.
6. Tell how hatching chicks communicate.

Student Objectives in Agricultural Science

As a result of studying this chapter, you should be able to

1. Summarize why the poultry industry is rapidly growing.
2. Define vertical integration.
3. Describe how broilers are produced in modern operations.
4. Describe how modern hatcheries operate.
5. Describe modern layer operations.
6. Describe modern turkey production.

Key Terms

broiler industry
cannibalism
layers
hybrid
heterosis
hybrid vigor
pigmentation
embryo
ovum
ovary
infundibulum
magnum
cells
mucin
albumen
yolk
chalazae
isthmus
uterus
shell gland
incubation
oxytocin
cloaca
fertilization
sperm
sperm nest
germinal disk
cage operations
metabolism
pullets
molting
candling
muscling

One of the fastest growing segments of the animal industry is that of the poultry industry. Worldwide consumption of poultry is increasing. Chickens, turkeys, ducks, geese, and other birds make up a large portion of the meat diet of people in most countries. In this country, the per capita consumption of broilers is almost 90 pounds. This amount has increased over 200 percent in the past thirty years, Figure 6–1.

Unlike some meats, poultry is generally accepted by most cultures. For instance, the Moslem and Jewish cultures do not eat pork, and Hindu culture does not allow the eating of beef. However, almost all cultures accept poultry as a wholesome meat for human consumption. Developing countries often begin to build a sound agricultural base with poultry. Birds are very efficient users of feed and are easily cared for in countries where a lot of human labor is available. The largest producers of poultry in the world are China, the countries of the former Soviet Union, and the United States.

THE BROILER INDUSTRY

At one time in the history of our country almost all families in rural areas had some type of poultry. Not only did chickens provide the family with fresh eggs but also with fresh meat. Today almost all of the poultry is raised in large operations. The term **broiler industry** refers to the raising of chickens for meat. This industry is concentrated in the Southeast where the mild winters provide an advantage to producers. The leading broiler-producing states are Arkansas, Georgia, and Alabama, Figure 6–2. A broiler is a bird that is grown out to about seven or eight weeks old and is dressed for market.

The vast majority of broilers produced in this country are raised on contract. In a typical grow-out contract, the company agrees to provide the producer with chicks, feed, medications, vaccines, and other supplies. The company also agrees to pay the producer a predetermined price per pound for the broilers produced, and it sometimes gives an added payment to the producer as an incentive for more efficient production. The producer supplies the house, feeding and watering equipment, utilities, litter material, waste disposal and labor. The company is usually *vertically integrated.* This means that the company owns the hatchery, feed mills, processing plants, and distribution centers.

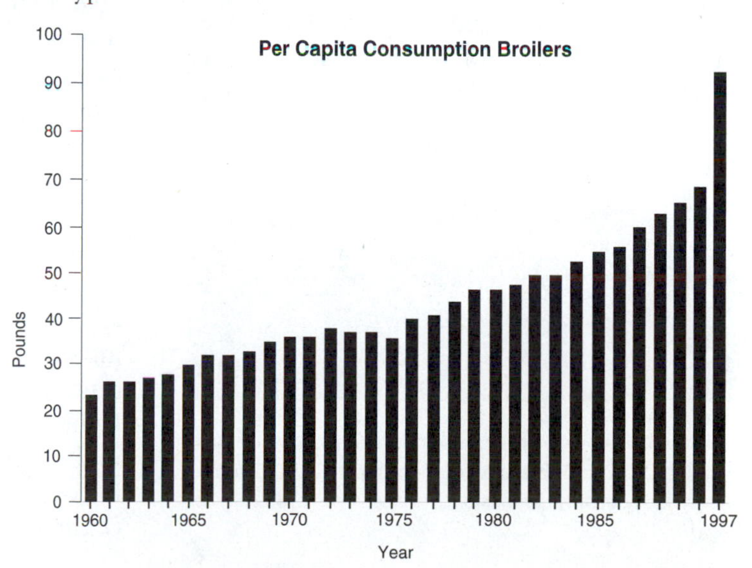

Figure 6-1. The per capita consumption of broilers is almost 70 pounds in the United States. *Charles Strong, Cooperative Extension Service, The University of Georgia.*

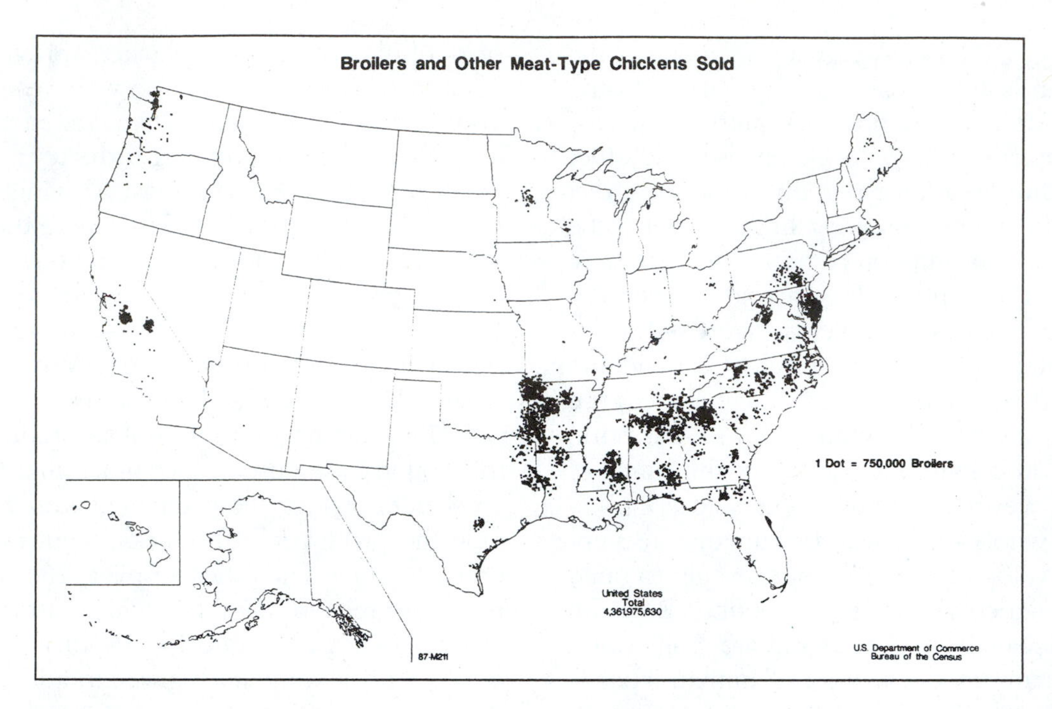

Figure 6-2. Much of the broiler industry is centered in the southeastern part of the United States. *Charles Strong, Cooperative Extension Service, The University of Georgia.*

Broiler Houses

Broilers are raised in large houses where the birds spend almost all of their lives. The broiler houses are designed to provide the birds with a clean, comfortable environment, Figure 6–3. The houses hold from 6 thousand to 40 thousand birds and are built to keep the animals warm in winter and cool in the summer. The houses are insulated and have ventilators that help remove the heat. When the birds are small, heat is provided by brooders that are powered by gas or electricity. As the birds grow larger, there is usually enough body heat generated by all of the birds in a well-insulated house to provide ample heat without any artificial heat.

Figure 6-3. Broilers are raised in large houses that provide a clean, comfortable environment for the birds. *James Strawser, Cooperative Extension Service, The University of Georgia.*

The houses are generally lighted almost around the clock. Research has shown that by leaving the lights on, incidents of **cannibalism** are greatly reduced. Cannibalism is the attacking of birds by other birds in the flock. It can be quite a serious problem if it is not regulated. Lights are routinely turned out for 1 hour each night. This is to help prevent the birds from becoming hyperactive should a power failure occur at night.

Broiler Production

The process of broiler production begins with the production of eggs to be hatched for young broilers. The parents are selected from breeds of chickens that grow rapidly and yield a large amount of breast meat. They are quite different in appearance from hens used only to produce eggs. **Layers** that produce eggs for consumption are selected for their egg-laying capacity—not for the amount of muscle they produce. Most broilers are **hybrid** birds. This means they are the result of the mating of different breeds of chickens to produce the type of meaty birds desired, Figure 6–4. Hens that lay eggs for hatching are bred either naturally or artificially. The resulting crossbred animals are generally healthier and grow faster than purebred animals. This is called **heterosis** or **hybrid vigor**.

Almost all of the broilers produced are white. Birds that are dark in color have spots of color or **pigmentation** left where the feathers were removed after slaughter. These spots do not lower the quality of the meat in any way, but consumers are reluctant to buy chicken with spots on the skin.

Figure 6-4. The bird on the left is the muscular type bred for meat production. The bird on the right is of the type bred for producing eggs.

Egg Production

Eggs produced by poultry serve the same purpose as eggs produced by other agricultural animals—reproduction. Unlike the eggs of most mammals, the eggs of poultry are produced in the body of the female and then expelled from the body. The development of the **embryo** takes place outside of the mother's body. The eggs of most mammals are microscopic in size and unprotected, whereas the egg of a chicken can weigh several ounces and is encased inside a hard shell.

The egg production process begins as it does in mammals with the release of the **ovum** from the **ovary**. The follicle in the ovary ruptures and the ovum is released. The ovum falls into a funnel-shaped structure called the **infundibulum** that surrounds the ovum and holds it for about twenty minutes.

If the hen has mated or has been artificially inseminated, the ovum will be fertilized here, Figure 6–5. The egg then moves into a tubelike tract called the **magnum**. **Cells** in the magnum secrete a substance called **mucin** that develops

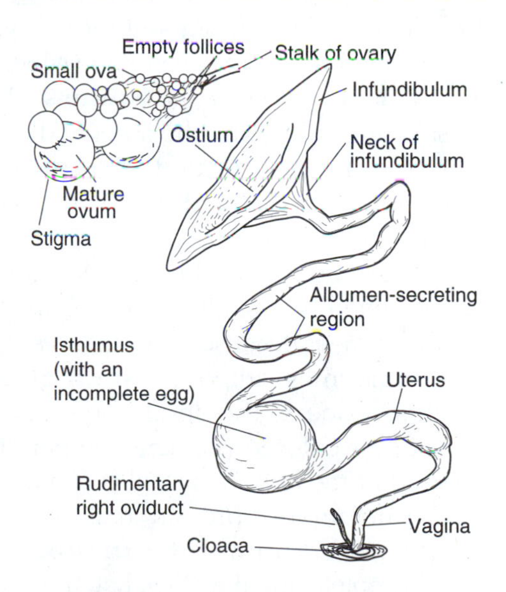

Figure 6-5. The reproductive system of a hen.

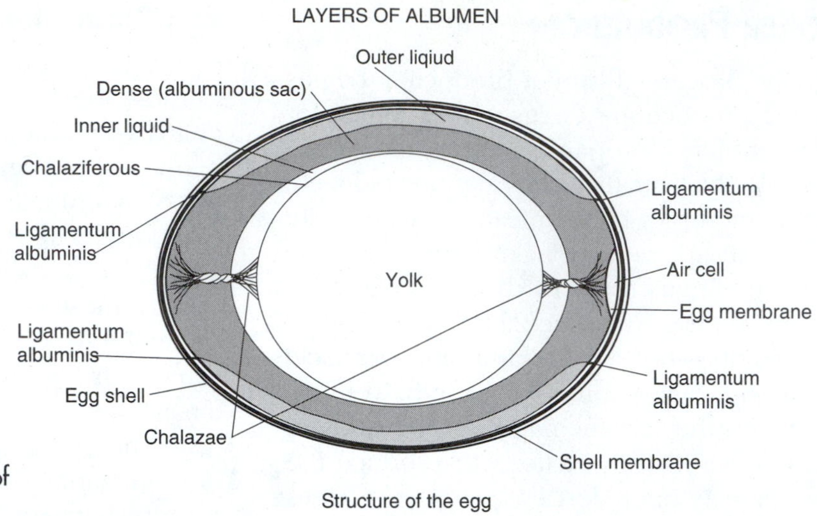

Figure 6-6. A cross section of an egg.

into the white or **albumen** of the egg. This substance is high in protein content and serves as nourishment for the developing ovum. The albumen (the white portion of the egg) surrounds the ovum, which is the **yolk** or the yellow part of the egg. Here a ropelike substance, called **chalazae**, is formed. Later as the egg is formed, these ropelike structures will serve to hold the yolk in position in the center of the egg.

In a process that takes about three hours, the egg moves through the magnum (which is over a foot long) and enters the **isthmus**. Here mineral salts are added, and the inner and outer shell membranes are formed. These membranes lie just inside the hard shell of the completed egg. After remaining in the isthmus for about an hour and a half, the egg moves into the **uterus** where the shell is formed around the egg, Figure 6–6. Because of this process, the uterus is sometimes called the **shell gland**. During this process, the shell may acquire brown or other colored pigments, depending on the breed of the hen. The shell is composed mainly of calcium and protein and serves to protect the embryo until the **incubation** process is complete and the chick hatches. Also, more water and minerals are passed into the egg white to fill out the egg, and a waxy substance is secreted that coats and seals the pores in the shell of the egg. In eighteen to twenty-two hours the egg shell is completed, hardened, and moved through the vagina and out of the hen's body.

Eggs are oval shaped, with one end smaller than the other end. In the process of development, the small end of the egg goes through the tract first. However, before the egg is laid, the ends are reversed and the large end emerges first. This turning helps the muscles of the tract to open and expel the egg. The process of laying is activated by a hormone called **oxytocin**. It causes the uterus to contract, forcing the egg through the vagina and out of the hen's body through an opening called the **cloaca**. In about a half hour the process begins all over again with the release of a new ovum from the ovary.

Hatching Eggs

Unlike the production of eggs for consumption, eggs produced for hatching are laid by the hens in nest boxes. A lot of scientific

Figure 6-7. Nesting boxes were once raised off the ground. *James Strawser, Cooperative Extension Service, The University of Georgia.*

Figure 6-8. The design of a nest box is based on research that determined what the hens prefer. *Steve Bolden, Seaboard Farms.*

research has gone into the design of the nest boxes to provide the type of environment the hens want. By providing the environment that makes the hens most comfortable, the producers ensure more efficiency in the laying operation. The nests consist of boxes that have concave bottoms and are filled with bedding or artificial turf to make a comfortable nest for the hen.

The hens naturally prefer nests that are enclosed because this gives them a feeling of security. This is perhaps a behavior passed down to the hens from their ancestors that lived in the wild and had to protect the nests from predators. Hens also prefer nests that are the gray color of galvanized metal. At one time, the nests were raised off the ground and had slatted floors that allowed manure to pass through to the ground where it could be removed, Figure 6–7. However, in modern breeder hen operations, mechanical nests are now used. The eggs roll onto a conveyor belt and an operator collects the eggs off the belt, Figure 6-8.

Hatching eggs must be kept as clean as possible. Any bacteria or other contamination can cause disease problems among the newly hatched chicks. As the egg comes from the hen, the surface of the egg is quite clean, but as the egg comes into contact with the surface of the nest or other areas within the house, the surface of the egg can become dirty and contaminated.

Even a small speck of foreign material on the shell can contain millions of microorganisms that can cause problems. Microorganisms must have an environment that allows their growth. Fecal material from the birds can provide an excellent place for the organisms to grow since there is moisture and material the microbes can feed on in the manure. Dirt or any foreign material that must be scrubbed from the eggs usually renders the eggs unfit for hatching. The washing or scrubbing of the eggs removes the protective coating from the eggs and presses the dirt into the pores of the eggs, Figure 6–9. Hatching eggs are not allowed to become wet.

Before they leave the farm where they are produced, the hatching eggs are sorted to remove dirty, undersized, oversized, misshapen, cracked, or defective eggs. They are then fumigated to kill harmful organisms on the surface of the eggs by a process that is precisely regulated to prevent harm to the eggs. The eggs are never allowed to become chilled and are

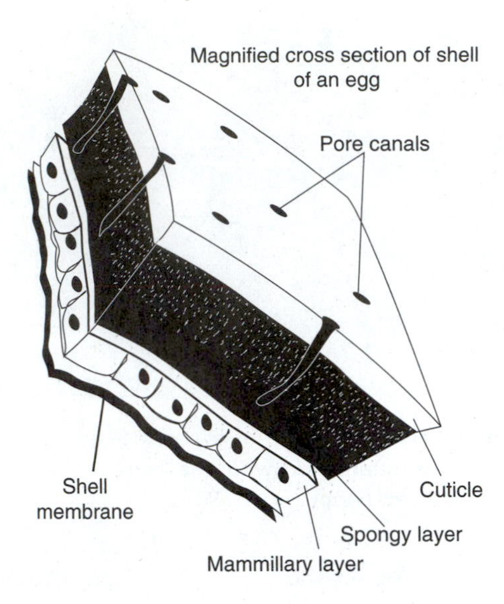

Figure 6-9. Dirt clogs the pores in the shell of the egg. *James Strawser, Cooperative Extension Service, The University of Georgia.*

stored at 70°–80° F until they are placed in the hatchery.

The eggs are carefully placed on racks that fit into carts designed to prevent damage to the eggs during transportation. The carts are then loaded into trucks for transportation to the hatchery, Figure 6–10. The carts, trays, and trucks are all kept clean and sanitized to prevent contamination of the eggs.

At the hatchery, the eggs are removed from the carts and placed in the incubator, Figure 6–11. The eggs are never allowed to be lowered in temperature, and the temperature is very gradually increased to prevent the eggs from *sweating*. Sweating is the condensing of water vapor on the surface of the eggs, and it occurs when the temperature of the eggs is raised too rapidly. Cold air holds more moisture than warm air. As the temperature of the air is raised, the water vapor in the air begins to condense. If moisture is allowed to collect on the surface of the egg, an environment is created that allows bacteria to grow and thrive. In a warm, moist environment a single bacterium can reproduce into two bacteria every 20 minutes. If this is allowed to continue, the single bacterium can become 16 million bacteria in eight hours.

Figure 6-10. Eggs are transported to the hatchery in carts loaded on a truck. *James Strawser, Cooperative Extension Service, The University of Georgia.*

Figure 6-11. At the hatchery, the eggs are removed from the shipping carts and are placed in the incubator. *James Strawser, Cooperative Extension Service, The University of Georgia.*

The relative humidity is also a factor in causing eggs to sweat. Relative humidity is the amount of moisture in the air relative to the amount possible at that temperature. The temperature and relative humidity are very carefully controlled in the hatchery, Figure 6–12.

Embryo Development

The development of the embryo begins before the egg is laid. As mentioned earlier, **fertilization** occurs very early in the formation of the completed egg. In contrast to most agricultural animals, **sperm** can remain viable in the hen's body for as long as 32 days, but highest fertility occurs if insemination occurs at least once a week. The sperm is stored in pockets inside the oviducts called **sperm nests**. The yolk portion of the egg contains a spot called the **germinal disk** that contains the genetic material from the female. The sperm fertilizes the egg within this germinal disk and the embryo begins to develop. If a newly laid egg is broken open, the germinal disk is visible to the naked eye and appears as a white spot in the yolk. After the egg is laid, the embryo remains dormant until it is stimulated by heat to grow. In nature, this heat is generated by the hen's body as she sits on the nest, but in modern operations the eggs are heated by artificial means in commercial incubators.

Within 48 hours after incubation begins, the embryo has developed a circulatory system that sustains life by carrying nourishment from the yolk to the embryo. At the end of the third day of incubation, three layers of membranes have developed. The first—the allantois—serves as a place to store the waste generated by the embryo. This membrane later merges with the second membrane—the chorion—to form a type of respiratory system until these organs are developed in the embryo. The third membrane—the amnion—is the membrane that is filled with the fluid that surrounds the embryo. It serves to protect the developing embryo from shock, Figure 6–13. To prevent the embryo from sticking to the outer membranes of the egg, the eggs must be turned several times a day. In nature, the hen turns the eggs in the nest. In the incubator, the eggs

Figure 6-12. In the hatchery, relative humidity and temperature are carefully monitored. *James Strawser, Cooperative Extension Service, The University of Georgia.*

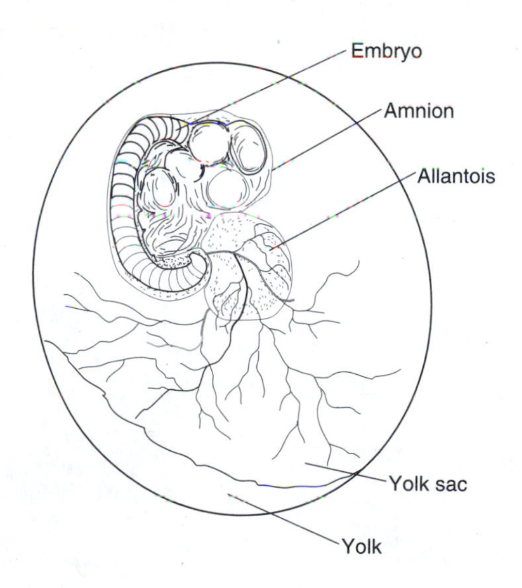

Figure 6-13. The chick embryo at the fifth day of development.

are either turned automatically by a time-controlled turning device or the eggs are turned by hand.

At the end of the first week of incubation, the embryo is recognizable as a chick embryo. Most of the chick's systems, such as the lungs, nervous system, muscles, and sensory systems are developed. By the end of the second week, the chick is covered with down. At the end of three weeks the chicks are fully developed. When the first chicks in the incubator begin to hatch, they make clicking sounds as they break the egg shell. This serves as a signal to the other chicks and stimulates them to begin hatching. This behavior is a carryover from the days in the wild when all the chicks needed to hatch at the same time for survival. The producer can mechanically provide slow clicking sounds to accelerate the hatching process.

In commercial hatcheries the eggs are incubated in two separate rooms: the setting room and the hatching room. The eggs are placed in the setting room incubator and are closely monitored for temperature and relative humidity. The eggs are turned every day to ensure that a high percentage of the eggs hatch. The eggs remain in the setting room incubator until one to two days before the eggs are ready to hatch. They are then placed in the hatching room where the temperature is lowered slightly and the chicks hatch into chick holding trays, Figure 6–14.

When the Chicks Hatch

The chicks are removed from the incubator and are cleaned, dried, and placed in a warm, dry environment, Figure 6–15. The chicks are sexed by examining their feathers. If the hatchery is producing chicks that are to become laying hens, it is essential that the females be separated from the males, Figure 6–16.

At one day of age, the chicks are vaccinated and their beaks are trimmed to help prevent cannibalism. The procedure is done with an electric knife and causes the chicks no harm since only a small portion of their beaks are removed. The day-old chicks are then placed in ventilated cardboard or plastic boxes and are transported to the broiler houses, Figure 6–17.

At the Broiler House

Prior to the arrival of the chicks, the producer has cleaned, disinfected, and placed clean litter in the house. Litter is the material placed on the floor to absorb moisture and to keep the birds clean and dry. This material is

Figure 6-14. One to two days before hatching, the eggs are placed in the hatching room. *James Strawser, Cooperative Extension Service, The University of Georgia.*

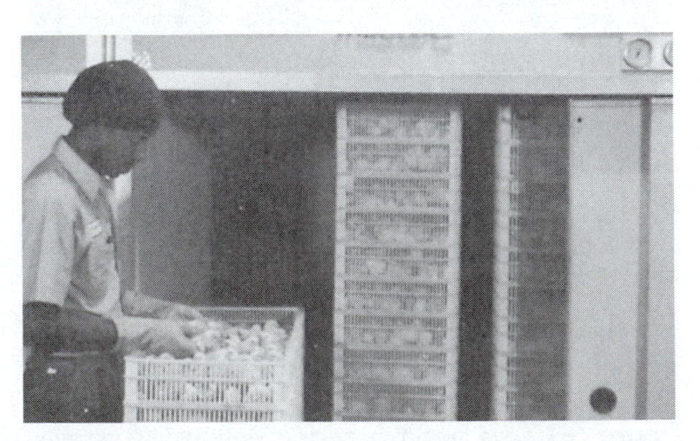

Figure 6-15. Chicks are removed from the hatchery trays and placed in a warm, dry environment. *James Strawser, Cooperative Extension Service, The University of Georgia.*

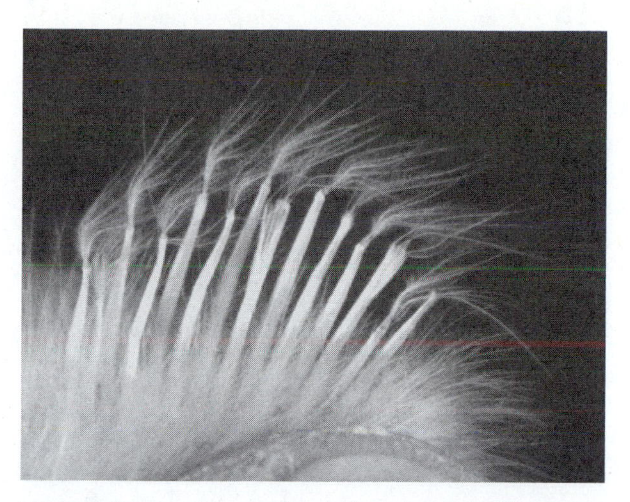

Figure 6-16. The sex of a chick is determined by examining the feathers. This chick is a male. *Joe Mauldin, Poultry Science Department, The University of Georgia.*

usually shavings or sawdust obtained from a sawmill.

Brooders are large, pan-shaped heaters that are used to keep the chicks warm during the first days in the house. Brooders are usually suspended from the ceiling and can be raised or lowered depending on the temperature and the size of the chicks, Figure 6–18.

Water is supplied by suspended waterers that the chicks quickly learn to use. They peck the nipple on the bottom of the waterer and obtain as much water as they need. Feed is given to the baby chicks by hand when they are small. But as they get older, they are fed through automatic feeders. The feed is brought to the birds by means of a conveyer chain in the bottom of a trough. The chain moves along the trough periodically throughout the day and night to ensure that the birds always have plenty of feed, Figure 6–19. Both the waterers and feeders are raised as the birds get larger. Every day the equipment and birds are continually checked to ensure that the equipment is functioning properly and that the birds are doing well.

The birds are generally kept in the broiler house from six to seven weeks. At this time, they weigh about 5 pounds and are ready for market. They are usually caught at night when they are less active. They are put into cages and loaded on a truck for transportation to the processing plant. The producer then begins to clean the house to get ready for the next batch of chicks. New litter consisting of sawdust or shavings is placed on the floor to absorb moisture from the manure, Figure 6–20. Since the litter must be removed periodically because it

Figure 6-17. Chicks are transported to the broiler houses in cardboard or plastic boxes. *James Strawser, Cooperative Extension Service, The University of Georgia.*

Figure 6-18. Young chicks are kept warm by brooders that are raised up as the chicks grow. *James Strawser, Cooperative Extension Service, The University of Georgia.*

Figure 6-19. The broilers are fed by means of a conveyor chain in the bottom of a long trough. *James Strawser, Cooperative Extension Service, The University of Georgia.*

becomes filled with manure, the disposal of the litter can be a large problem. A broiler house that holds 20 thousand broilers produces about 180 tons of litter per year. Since the manure in the litter has a very high concentration of nitrogen and other elements necessary for plant growth, the litter is a valuable source of fertilizer, Figure 6–21. The litter can also be processed and used as a source of protein in cattle rations.

At the Processing Plant

When the birds reach the processing plant, they are slaughtered and prepared for market. Some plants process the chickens to be sold whole, some cut the broilers into parts, such as breasts, thighs, and drumsticks, Figure 6–22. Other plants may process the chicken further into more complex prepared food such as chicken franks, chicken bologna, or complete frozen dinners, Figure 6–23.

Figure 6-20. When the broilers are sent to market, fresh litter is spread on the floor. *James Strawser, Cooperative Extension Service, The University of Georgia.*

Figure 6-21. The manure in broiler litter is a valuable source of fertilizer. *James Strawser, Cooperative Extension Service, The University of Georgia.*

Layer Industry

The per capita egg consumption in the United States has decreased sharply over the past thirty years. As Figure 6–24 shows, the consumption of whole eggs in the shell has greatly decreased while the consumption of egg products has increased. This is a reflection of changing dietary habits and consumer pref-

Figure 6-22. Chickens may be cut up or packaged whole at the processing plant. *James Strawser, Cooperative Extension Service, The University of Georgia.*

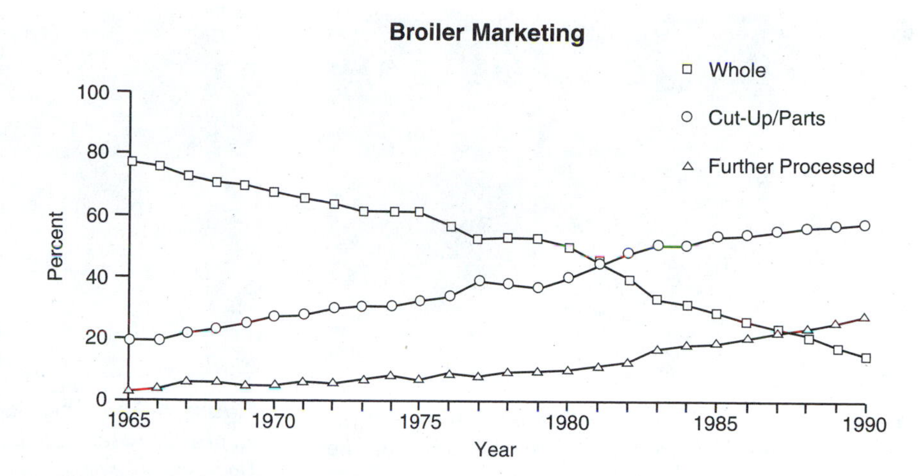

Figure 6-23. The trend is toward more processing of broiler products. *Charles Strong, Cooperative Extension Service, The University of Georgia.*

erence for processed foods. Even with the decrease in demand, the layer industry in this country is still quite large.

Cage Operations

Over 90 percent of eggs are produced by layers in cages, Figure 6–25. The hens live in the cages in groups of two to twelve hens, depending on the particular operation. The most common grouping is that of four hens per cage. The birds used for **cage operations** have been developed to tolerate the confinement operation and to efficiently produce eggs. As mentioned earlier, the type of hen used to produce eggs is quite different from those used to produce meat. They are smaller and are much less muscular. The smaller, trimmer birds use a large portion of their **metabolism** in producing eggs instead of developing muscles and body size. Some layers produce brown eggs and some produce white eggs. The vast majority of the eggs sold in the United States are white eggs. Consumers simply prefer to buy white eggs.

Modern cage operations are scientifically designed to provide the hens with adequate

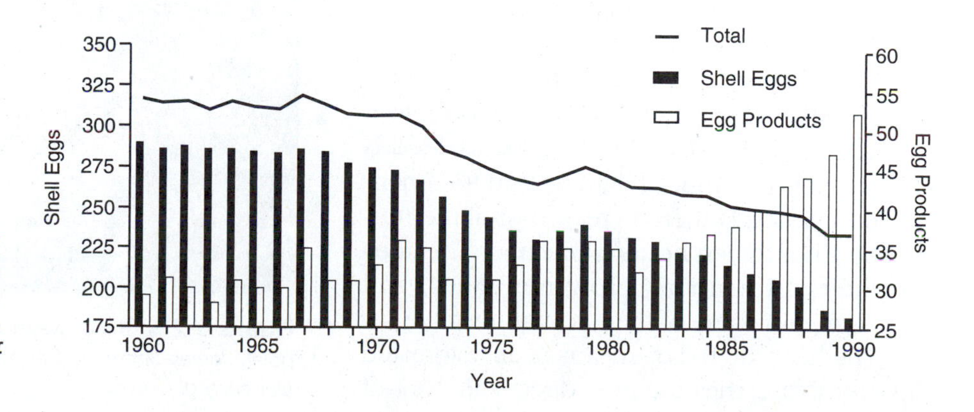

Figure 6-24. Overall consumption of eggs in the shell has declined. *Charles Strong, Cooperative Extension Service, The University of Georgia.*

Figure 6-25. Most eggs are produced by hens in cages. *James Strawser, Cooperative Extension Service, The University of Georgia.*

Figure 6-26. When laid, the eggs roll onto a conveyer. *James Strawser, Cooperative Extension Service, The University of Georgia.*

room, proper ventilation, correct temperature, plenty of food, and fresh water. In addition, lighting is carefully controlled. Hens naturally lay eggs in the spring and summer months. In the wild the chicks would have a much greater chance of survival if the eggs were laid and hatched during the warmer months. As spring approaches, the days have more hours of light and fewer hours of dark. The longer periods of light stimulate the hen's hormonal system into producing eggs. In a commercial cage operation, the lighting is carefully controlled to allow fourteen to fifteen hours of light every day. As the young hens (called **pullets**) are reaching maturity, the light is gradually increased until they have fifteen hours of light per day. This causes the hens to be in full production.

As the hens grow older, they lay fewer eggs. Some producers sell the hens when production decreases; others submit the hens to a process called **molting**. In this process, feed is withheld and no artificial light is used for a period. The lighting is then increased until the normal length of fifteen hours is reached. The hens lose their old feathers, grow new ones, and seem to regain some of their youthful vitality.

The hens are fed by means of an automated conveyer that carries the feed directly in front of them, Figure 6–26. Water is supplied by means of a narrow, free-running trough or a nipple waterer that the hens learn to peck to obtain water. As the eggs are laid, they roll onto a conveyer that periodically moves the eggs to a collection point where a worker gathers them and places them in flats, Figure 6–27. Dirty eggs are separated out and the clean eggs are refrigerated.

At the processing and packing plant, the eggs are coated with a thin coat of light mineral oil to prevent carbon dioxide from escaping from within the egg. The eggs are graded according to shape and size and are checked for cracks and interior spots in a process called **candling**.

Figure 6-27. The conveyer moves the eggs to a collection point. *James Strawser, Cooperative Extension Service, The University of Georgia.*

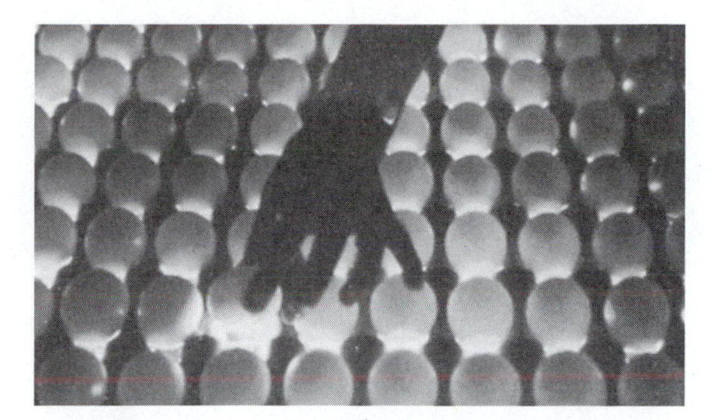

Figure 6-28. The eggs pass over an intense light to reveal blood spots in the eggs or cracks in the shell. *James Strawser, Cooperative Extension Service, The University of Georgia.*

The eggs are passed over an intense light in a dark room and any blood spots or cracks in the shell will show up in the light, Figure 6–28. The eggs are packaged and sent to the retail market or are sent to a processing plant where they are broken and processed.

THE TURKEY INDUSTRY

The production of turkeys for meat is a rapidly expanding segment of animal agriculture. The sale of turkeys is second only to that of chicken in the overall sale of poultry meat. In 1997, the per capita consumption of turkey was almost 19 pounds. This is an increase from a little over 10 pounds in 1980, Figure 6–29. Turkey represents a high-quality, low-cost, nutritious source of food protein. Although one-third of all turkey sales still occur in the weeks surrounding the Thanksgiving and Christmas holidays, the trend is toward a steady year-round sale.

The turkeys that are produced in this country are the descendants of wild turkeys native to the Americas. The wild turkey is a bronze-colored bird that lacks the broad breast and overall **muscling** of the commercial turkeys. Just as in broilers, consumers demand

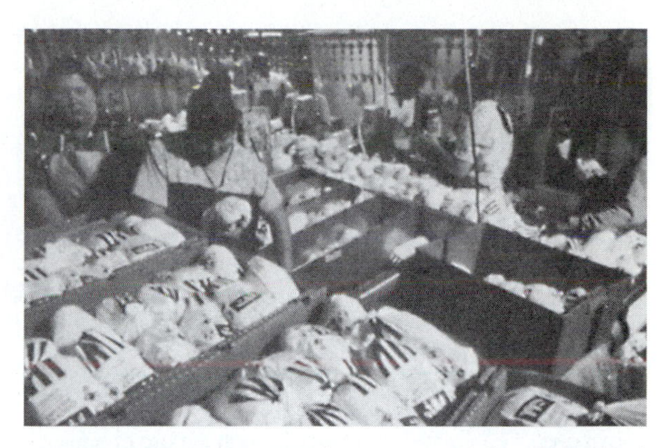

Figure 6-29. Americans buy over 4.5 billion pounds of ready-to-cook turkey each year. *James Strawser, Cooperative Extension Service, The University of Georgia.*

that the turkeys they buy be white. As mentioned in the section on broilers, a colored bird will have specks of pigmentation left in the skin when the feathers are removed. The carcasses of white birds look a lot cleaner. The modern white turkey is the result of a mutation or accident of heredity that left out the gene for the feather and skin pigmentation, Figure 6–30.

From these mutated white turkeys, a heavy-muscled, broad-breasted bird was developed. A problem with this highly-developed bird is that they are not efficient breeders. The physical act of mating is difficult because of the heavy muscling, and the birds seem more reluctant to breed. This problem

Figure 6-30. The modern white turkey is the result of a mutation. *James Strawser, Cooperative Extension Service, The University of Georgia.*

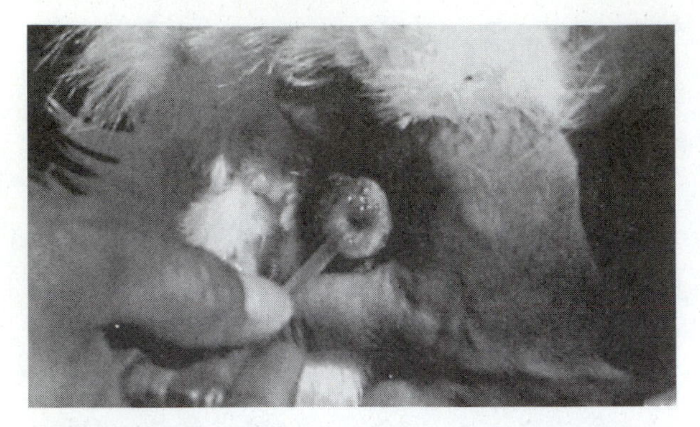

Figure 6-31. Turkeys are bred through the use of artificial insemination. *James Strawser, Cooperative Extension Service, The University of Georgia.*

has been solved through the use of artificial insemination, Figure 6–31.

The majority of the turkeys in the United States are grown in the western part of the northern central region, the south Atlantic region, and the Pacific region, Figure 6–32. Turkeys seem better able to tolerate cold weather than hot weather. Most are produced by small operations of 30 thousand birds or less. There are two ways of growing turkeys: in confinement houses and on open ranges. Confinement houses offer the advantages of environmental control of temperature and humidity. Most turkeys are raised in confinement. Although the open range offers the advantage of being less expensive, few turkeys are raised this way. Turkeys on the range can stay outdoors completely, or they are provided with housing where the birds can get shelter when they want, Figure 6–33. Producers usually move the range every three to four years to help keep down problems with disease and parasites.

Other Poultry

In some parts of the world, poultry such as ducks and geese make up a major portion

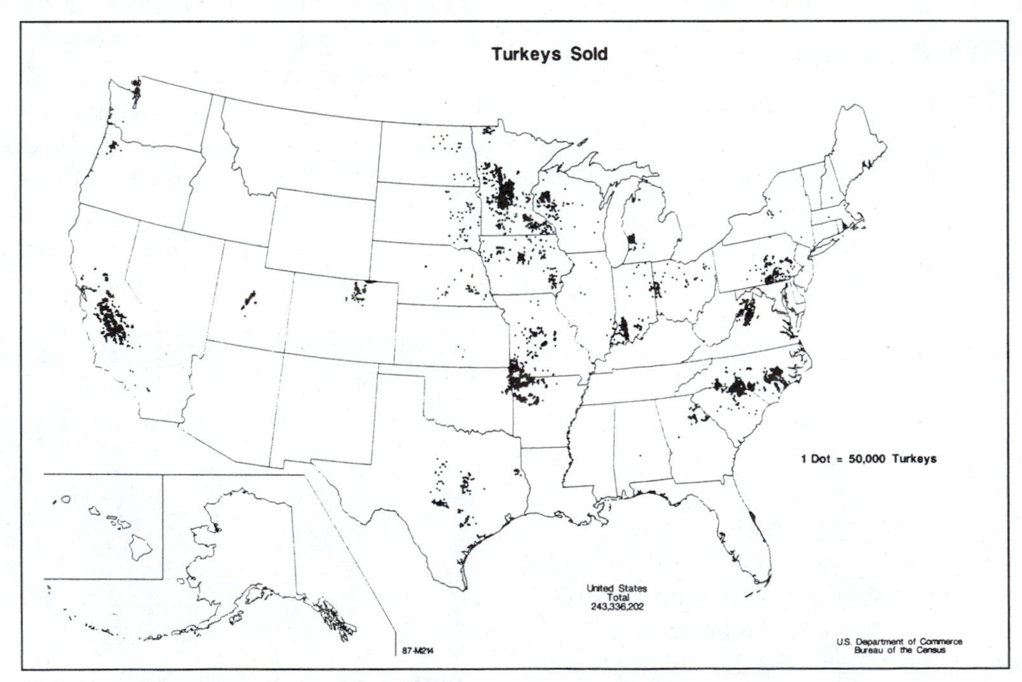

Figure 6-32. The turkey production industry in the United States. *James Strawser, Cooperative Extension Service, The University of Georgia.*

Figure 6-33. Many turkeys are raised on the range. *James Strawser, Cooperative Extension Service, The University of Georgia.*

Figure 6-34. In this country most ducks and geese are raised by hobbyists.

of the total poultry output. In China and Southeast Asia, ducks and geese are a major part of the overall diet. These birds are more hardy than chickens because they are not as susceptible to diseases and can forage for themselves better. Also the feathers are used to make bedding and other goods. In this country, most ducks and geese are raised in small flocks by hobbyists or part-time producers, Figure 6–34. Most of the meat sold goes to the restaurant trade or gourmet food market.

About the only other poultry production of any significance is that of growing quail and pheasant. Both of these birds are grown for the restaurant trade and the gourmet food market, Figure 6-35. In addition, they are raised for restocking wildlife areas. Each year, thousands of quail and pheasants are released in the wild to provide birds for hunting. This helps replenish the areas with game that is difficult to produce in the wild.

Figure 6-35. Quail are grown for the restaurant trade. *James Strawser, Cooperative Extension Service, The University of Georgia.*

SUMMARY

The poultry industry is one of the most dynamic industries in agriculture. Few areas have come close to the progress made in the growing of poultry and poultry products. Most of this progress is directly due to the discoveries found through scientific research. The amount of poultry eaten by humans is on the increase. It represents a relatively inexpensive, healthy alternative to other foods. This industry has made a gigantic contribution to our food supply system.

◆ Review Exercises

TRUE OR FALSE

1. Poultry, although accepted as wholesome meat for human consumption in most countries, is not consumed to any great extent in China and Russia. ______.
2. No matter how well a broiler house is insulated, the birds must be kept warm by gas or electric heaters. ______.
3. The development of poultry embryos takes place outside the mother's body. ______.
4. Providing hens with a comfortable environment encourages them to lay eggs more efficiently. ______.
5. Washing or scrubbing the eggs removes the protective coating and presses the dirt into the pores of the eggs. ______.
6. Moisture on the surface of an egg allows bacteria to grow and thrive. ______.
7. At the end of the first week of incubation most of the chick's systems—such as the lungs, nervous system, muscles, and sensory systems—are developed. ______.
8. The eggs are placed in a setting room in commercial operations until one week before hatching when they are removed to the hatching room. ______.
9. Brooders are large pan-shaped heaters that are used to keep the birds warm during the latter stages of growth. ______.
10. Chicken litter can be fed to cattle as a source of protein. ______.
11. The consumption of whole eggs in the shell has greatly decreased, while the consumption of egg products has increased. ______.
12. Shorter periods of light stimulate the hen's hormonal system into producing eggs. ______.
13. At the processing and packing plant, the eggs are coated with a thin coat of mineral oil to prevent cracking. ______.
14. The turkeys that are produced in this country are descendants from turkeys brought to this country by the early settlers. ______.
15. Quail and pheasant are grown for the restaurant trade and the gourmet food market as well as for restocking wildlife areas. ______.

FILL IN THE BLANKS

1. A broiler is a ____________ that is kept in the ____________ ____________ to about six or seven ____________, is __________, and is dressed for ______________.

2. Layers that produce eggs for ____________ are selected for their ____________ ____________ capacity and not for the amount of ____________ they produce.

3. Cells in the magnum secrete a ____________ called ____________ that develops into the ____________ or ____________ of the egg.

4. The shell of the egg is composed mainly of ____________ and ____________ and serves to ____________ the embryo until the ____________ process is complete and the chick ____________.

5. Within forty-eight hours after the ____________ begins, the embryo has developed a ____________ system that sustains ____________ by carrying ____________ from the yolk to the ____________.

6. To prevent the ____________ from sticking to the ____________ membrane of the egg, the egg must be ____________ ________several times each ____________.

7. At one day of age the chicks are ____________ and their beaks are ____________ to help prevent ____________.

8. Every day the ____________ and the birds are continually checked to ensure that the ____________ is functioning ____________ and that the birds are doing ____________.

9. The manure in chicken litter has a very high concentration of ____________ and other ____________ necessary for plant ____________, it is a valuable source of ____________.

10. The smaller, ____________ birds use a large portion of their ____________ in producing eggs instead of developing ____________ and ____________ ____________.

11. Modern cage operations are ____________ designed to provide the hens with adequate ____________, proper ____________, correct ____________, plenty of ____________, and fresh ____________.

12. Candling is a process by which eggs are passed over an ____________ ____________ in a dark room and any ____________ ____________ in the eggs or ____________ in the shells will show up.

13. The majority of turkeys in the United States are grown in the ____________ part of the northern ____________ region, the ____________ Atlantic region, and the ____________ region.

14. Ducks and geese are more hardy than chickens because they are not as susceptible to ____________ and can ____________ for themselves better.

DISCUSSION QUESTIONS

1. Why is poultry production so popular around the world?
2. What is a vertically integrated company?
3. How is cannibalism reduced in broiler flocks?
4. Why are white broilers preferred over broilers that are colored?
5. In what ways are the eggs of poultry different from eggs of mammals?
6. List the parts of the egg-producing tract of the hen.
7. Describe the type of nests that are preferred by laying hens.
8. Explain why dirty eggs are not used for hatching.
9. Why is it important that hatching eggs not get wet?

10. Describe the characteristics of a chick embryo at the end of each week of incubation.
11. Indicate two uses for chicken litter.
12. What effect does light have on layers?
13. What is meant by the molting process?
14. Why are most domesticated turkeys produced through artificial insemination?
15. What are two methods used to produce turkeys?

STUDENT LEARNING ACTIVITIES

1. Visit the meat counter in a large grocery store and make a list of all the different ways in which poultry meat is offered for sale (e. g., whole fryers, cut up fryers, breasts only, thighs only, processed, etc.). Interview the manager and determine which method of packaging and processing sells the best.
2. Break open some eggs and locate the germinal disk. Also take note of the separation of the albumen and the yolk.
3. Using a small incubator, place eggs in the incubator for one, two, and three weeks. At the end of each day during the first week open an egg and take note of the development. Write a detailed account of the differences noted from one day to the next.
4. During a two-week period keep track of the type and amount of meat consumed by your family. Determine the percentage of poultry meat in the total amount. Discuss with your parents the economics of buying poultry as compared to other types of meat. Report to the class.

Chapter 7

The Sheep Industry

Student Objectives in Basic Science

As a result of studying this chapter, you should be able to

1. Explain why sheep have been important to humans throughout history.
2. List the characteristics of sheep that allow them to make good domesticated animals.
3. Discuss the controversy over predators of sheep.
4. Explain the characteristics of wool that makes it useful to humans.

Student Objectives in Agricultural Science

As a result of studying this chapter, you should be able to

1. Discuss the importance of lamb and mutton in the diets of Americans.
2. Explain the importance of the sheep industry to our economy.
3. List the ways wool is used by humans.
4. List the uses for by-products of wool.
5. Discuss how wool is made into clothing.

Key Terms

ecological balance
mutton
cuticle
cortex
felting
crimp
suint
yolk
grease wool
scouring
mohair

Humans have raised sheep for at least the past 10,000 years, Figure 7–1. During this time, sheep have supplied food, shelter, and clothing for human use. The meat from the carcasses and milk from the females have provided a protein-rich diet for even the poorest of societies. Since sheep can live and thrive in areas where other agricultural animals can not, they have played an important role in the feeding of people all over the world. These animals eat plants ranging from grasses and legumes to brush. Often they may even eat plants that are toxic to other animals.

Compared to other agricultural animals, sheep are unique in that they are very docile and easy to handle. This may be due to the fact that they were one of the very first animals domesticated and have been continuously raised and bred for human use for the past 10,000 years. In fact, they are so tame that they have very little defense from predators.

Compared to beef and pork, Americans eat relatively little lamb and mutton. (Lamb refers to meat from a sheep that is less than a year old; mutton refers to meat from a sheep that is over a year old.) In many parts of the world, lamb and mutton are a basic part of people's diet. The United States per capita consumption of mutton and lamb is only about two and a half pounds. Of this consumption about 95 percent is lamb and only about 5 percent is mutton. Americans seem to have never developed a taste for the stronger flavored mutton.

Figure 7-1. Humans have raised sheep for thousands of years. *USDA.*

However, in certain areas of the country, lamb is a favored food. The large cities along the Eastern Seaboard account for almost half of the market for lamb and mutton in this country. This presents a problem since most of the lambs are produced west of the Mississippi River, Figure 7–2. The leading states in sheep production are Texas, California, Wyoming, Colorado, South Dakota, Montana, New Mexico, Utah, and Oregon. Modern refrigerated trucks and railroad cars have helped alleviate the spoilage problem, so most of the problem is the economics of transportation.

One advantage of producing lambs for market is that good quality lambs can be produced on grass and do not have to be fed a lot of expensive grain. Although an increasing number of lambs are being fed on grain in the feedlot, roughages still make up about 90 percent of all the feed consumed by sheep.

In the Willamette Valley of Oregon, lamb production fits in well with the production of rye grass seed. Rye grass bunches and spreads better if it is closely grazed for a period at a certain time of the year. Lambs are used to graze the rich green grass down and, in turn, the animals are fattened by the nutrients in the grass, Figure 7–3.

Sheep also can make better use of lower quality forage than can cattle. For this reason they can be successfully grazed on poorer grazing lands of the desert areas of the West. In addition, a drier climate helps reduce parasite and disease problems associated with sheep that are grown in the more humid areas. For example, very few large herds of

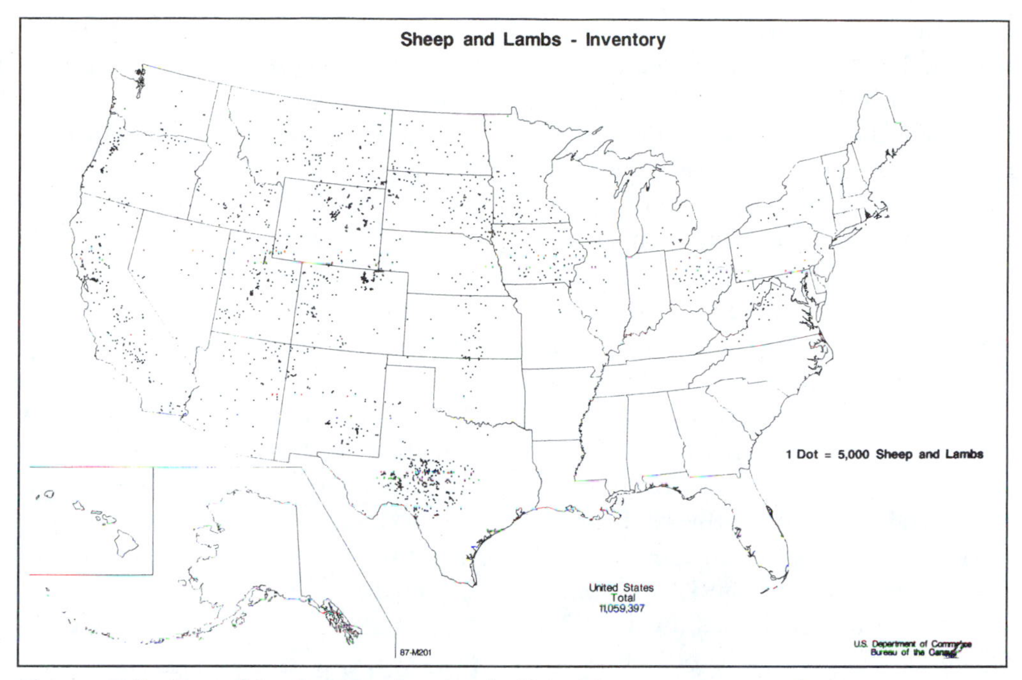

Figure 7-2. Most of the sheep produced in the United States are grown in the western states. *Bureau of the Census.*

Figure 7-3. Sheep are used to graze down grass raised for seed. *Calvin Alford, Cooperative Extension Service, The University of Georgia.*

Figure 7-4. Many lambs are now being raised in the Southeast as FFA and 4-H show animals. *Calvin Alford, Cooperative Extension Service, The University of Georgia.*

sheep are grown in the Southeast because of problems with heat and humidity. Hot humid weather makes a good environment for parasites and disease organisms such as foot rot; nevertheless, many lambs are now being raised in the Southeast as FFA and 4-H show animals, Figure 7–4.

Sheep breeds are generally grouped according to use. The use is determined by the type wool the animals grow. Wool type is broadly classified as fine wool, long wool, medium wool, hair, and fur. The medium wool breeds are used most often to produce lambs for slaughter. The other types will be discussed under the heading of *Animal Fiber Production*. Medium wool breeds commonly used to produce slaughter lambs are Suffolk, Hampshire, Southdown, and Dorset, Figure 7–5.

A large problem facing sheep producers is that of predators. Animals such as coyote and both wild and domesticated dogs kill large numbers of sheep each year. Although it is difficult to obtain precise data on losses due to predators, research indicates that each year around four to eight percent of the lambs and 1.5 to 2.5 percent of the ewes in the western 17 states are lost annually to predators. Some producers have reported losing as much as 29 percent of their lambs in a year to predators.

In the past, sheep producers have used measures such as trapping and poisoning to rid the area of predators. Today, such measures are closely regulated because of damage that may occur to the **ecological balance** of an area. The ecological balance refers to the balance nature has regarding the number of living things in a given area—too many or too little of a certain type of animal can upset the balance. If too many animals that are not guilty of preying on sheep are killed, the balance of nature can be upset.

Figure 7-5. The Hampshire is a medium wool breed raised primarily for meat. *American Hampshire Sheep Association.*

Government programs are now in effect to help producers with losses incurred by wild predators. The use of guard dogs and improved fencing has also helped with the problem. Difficulties still remain for the sheepherder whose animals are raised on open government lands. Controversy exists between the environmentalists and sheep producers over the amount of control necessary to prevent predation. In remote areas, the reintroduction of native animals such as the timber wolf and the mountain lion has prompted protest from producers who are concerned that further loss to predators will be incurred.

The Wool Industry

Wool is made of the fibers from the hair coat of sheep and has been used as a material for making clothing for thousands of years, Figure 7–6. The spinning of wool is probably one of the oldest industries in which people have been engaged. In many places throughout

Figure 7-6. Wool is the fiber from the hair coat of sheep. *The Wool Bureau Inc.*

the world archaeologists have uncovered clothing made from wool that is well over 10,000 years old. Records of the ancient Greeks, Romans, Egyptians and Hebrews indicated that they used wool for clothing. Several accounts of the production and use of sheep are written in the Bible and other ancient writings.

Throughout all history, wool is mentioned as being the standard material from which cloth was made. The wool was grown by family-owned flocks of sheep and harvested periodically by shearing the animals. The thread was made by spinning the wool on a hand-operated machine. The thread was then woven into cloth on a hand loom that was operated right in the home. This cloth, referred to as homespun, was rather coarse and plain. Almost all of the poorer people wore clothing made from this cloth.

Up until the early- to mid-19th century, almost all clothing was made from wool. During the first half of the 1800s, the cotton industry began to flourish in the southern portion of the United States, and it provided the world with a cheap alternative to wool clothing. In modern times, synthetic fibers from the petroleum industry have competed with both cotton and wool in the manufacture of textiles.

Even though wool is one of the oldest materials for clothing, it is still very popular today. Many of the very finest, most expensive garments worn today are made from wool. Also, fine carpets and tapestries are made from wool. Countries in the Middle East have always been famous for beautiful carpets made from wool.

Wool has certain characteristics that make it desirable over cotton and synthetics as a fabric. Wool can be worn in the winter or the summer. Because it can absorb up to 30 percent of its weight in moisture and still feel dry, the fabric makes a good insulator from both the heat and the cold. Wool is also very strong. A fiber of wool is stronger than a fiber of steel the same size. This characteristic makes wool fabric very durable, a trait desirable in the manufacture of both clothing and carpets. In the manufacture of woolens, the material takes and holds dye very well and, as a result, many beautiful patterns and colors can be made from wool.

Wool at one time was known as the material from which soldiers' uniforms were made. The reason for this was not only wool's ability to shed water and provide insulation from heat and cold but also its resistance to burning. During the Civil War, both Confederate and Union soldiers wore wool clothing. The weapons used by the troops were mainly black powder rifles and artillery, and this type of gunpowder often propelled bits of burning embers when the guns were fired. If an ember landed on the soldier's clothing, he might not be aware of it for a while. A cotton uniform would blaze up in a short time and cause a painful burn, whereas a wool uniform would only smolder and not burn. This allowed for a greater degree of safety. For this reason, even the soldiers of the South who had a ready supply of the cheaper cotton, preferred wool for their uniforms.

Wool fibers are made up of two distinct layers of cells, the **cuticle** on the outside and the **cortex** on the inside. The cuticle cells of the outer layer are arranged together into scales that overlap each other very much like the scales on a pine cone, Figure 7–7. Because of this characteristic, the fibers of wool lock together and bond into a solid mass when they are put under pressure. The fibers form a strong, thick bond of solidly matted fibers. This process is called **felting** and is used to make such things as hats and other objects requiring a thick layer of matting. The cortex layer that

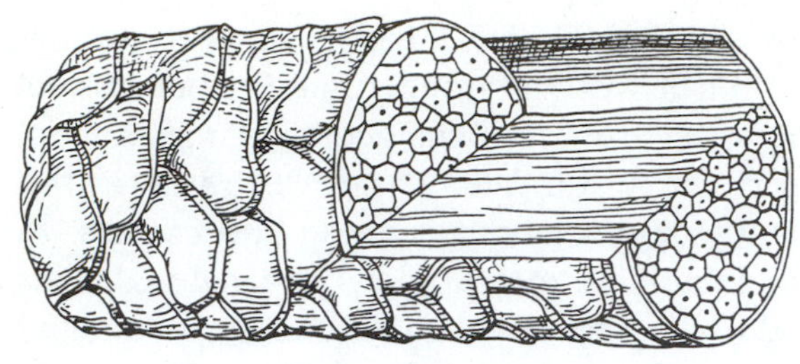

Figure 7-7. Close-up of a cross section of a wool fiber. *Sheep Industry Development Program.*

makes up the majority of the fiber gives it strength and elasticity. Elasticity is the ability to return to its original shape after being stretched.

The fibers are never perfectly straight, Figure 7–8. In fact, most are quite wavy and may be stretched up to 50 percent of its length and then return to its original length. This effect is called **crimp** and is used to help determine the quality of the wool. Usually the more crimp (the wavier) and the more uniform the crimp, the higher the quality of the wool. Wool is graded according to the diameter of the fiber with the wool having the finer or smaller diameter receiving the highest grade. Higher quality wool is free of foreign material, and is bright and white in color with no black or off-color fibers.

Wool contains oils, grease, and the salts from the perspiration of the sheep **(suint)**. This material is referred to as the **yolk**. The yolk helps to hold the scales on the outer layer of the fiber together and provides the fiber with the ability to shed water. It also helps the fibers from becoming matted together in the felting process, Figure 7–9.

Wool as it comes from the sheep is called **grease wool**. The first step in the processing of the wool is to remove the loose dirt and other particles from the fibers. This is done by a machine that opens up the fibers and dusts foreign particles from the wool, Figure 7–10.

The second step is called **scouring**. In this process, the wool is gently washed in detergent to remove yolk, suint, and other materials not removed in the dusting process. Almost half the weight of grease wool is removed in scouring. After the fibers have been cleaned, the wool is called *scoured wool.*

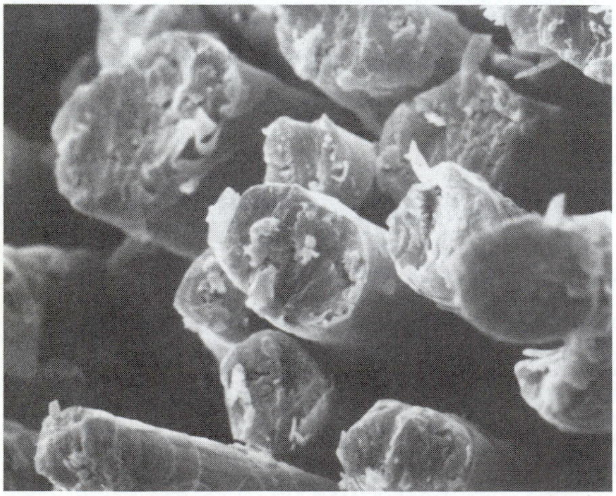

Figure 7-8. A microscopic view of wool fibers. Note the scales on the fibers. *American Wool Council.*

Figure 7-9. Scouring removes the yolk, suint, and other materials from the wool. *American Wool Council.*

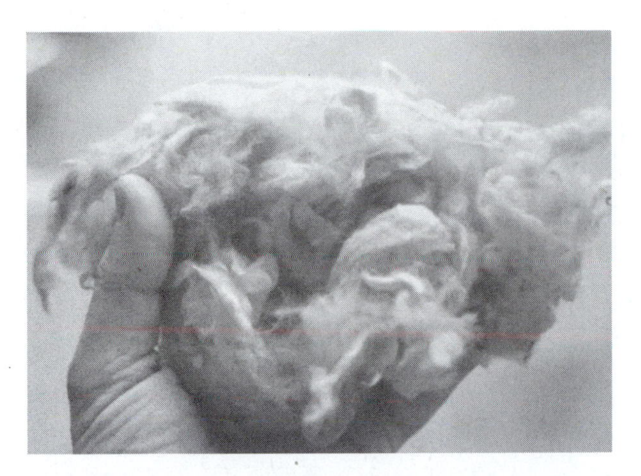

Figure 7-10. After the wool has been cleaned, it is called scoured wool. *American Wool Council.*

Figure 7-11. Wool dyed before spinning is called stock dyed wool. *American Wool Council.*

The water used in the scouring process is retained in order to remove the oils extracted from the wool. These oils are used to produce lanolin, an important ingredient in soaps and lotions.

After the wool is scoured, the wool is dried and treated to remove any remaining vegetable matter. This is done by mechanical or chemical means. The removal of the matter by chemical means (carbonizing) is the method most often used. This consists of treating the wool with acids or other chemicals in order to dissolve the vegetable matter.

Once the wool is clean, it is blended. This means that wool fibers of different types are mixed together mechanically to achieve a particular type of fabric or product. It is estimated that wool fibers can be blended into 2,000 combinations to produce a very wide assortment of products.

At this stage the wool can be dyed. If so, the wool is called *stock dyed* wool, Figure 7–11. If the wool is dyed after it is spun into yarn, it is called *yarn dyed.* Wool dyed after it is woven into cloth is called *piece dyed.*

The wool fibers are untangled and laid out parallel to each other in a process called *carding.* If the wool is to be made into a type of fabric called *worsted wool,* the fibers are further untangled and smoothed by combing and carding, Figure 7–12. If the wool is to be made into woolen fabrics, the wool goes from carding to spinning.

After the fibers are smoothed out and laid parallel to each other, they are processed into

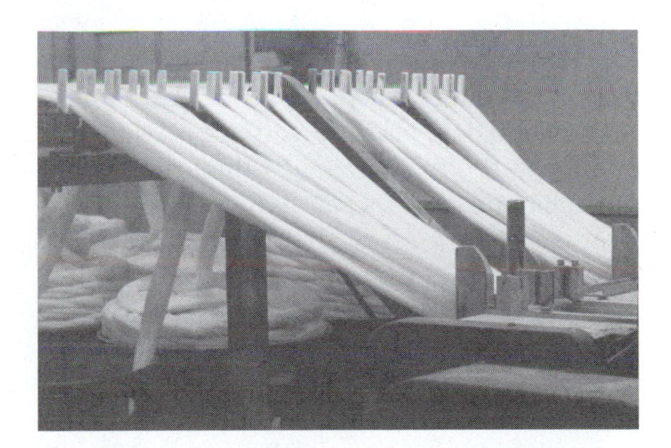

Figure 7-12. In the combing process, the fibers are untangled and smoothed. *American Wool Council.*

Figure 7-13. The spinning process creates long continuous threads from the fibers. *American Wool Council.*

Figure 7-14. The weaving of yarn creates fabrics in many patterns and designs. *American Wool Council.*

yarn by the spinning process. In the spinning process, the fibers are spun around and twisted into a long continuous thread that is used for weaving, Figure 7–13.

The wool cloth is made by weaving the yarns together. The yarn can be woven together into a variety of patterns and designs that go to make up the clothing or tapestries that are so popular with consumers, Figure 7–14.

Each year, Americans use about one pound of wool per person. This means that about 100 million pounds of grease wool is used in the United States per year. Of this amount, about half is produced in this country and about half is imported. The countries exporting the most wool to this country are Australia and New Zealand. Fine wool comes from such breeds of sheep as the Merino, Debouillet, Delaine, and Rambouillet.

Karakul sheep are raised for their pelts, which have a fur-like quality, Figure 7–15. Pelts are the skins of the animals with the hair left on. Most of these sheep are raised in the countries that made up the former Soviet Union, Afghanistan, and Iraq; however, some of these sheep are raised in the United States. These high-quality pelts are valuable in making coats and jackets.

Mohair is a fiber from the fleece of the Angora goat, Figure 7–16. This fiber is used to make a fabric that is resistant to wrinkles, is very soft and lustrous, and is unequaled in its ability to retain rich colors. This fiber differs from sheep wool in that the fibers are smooth and have less crimp. In addition, the fibers are long, often reaching a length of one foot. This quality makes the fibers easier to weave and provides a wider range of usages.

Figure 7-15. Karakul sheep are grown for their fur-like pelts. *American Karakul Sheep Registry.*

Figure 7-16. Mohair is the fiber from the fleece of Angora goats. *Mohair Council of America.*

The United States produces about one third of the world's mohair; most of which is exported to England where the fibers are processed into fabric. Most of the Angora goats raised in this country are produced in Texas, Figure 7–17. The goats make good use of the sparse browse found in the western part of the state. Much of the production of mohair comes from castrated males that are raised exclusively for their hair. The goats produce about one inch of hair per month and are sheared twice a year. The freshly shorn goats must be protected from the elements until their hair can grow out enough to protect them from the cold and rain.

SUMMARY

Sheep are among the oldest animals domesticated for human use. In the United States, sheep are not as widely grown as some of the other agricultural animals. Much of our wool is imported from New Zealand and Australia; however, a significant industry exists around the production of lamb and wool in this country. The popularity of wool continues to grow because of the unique characteristics of the fiber. Because of this, the production of wool will be with us for years to come.

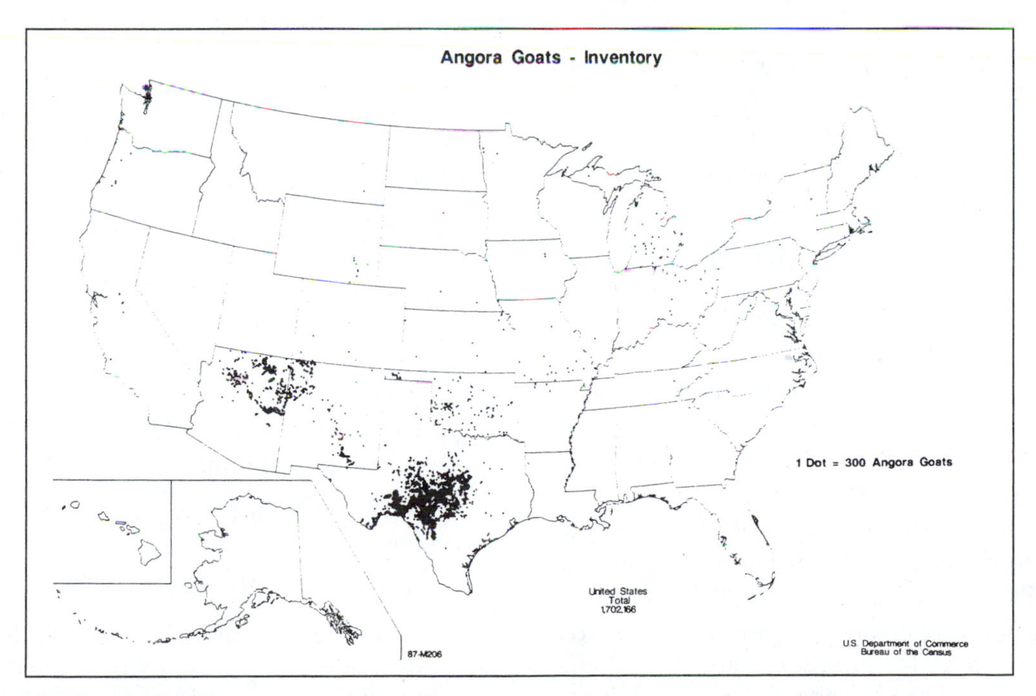

Figure 7-17. Texas is the largest producer of mohair in the United States. *Bureau of the Census.*

◆ Review Exercises

TRUE/FALSE

1. Compared to beef and pork, Americans consume less lamb and mutton._____.
2. The biggest market for lamb and mutton in the United States is along the West Coast. _____.
3. Very few sheep are grown in the southeast due to the heat and humidity. _____.
4. The medium wool breeds are used most often for slaughter. _____.
5. Lambs are usually weaned around nine months of age. _____.
6. A fiber of wool is stronger than a fiber of steel the same size. _____.
7. Usually, the wavier the crimp, the lower the quality of the wool. _____.
8. Wool as it comes from the sheep is called *suint* wool. _____.
9. Most of the Angora goats raised in this country are raised in Florida and Georgia. _____.
10. Most of the production of mohair comes from castrated males that are raised exclusively for their hair. ______.

FILL IN THE BLANKS

1. The top three sheep-producing states are ________, ________, and ________.
2. Meat from a sheep that is less than one year of age is referred to as ________, whereas meat from a sheep that is over a year old is called ________.
3. The sheep breeds are generally grouped according to _______, which is determined by the type of _______ the animals grow.
4. Medium wool breeds commonly used to produce lambs for slaughter are ________, ________, ________, and ________.
5. Sheep are naturally _________ breeders, meaning that breeding takes place during a certain time of the year.
6. Most lambs are born in the early ________, and are called ________ lambs.
7. Because it can absorb up to 30 percent of its weight in ________, the fabric makes a good insulator from both the ________and ________.
8. Wool fibers are made up of two distinct layers of cells, the ________on the outside, and the ________ on the inside.
9. Wool is graded according to the ________ of the fiber, with the smaller receiving the highest grade.
10. The ability of wool fibers to return to their original shape after being stretched is called ________.
11. Wool dyed after it is woven into cloth is called _______ _______.

DISCUSSION QUESTIONS

1. Explain why growing lambs and rye grass is a good combination.
2. What are some characteristics of wool that make it more desirable than cotton and synthetics? Why did the soldiers in the Civil War prefer wool uniforms?
3. Why is the Southeast an undesirable place to raise sheep?
4. What is yolk? What purpose does it serve?
5. Describe ecological balance. How does this affect sheep producers when dealing with the problem of predators?

STUDENT LEARNING ACTIVITIES

1. Determine how many different materials in your house are made from wool. List the characteristics of wool that makes wool a good material for each item.
2. Conduct a survey among students in your school to determine how many of them ate lamb or mutton in the past month. Also ask how many have never eaten lamb or mutton. Determine why.
3. Go to the internet and locate information about a particular breed of sheep. Report to the class.

Chapter 8

The Horse Industry

Student Objectives in Basic Science

As a result of studying this chapter, you should be able to

1. Explain how the anatomy of the horse makes it ideal for carrying and pulling loads.
2. Discuss the different ways of classifying horses.
3. Tell how horse behavior affects management practices.
4. Describe the scientific classification of the horse.
5. Explain the process of mating horses.

Student Objectives in Agricultural Science

As a result of studying this chapter, you should be able to

1. Discuss the importance of the horse industry.
2. List the various uses for horses in this country.
3. Explain several advantages mules have over horses.
4. Discuss how horses are raised.

Key Terms

mules
light horses
draft horses
ponies
perissodactyl
cecum
pasture breeding
hand breeding

Humans have used horses for transportation, work, and war from the beginning of recorded history. At one time, almost all civilizations relied on horses or donkeys to provide these services. Up until about 60 years ago, military history was written around the horse. From the time the ancient Assyrians used horse-drawn chariots to transport soldiers until horses and mules were used to transport supplies and to pull artillery during World War II, horses and mules have been used to wage war. From the ancient Romans to our American Civil War, generals have used cavalry to increase the efficiency of their fighting forces. Only with the advent of modern weapons and the use of self-powered machines, have horses become obsolete in warfare.

In the United States, much of our history has been built around power supplied by horses and **mules**, Figure 8–1. The very first explorers and settlers brought these animals to help tend fields and to provide the power necessary to build an agricultural base. As settlers moved westward, it was horses and mules that took them there and worked the farms once they were settled.

Figure 8-1. Much of our history has been built around power supplied by horses and mules.

The number of horses and mules in this country grew until the 1920s when the rapid increase in cars, trucks, and tractors caused a sharp decline in their numbers. From then until 1960, their numbers steadily declined. Since the 1960s the number of both horses and mules have increased dramatically. Although they no longer serve as the basis of agricultural power and transportation, they serve an important role in the agricultural sector of the United States. In terms of world production, the United States is the second largest producer of horses with 2.5 million head of horses, Figure 8–2. China leads the world with 11 million head; the majority of these animals are used as work animals.

CLASSIFICATION

The horse belongs to the genus *Equis*. Within this genus there are three groups of species. The domesticated horse belongs to the species *E. caballus* and includes the animals we generally associate with working and riding. Another group of species includes the zebras and yet another group includes the asses or donkeys. The only true wild horse is the *E. przhevalskii* that now exists only in parts of Mongolia. Wild horses in other parts of the world are the descendents of the domestic horse (*E. caballus*) that have escaped into the wild. Most of the species of Equis will interbreed to some degree, some much more successfully than others.

While some horses and mules in the United States are used for work, the majority are used for recreational purposes, Figure 8–3. In modern times, horses are generally categorized into one of three classes; **light horses**, **draft horses**, and **ponies**. Light horses refer to animals that weigh 900 to 1400 pounds. These

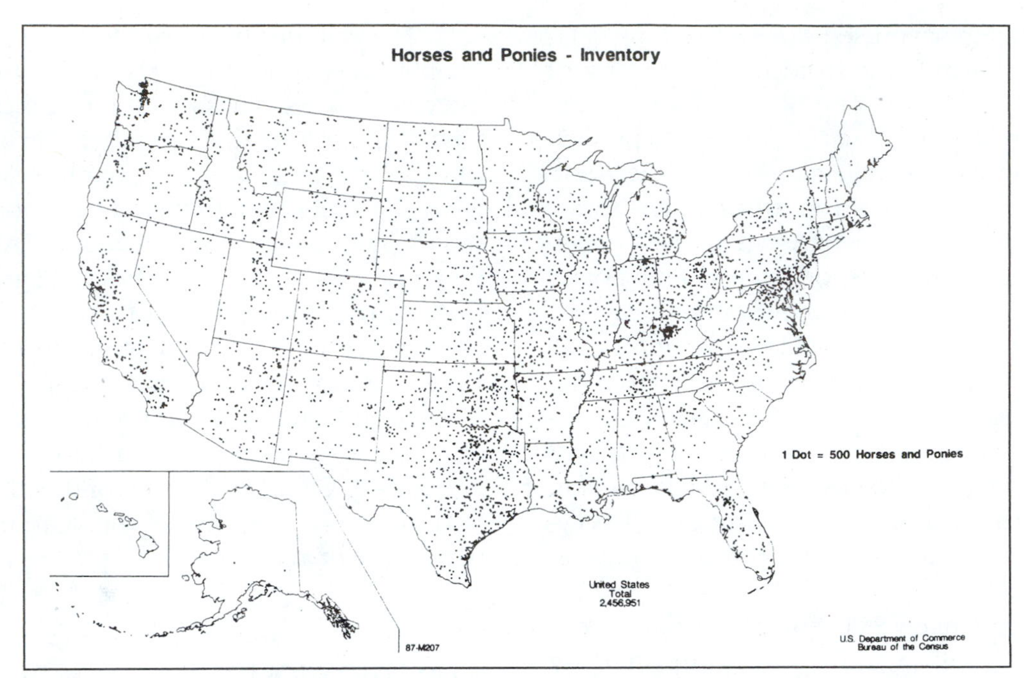

Figure 8-2. There are almost 2.5 million horses in this country. *Bureau of the Census.*

horses are then further divided according to use. Horses can be used for many events such as pleasure riding, trail riding, rodeos, racing, endurance riding, horse shows, fox hunting, dressage, combined training, polo, and driving. While some breeds are better suited for certain events, the majority of horse breeds in the United States are very versatile, Figure 8–4.

Figure 8-3. In this country, most horses are used for recreational purposes. *Kentucky Horse Park*

Figure 8-4. Gaited saddle horses such as the American Saddlebred are used for pleasure riding. *Jamie Donaldson.*

Figure 8-5. Draft horses are bred to pull heavy loads. *Russ Reagan, Missouri Department of Conservation.*

Figure 8-6. Driving horses are used to pull carts. *James Strawser, Cooperative Extension Service, The University of Georgia.*

Draft horses are those breeds that weigh over 1400 pounds. At one time, these were the animals that provided the power for pulling heavy loads such as wagons, plows, and other agricultural implements, Figure 8–5. Today, these animals are used in pulling competitions, shows, and parades.

Ponies are breeds of horses that weigh 500 to 900 pounds. While some are used to pull carriages and for show, the majority of ponies are used as horses for children.

Altogether, the horse industry in this country is a $15 billion industry. Surprisingly, horse racing ranks third behind baseball and auto racing as the largest spectator sport in the United States. Each year over 75 million people attend thoroughbred and harness races, Figure 8–6. Each year there are around 7,000 horse shows in this country where young people and adults compete, with their horses, in a variety of events. Many horses are owned as individual saddle horses that are used for recreation and are never raced or shown.

There are more registered quarter horses than any other breed in this country. The American Quarter Horse Association has a current count of 1.8 million registered horses compared to the next most numerous breed, the Arabians with 620,000 head. Quarter horses are still used to help herd cattle. No mechanized substitute has ever been developed that is more effective in working cattle over the rough terrain of the open range, Figure 8–7. Personnel who work in remote wilderness areas rely on horses for transportation and the packing of gear to areas inaccessible by car.

MULES

Mules have been bred and raised since ancient times when humans recognized the special characteristics that make them so valuable. By combining the size, speed, and strength of the horse with the patience, perseverance, toughness, and agility of the donkey,

Figure 8-7. Horses are still used to work cattle.

a very unique animal was created. The mule is a true hybrid, a cross between a male ass (jack) and a female horse (mare). Because of this, the mule rarely can reproduce (*see* the section on genetics).

The mule owns a particular place in American history. Around the time of the Revolutionary War, mules began to be bred and raised in this country to work the farms and plantations. They were particularly popular in the South where they more easily adapted to working in the hot, humid weather than did the horse, Figure 8–8. Mules have several other advantages over horses.They usually have sounder feet and legs than horses. This means that mules have fewer problems with lameness, split hooves, splints, and other leg problems associated with horses. In rocky, hilly terrain, mules are more sure-footed and are less likely to stumble than horses. For this reason, mules are used for transporting people up and down rough trails such as those found in the Grand Canyon. Tourists who make the journey down to the bottom of the Grand Canyon usually travel on the back of a mule.

If given the opportunity, horses will often eat so much grain that they do themselves harm, but a mule will seldom overeat, even if given free access to all the grain it wants. Mules also have the reputation for being stubborn, but most of the stubbornness results from the mule's refusing to overwork themselves. When

Figure 8-8. Mules are better adapted to working in the hot, humid areas of the South.

they become tired, they may balk and refuse to do any more until they are rested.

Mules are enjoying an increase in popularity. Each year, mules are shown in various shows across the country. Mules are bred specifically for purposes such as pleasure riding, hunting, packing, and pulling wagons. Many parades in all parts of the country feature mule-drawn wagons as a part of the history of the United States.

Anatomy of the Horse

The horse has certain anatomical features that make it suitable for use by humans, Figure 8-9. The skeletal system is composed of strong bones that are connected with ligaments, giving the horse a fluid, gliding movement that allows a rider to sit atop in comfort. Long bones in the hip and legs aid in the long stride of the horse; these bones act somewhat like a lever in propelling the horse forward.

The horse's muscular system is well advanced and is adapted to carrying heavy loads. Massive muscles down the back, over the croup, and down the legs give the horse the ability to pull loads and to sustain hard work for long periods. Horses with relatively short backs are better equipped to carry heavy loads than horses with long backs because the muscles are concentrated in a shorter span.

The feet of horses are especially well suited for carrying loads. The horse is classified as a **perissodactyl** or an animal with

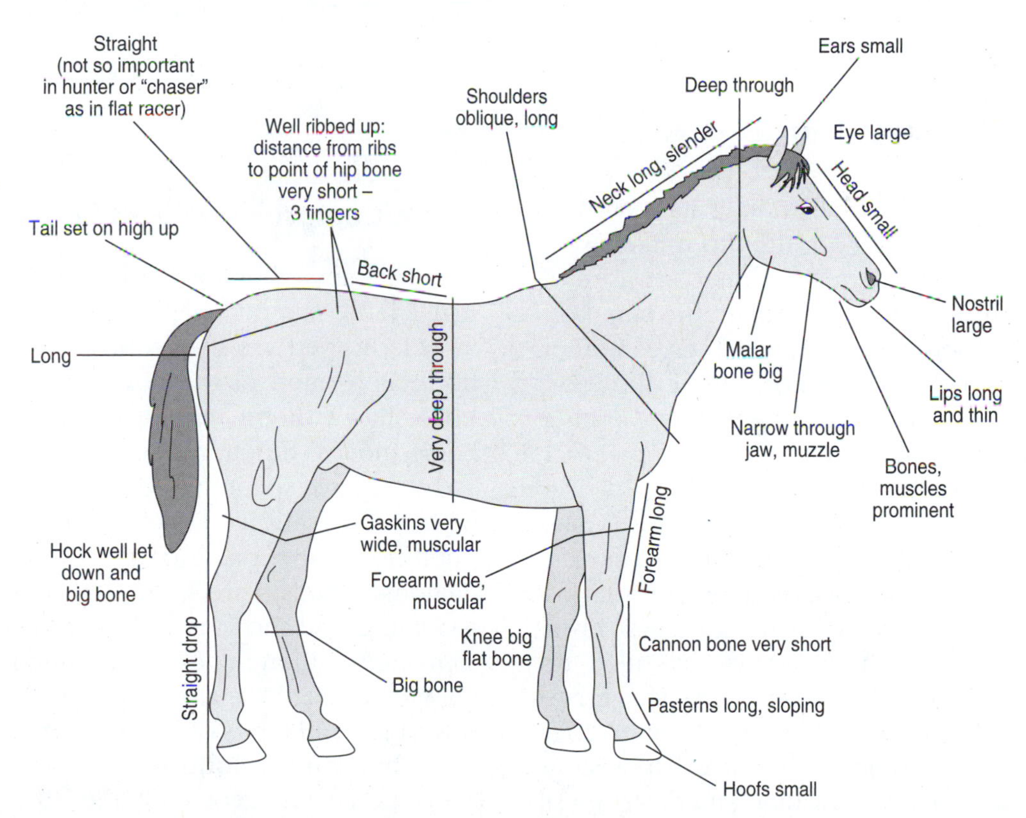

Figure 8-9. The horse has certain anatomical features that make it suitable for use by humans.

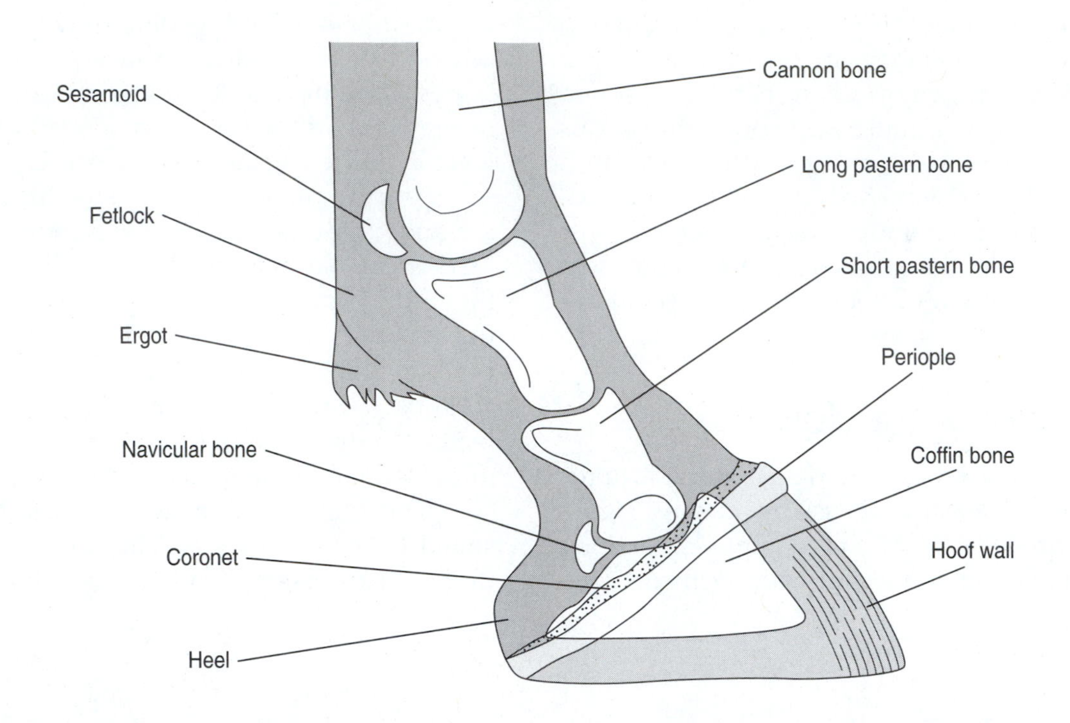

Figure 8-10. The feet of horses are well designed to carry heavy loads.

only a single toe on all four of its feet. The foot is enclosed with a tough horn-like structure that protects the tender inner structure of the hoof. Inside the hoof wall on the very bottom of the foot is the sole, which gives further protection to the inner portions of the foot. The heel provides a flexible weight-bearing structure that also serves as a shock absorber for the foot and leg, Figure 8–10.

The digestive system of the horse is also highly specialized. Unlike many large animals, horses are nonruminants; therefore, they lack a rumen that produces enzymes to digest fiber. An enlargement in the digestive tract called the **cecum** provides a repository for large amounts of microbes that break down fiber into a form digestible to the horse. The cecum allows the animal to consume and digest grass and hay.

Horse Conformation and Body Type

Like any other species of agricultural animal, a horse should have the proper body type in order to perform the required tasks. How a horse is formed or the conformation of the horse has a direct impact on how well the horse moves, functions, or performs. Animal scientists have spent untold hours studying the best conformation for horses,and they have published many, many papers on the subject. Millions of dollars are spent each year buying, grooming, and showing horses with good conformation. Although horses are used for a variety of reasons, there are characteristics that are desirable in all types of horses; For example, a short back and a long level croup (the hip area) are advantageous whether the animal is carrying a rider or pulling a heavy load. In addition, the neck should be long and slender in order to give the horse balance. Long smooth muscles allow the animal to move freely and to work for long periods of time with less fatigue than an animal with short "bunchy" muscles.

A horse should be able to move freely on all its legs. For a horse to function properly, its feet and legs must be structurally sound. Many horses are born with defects that make the feet and legs less than perfect. These defects can cause problems as the horse walks or runs. If the legs are too straight, the bones will be jarred as the animal moves and a rider will not experience a smooth ride. If there is too much curve to the legs, undue strain will be placed on the muscles of the legs and the stifle. Figure 8–11 shows the proper placing of the legs and some common defects that should be avoided. Properly shaped and conformed bones and muscles allows the animal to function in the way it is expected to perform.

Raising Horses

Horses are generally bred using one of two methods—**pasture breeding** or **hand breeding**. Pasture breeding simply means that

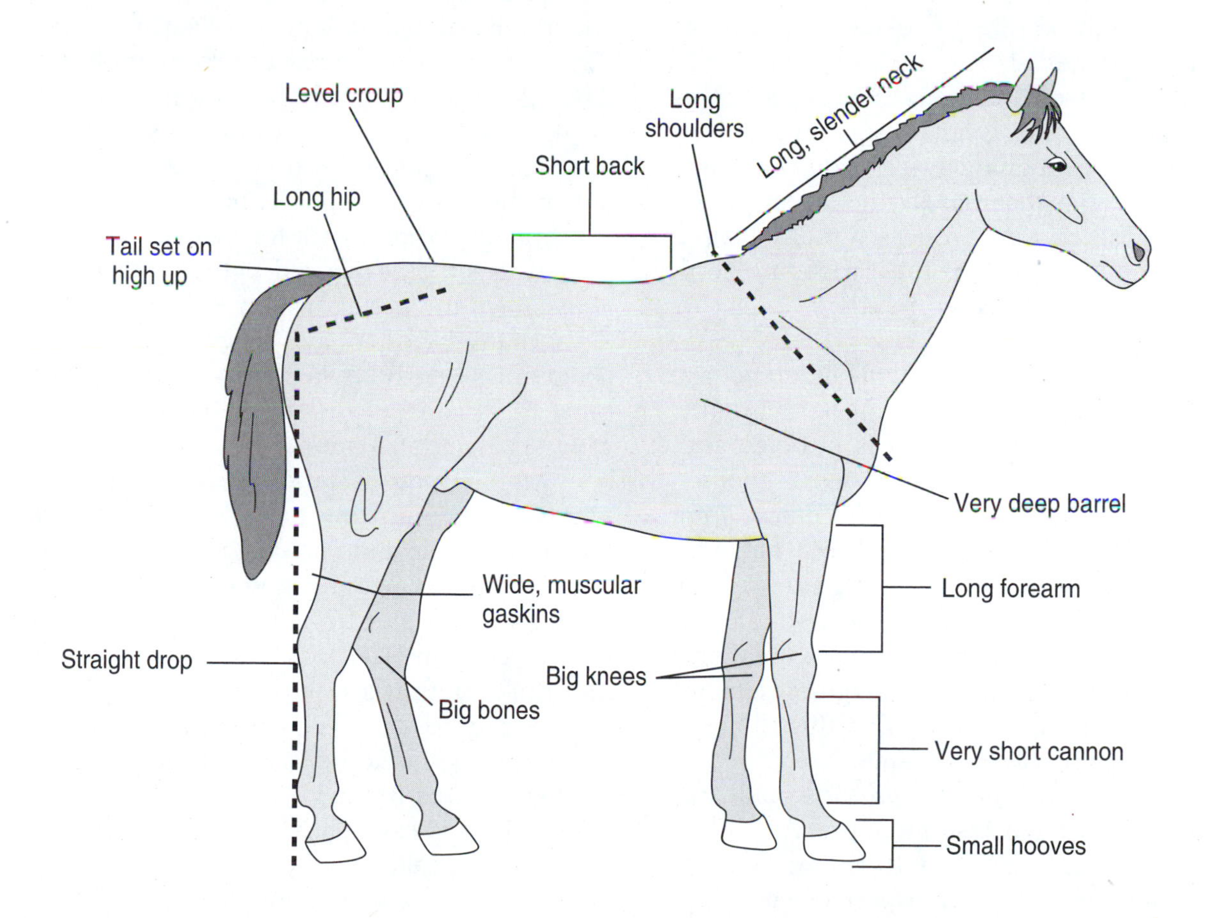

Figure 8-11. Desired horse traits.

Figure 8-12. Pasture breeding is less labor intensive and results in a greater pregnancy rate. *Kentucky Horse Park.*

stallions are turned into a pasture with mares, and mating takes place naturally, Figure 8–12. The advantage of this method is that it is less labor intensive and usually results in a greater percentage of pregnancies. The disadvantage of this method is that mares can sometimes get rough in the mating process. Often kicking and biting are a part of the procedure, and may result in blemishes in the skin of the mares or stallion.

Hand mating can be performed under a variety of conditions. The stallion is usually brought to the mare, which is hobbled and restrained during the mating process. Another alternative is for the mare to be placed in a breeding chute, a rectangular box with short walls in the front and on both sides. With either method, extreme caution should be used to avoid injury to both humans and horses. This method has the advantage of certainty about the date of breeding and this allows a closer estimation of foaling time.

Many breed registries are now allowing the use of artificial insemination (AI) in the equine industry. When performing AI, only fresh semen or cooled, transported semen can be used; shipping frozen semen, as is common in the cattle industry, is not allowed. On the other hand, embryo transfer, the process of removing an embryo from one mare and transplanting it into another mare, is also becoming fairly common in the equine industry. Before deciding on AI or embryo transfer, the breeder should contact the specific breed registry to obtain specific rules and regulations.

Equine reproduction is very dependent upon photoperiod and hormones. Horses are seasonal breeders, meaning they are only receptive to mating during specific times of the year. Mares begin to cycle when the days become longer, in the spring, and stop cycling during the fall months. Mares can be placed under artificial light to induce follicular activity sooner. A mare requires approximately 60 days of artificial light before ovulation will occur. The light cycle should consist of 16 hours of daylight and 8 hours of darkness to obtain the correct artificial photoperiod. This concept is very important for many breeds, as January 1 is the designated birthday for many foals. Since the gestation period is about 340 days, many breeders aim to get their mares pregnant as early as February. Having late born foals can be a huge setback in many show and racing situations.

After a long gestation period, the new foal is born. Following parturition, the navel cord is treated in a ten percent iodine solution, and the foal is usually given an enema of warm, soapy water to help the foal pass the fetal meconium. The foal should try to stand and nurse within a few minutes. If the foal does not try to stand and nurse within one to two hours, it should be helped to its feet and guided towards the mare's udder. As with all mammals, it is essential that the baby receive the first milk, called colostrum, from the mother. This milk is rich in antibiotics and nutrients needed by the newly born animal.

Figure 8-13. Foals are generally weaned between four to six months of age. *Courtesy of Progressive Farmer.*

Foals are generally weaned between four and six months of age, Figure 8–13. Males may be castrated any time from birth to two years of age. This is done to prevent the very aggressive behavior of a mature intact male. Geldings (a castrated horse) is easier to handle and has a better disposition than a stallion.

Training usually begins before the foals are weaned because very young horses are easier to handle and train. Training begins by teaching the foal to lead with a halter and accustoming it to humans. The age at which a horse is trained to accept the saddle and be ridden depends on the breed of horse and its intended usage. Most horses are taught this at age two, but some breeds are taught at a later age. The amount of training and the overall time it takes to train a horse depends on what the horse is going to be used for. Care and patience are required to train the animals to respond properly to humans.

Like all animals, there are many management practices that must be observed when owning a horse; for example, horses have to be dewormed and vaccinated regularly. The services of both a veterinarian and farrier are required periodically. A properly trained and well-managed horse will give many years of faithful service, Figure 8–14.

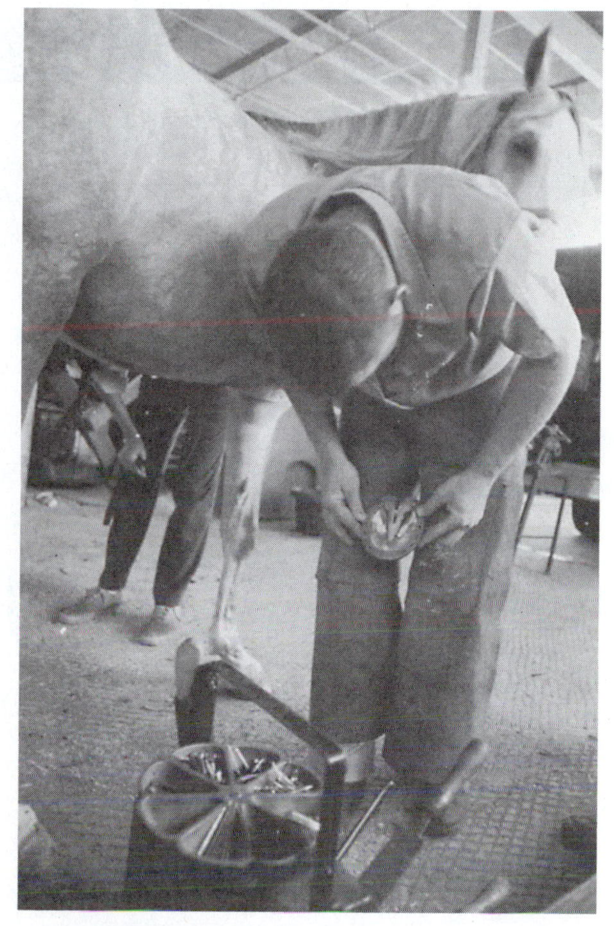

Figure 8-14. Horses require a lot of management and service. Care has to be given to the feet. *James Strawser, Cooperative Extension Service, The University of Georgia.*

Summary

Horses are a vital part of the history of this country. Over the years they have served well as workers and companion animals. These animals are uniquely designed to provide service to human beings. Currently, the numbers of horses are on the increase for use as pleasure animals. This trend is likely to continue as more people become involved in the horse industry.

◆ Review Exercises

TRUE OR FALSE

1. Horses and mules have been relatively unimportant in the building of this country. ______.
2. The number of horses and mules is increasing in the United States. ______.
3. France leads the world in the number of horses. ______.
4. Gaited horses are used primarily as draft animals. ______.
5. As a spectator sport, horse racing is relatively small compared to football. ______.
6. Mules are generally incapable of reproducing. ______.
7. Horses are much more intelligent than mules. ______.
8. The gestation period for a horse is about 340 days. ______.
9. Horses are seasonal breeders and can only be bred during the winter months. ______.
10. The Arabian breed has the largest number of registered horses in the country. ______.
11. The cecum is the digestive organ in the horse that allows for fiber digestion. ______.

FILL IN THE BLANKS

1. Horses have been used for _________, _________, and _________ since the beginning of recorded history.
2. The horse belongs to the genus _________.
3. While some horses and mules are used for work in the United States, most are used for _________ purposes.
4. _________ horses are those that weigh over 1,400 pounds.
5. Mules are crosses between _________ and _________.
6. The horse is classified as a _________ or an animal with only a single toe on all four of its feet.
7. Horses are generally bred using one of two methods—_________ breeding or _________ breeding
8. It is very important for foals to obtain _________, which is present in the mare's milk, to aid in boosting their immune system.
9. Foals are generally weaned between _________ and _________ months of age.
10. A castrated horse is called a _________.

DISCUSSION QUESTIONS

1. Discuss some of the uses humans have made of horses in the past.
2. What are the three classes of horses?
3. Explain why mules make such good work animals.
4. Discuss the anatomical features of the horse that make it suitable for use by humans.
5. List at least five ways that horses are used presently in the United States.
6. Explain why horses are able to digest roughages.
7. Explain the advantages of hand breeding over pasture breeding.
8. How are foals cared for at birth?
9. Why are male horses castrated?
10. When does the training of a horse usually begin?

STUDENT LEARNING ACTIVITIES

1. Log onto the internet and research a particular breed of horse. Find out about its origin, characteristics, uses, and popularity. Report to the class on your findings.
2. Take a field trip to a horse show or training facility. Make notes on how the behavior and nature of the horses are used in the training. Report to the class.
3. Invite a veterinarian to visit the class to discuss good management procedures and basic first aid in caring for horses.
4. Go to the Agricultural Census site on the internet and find out the number of horses in your state. How does your state compare to other states? Think of some reasons why there are relatively many or few horses in your state. Share your reasons with the class.

Chapter 9

The Aquaculture Industry

Student Objectives in Basic Science

As a result of studying this chapter, you should be able to

1. Explain why fish gain more on less feed than other animals.
2. Discuss how fish obtain oxygen.
3. Explain how oxygen is dissolved in water.
4. Explain how oxygen is depleted from the water.
5. Distinguish between cool water fish and warm water fish.
6. Discuss the behavioral characteristics of bullfrogs that make them difficult to raise.

Student Objectives in Agricultural Science

As a result of studying this chapter, you should be able to

1. List the reasons why aquaculture is a rapidly growing industry.
2. List the advantages fish have over other agricultural animals in a production operation.
3. Discuss the problems associated with fish production.
4. Explain why catfish are the most widely produced aquatic animal in the United States.
5. Describe the methods used in the production of various aquatic animals.

Key Terms

aquaculture
crustaceans
steer
cold-blooded
photosynthesis
warm-water fish
cold-water fish
fry
fingerlings
seines
larvae

Aquaculture is the growing of animals that normally live in water. This production includes freshwater and saltwater fin fish; **crustaceans** (shrimp, prawns, and crayfish); mollusks (clams and oysters); amphibians (bullfrogs) and reptiles (alligators). Throughout recorded history, humans have eaten fish and other animals that live in streams, lakes, ponds, and the ocean. A ready supply of high quality, protein-rich food could be obtained by harvesting organisms from the waters. As with other agricultural animals, humans soon discovered that by producing their own aquatic animals, the supply would be more dependable and easier to harvest.

Although it is difficult to determine just when aquaculture began, archaeologists know that people have raised fish for at least three thousand years. The ancient Chinese and Egyptians kept fish in captivity for use as food. This is evidenced by the paintings and drawings on the walls of the tombs of the ancient Egyptians and by the writings of ancient Chinese scholars. Later the Romans grew such aquatic animals as fish and eels.

The commercial growing of fish has increased in recent years. Each year over 5 million tons of fish are produced by fish farmers throughout the world. Asian countries grow more tonnage of aquatic animals than any other region. Europe follows Asia in fish production; North America ranks third. In the United States, fish culture is one of the fastest-growing agricultural enterprises. There are several reasons for the rapid increase in the amount of fish raised. The demand by consumers for seafood (a term for all of the aquatic animals) has increased to almost 15 pounds per capita. Although this amount is far behind countries such as Japan where the per capita consumption is 148 pounds, seafood still accounts for over 12 percent of the meat consumed by Americans.

Until a few years ago, the demand for fish and seafood had been easily met by the commercial fishing industry that harvested wild fish from the sea and from freshwater sources. As the world demand increased due to population increases, the commercial fishing industry had trouble meeting the demand because of overfishing in certain areas of the world. As a result of scientific research, aquatic animals are understood much better now than they were in the past. This allows producers to use new knowledge in order to provide the type of environment that allows the animals to be produced efficiently and economically.

FISH PRODUCTION

Fish have several advantages over other agricultural animals. While it takes about nine pounds of concentrate feed for a **steer** to put on a pound of gain, a fish will gain a pound on about two pounds of feed. This is because fish are ectothermic (once called **cold-blooded**) animals that do not require a large portion of their nutrient intake to go into maintaining body temperature. The bodies of ectothermic animals adjust to the temperature of their environment, so their body temperature is regulated by the surroundings. An endothermic animal—such as a mammal—regulates its own body temperature, and the internal temperature remains relatively constant. In addition, the natural buoyancy of their bodies in water helps fish move and gives their bodies support; therefore, they use less energy in moving about.

Fish have a higher percentage of edible meat than other animals. A steer generally has only about 35–40 percent of its body weight in edible meat. A catfish is about 55 percent edible meat; a trout may be as high as 85 percent edible meat. Because of this factor, much more meat can be produced on an acre devoted to

Figure 9-1. A well-managed pond can produce as much as six thousand pounds of catfish per acre. *USDA photo.*

fish production than with any other agricultural animal. A well-managed pond can produce as much as six thousand pounds of catfish per acre, Figure 9–1. With an ever-increasing world population, this could prove to be quite substantial in the future.

Fish producers face problems not encountered with the production of other agricultural animals. The grower must make sure that the dissolved oxygen level in the fish ponds is adequate for the fish. Like all animals, fish must have oxygen to live. While land animals obtain oxygen from the air they breathe, fish get oxygen from the water. Fish have gills that serve much the same purpose as lungs in land animals. The lungs in air-breathing animals separate oxygen from the other gases in the air and pass the oxygen into the bloodstream. In fish, the gills take oxygen from the water and pass the oxygen to the bloodstream, Figure 9–2. This oxygen comes in the form of oxygen that is dissolved in water by green aquatic plants that release oxygen in the process of **photosynthesis**. Since photosynthesis occurs only when the sun is shining, oxygen is released from these plants only during the daylight hours.

Oxygen is also passed into the water directly from the atmosphere through waves blown by the wind, ripples produced by a moving stream, and waterfalls that allow water to drop through the air. Most human-made ponds are static and get very little movement like we see in flowing streams or waterfalls.

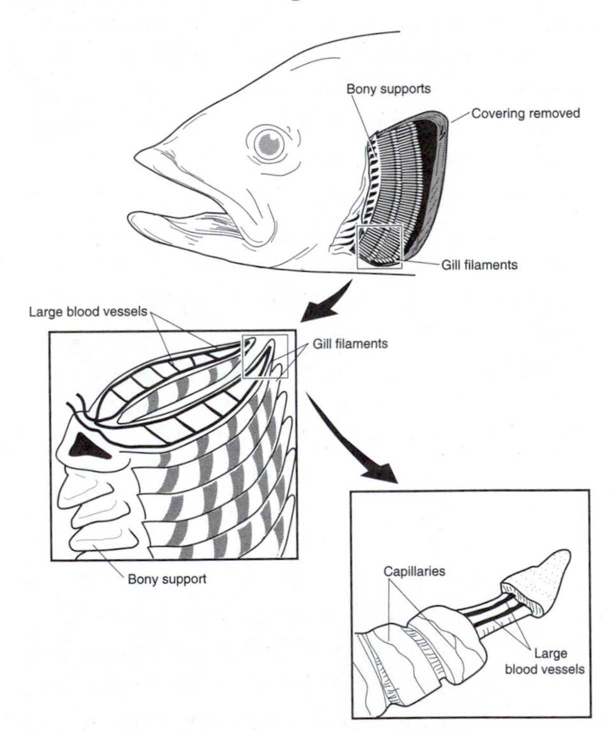

Figure 9-2. The gills of a fish take oxygen from the water.

Fish producers must rely on power-driven devices to fling the water into the air in order to absorb oxygen when the oxygen level of the water is low due to atmospheric conditions. Calm, cloudy days that have a very high temperature may cause the oxygen level in the ponds to fall below the level needed by the fish. Under these conditions, nights are particularly bad because aquatic plants are not undergoing the photosynthesis process that releases oxygen into the water. Without aeration, the entire population of a pond may die on a hot night. To prevent this from happening, the water is periodically monitored throughout the day and night with an oxygen meter that tells the operator how much dissolved oxygen is in the water. When the oxygen falls below an acceptable level, large aerators are turned on to throw the water high into the air in order to absorb more oxygen, Figure 9–3.

Shipping the fish presents another problem. After fish die, the meat spoils very quickly; therefore, the fish have to reach the processing plant alive in order to produce the highest-quality product. Specially equipped tank trucks are used to deliver the fish; they are equipped with gauges that closely monitor the water for temperature and oxygen levels, Figure 9–4. The fish are loaded and unloaded through large tubes that pump water containing fish to and from the truck tank, Figure 9–5.

Figure 9-3. Large mechanically powered aerators are used to replace depleted oxygen. *George Lewis, Cooperative Extension Service, The University of Georgia.*

Figure 9-4. To deliver fish live to the processor, they are transported in special trucks that monitor the oxygen. *George Lewis, Cooperative Extension Service, The University of Georgia.*

Because fish have to be so closely monitored in all phases of their production, the operations are said to be labor-intensive. This means that people have to spend a lot of time on the job in order to produce the fish. The operations are also considered to be relatively high risk because the fish can be lost so rapidly.

Figure 9-5. The fish are loaded and unloaded through large tubes. *George Lewis, Cooperative Extension Service, The University of Georgia.*

Commercially grown fish are grouped into two broad categories: **warm-water fish** and **cold-water fish**. Warm-water fish will not thrive in water temperatures below 60°F and cold water fish will not thrive in water warmer than 70°F. In the United States, the most popular commercially raised fish are the catfish and tilapia (warm-water fish) and the trout and salmon (cold-water fish).

Catfish Production

The most widely grown fish in this country is the catfish. Each year producers in the United States harvest and market almost 100,000 metric tons of catfish, amounting to a farm value of catfish that exceeds $380 million. These fish are gaining an increasingly wide market in all areas of the nation. Consumers are beginning to recognize farm-raised catfish as a tasty alternative to other forms of fish and seafood.

Catfish are different from most fresh water fish in that they have a smooth skin with no scales, Figure 9–6. They are hardy fish that produce well in small ponds and survive in lower levels of oxygen than most fish. Although there are many different varieties of catfish, the only variety of economic importance is the channel catfish. If left to grow for many years, these fish can grow to weigh over a hundred pounds. Huge channel catfish are caught in large lake reservoirs every year in the deep South.

Figure 9-6. Catfish have a smooth skin with no scales.

Since these fish do best when the water temperature is around 85°F, most are produced in the South. Mississippi is the leading producer of catfish with about 80 percent of production followed by Arkansas and Alabama. Most catfish are grown in open ponds in not more than six feet of water. In Mississippi, the acreage of ponds can be quite large. Farms with 250 acres under water can commonly be found, and a number of farms in the state have as much as 1,000 acres in ponds.

Eggs are collected from female catfish by allowing them to lay in nests provided by the producers. The eggs are then collected and placed in tanks or jars in the hatchery. The eggs are gently moved back and forth by means of paddles in the tanks that slowly move the water in a wavelike action, Figure 9–7; or the bottles containing the eggs are moved in a slow rocking action.

Figure 9-7. Fish eggs are hatched in tanks with paddles that keep the water moving. *George Lewis, Cooperative Extension Service, The University of Georgia.*

Just as with bird eggs, catfish eggs have to be turned in order for the embryo to grow properly. The rocking motion involved in the turning provides the movement necessary for the embryos to develop.

When the small fish (called **fry**) hatch, they are placed in a tank until they are about two inches long. These young fish, called **fingerlings**, are then placed in ponds or put in cages where they will remain until they reach a weight of one to two pounds. The fish are fed a commercially processed food that is compressed into small pellets. The fish are fed twice a day by spreading the pellets on the water. In larger operations, labor is reduced by using a feed truck that drives to the edge of the pond and blows feed into the water.

The ponds are constructed so that producers can move through the ponds with large nets called **seines** to harvest the fish. Several passes with the seines are usually necessary to get most of the fish out of the pond, Figure 9–8. The fish are then placed in holding tanks until they are pumped into trucks for transport to market. Salt is added to the water of the transport truck in order to keep the fish alive and well during the trip to the market. The salt water has a calming effect on the fish, and therefore the stress is lessened.

Figure 9-8. Catfish ponds are designed so the fish can be caught using seines. *USDA.*

Figure 9-9. Fish are sometimes raised in cages that are submerged in the water. *George Lewis, Cooperative Extension Service, The University of Georgia.*

Another method is that of growing the fish in cages that are submerged in water, Figure 9–9. This method has several advantages to growing fish in open ponds. The fish are kept in a confined area where they may be inspected more closely. Less feed is wasted since the feed is only spread out within the cage. Cage raising helps solve problems of predators—such as turtles, cranes, and herons—that feed on fish in an open pond. Also, fish raised in a cage are much more easily harvested, Figure 9–10.

Tilapia Production

Tilapia, Figure 9–11, are fish native to Africa that are grown commercially all over the world. These fish resemble our native sunfish,

Figure 9-10. Fish raised in cages are much easier to harvest. *George Lewis, Cooperative Extension Service, The University of Georgia.*

reproduce prolifically, grow rapidly, and are considered to be a good-quality food fish. They are very hardy fish that survive high temperatures, low oxygen levels, and overcrowded conditions. Tilapia are a widely cultured fish and are second only to carp world wide. Tilapia have been raised in the United States only within the past few years, but they are gaining in importance. Fish biologists consider them to have high potential as a commercially raised food fish in the United States. Since it is a warm-water fish, it grows best in the southern region of the country. If the water temperature falls much below 50°F, tilapia cannot survive.

Figure 9-11. Tilapia are native to Africa and have potential as a food fish in this country. *George Lewis, Cooperative Extension Service, The University of Georgia.*

Trout Production

Trout is considered to be among the best-tasting fish; it is served on the menus of restaurants all across the country. They are highly desirable as a food fish not only for eating quality, but also because such a high percentage of their body is edible meat. These fish are cold-water fish that cannot survive in climates where the water temperature gets above 75°F. Trout are raised in smaller quantities than are catfish and are produced in the northern part of the country.

Most trout are raised in concrete raceways where the water is kept clean and moving, Figure 9–12. The moving water helps keep the temperature low and oxygen in the water at an acceptable level. Diseases can be more easily controlled in concrete raceways; whereas disease organisms can be harbored in the soil of a regular pond. Harvesting in concrete spillways is much easier than in an open pond with an irregular bottom surface.

Figure 9-12. Trout are raised in concrete raceways. *George Lewis, Cooperative Extension Service, The University of Georgia.*

Salmon Production

Another cool-water fish that is gaining popularity as a cultured fish is the Atlantic salmon. In the coastal states of Washington and Maine, salmon are stocked in floating net cage enclosures that are anchored in coves and bays. The salmon are fed and cared for during an eighteen to twenty-four month period. The fish are then harvested at around 9–11 pounds. These fish are very meaty and have a flavorful taste. Although most of the salmon consumed in the United States comes from ocean fishing or from Norway, many authorities feel that the culture of salmon has a bright future in this country.

Sport Fishing

Fish are also grown for sport fishing. Hatcheries all over the country raise small fish for stocking ponds, lakes, and streams for the benefit of the sport fisher. Almost every state in the country has large human-made reservoirs that are kept stocked with game fish for recreational purposes. Many of the fish caught were hatched in commercially operated fish hatcheries or in hatcheries run by the government. Each year these hatcheries stock lakes and streams with bass, bream, crappie, muskie, trout, and several other species of fish. This use of taxpayers' money can be easily justified since recreational fishing is a big industry in the United States. This agricultural enterprise spurs offshoot industries such as fishing tackle, boats, and guiding services. People who run restaurants, hotels, and other stores all benefit from the people who come to the lake to enjoy a weekend of fishing.

Some people grow fish in their privately owned ponds and make money by charging people to fish. Charges are made either by the day or by the pounds of fish that are caught. As in most of the other types of fish-production operations, the fish are hatched, cared for, and harvested for a profit.

Bullfrogs

Frog legs are considered a gourmet food that is served in many restaurants. Although most of the frog supply comes from the wild, some cultivation of frogs occurs in Taiwan and Japan. In this country, frogs have not been successfully grown in large numbers. However, because the demand far exceeds the supply, small profits have been made as a sideline crop from fish ponds, Figure 9–13. The frogs are harvested much as they would be in the wild.

Figure 9-13. Bullfrogs are raised commercially in some parts of the world. *George Lewis, Cooperative Extension Service, The University of Georgia.*

Attempts have been made in the United States to produce frogs, but because of the problems faced with production, few have been successful. One of the major problems is that these animals are so territorial. This means that the animals claim a certain area and do not allow other frogs into their space. A second problem is that frogs will only eat food that is alive so they cannot be fed processed feed. Their diet includes their own young, which can cause a real problem when they are raised in captivity. A third problem is predators. Almost all of the predators that inhabit areas near water feed on frogs, including raccoons, fish, herons, snakes, ducks, and cranes. As with the development of all the other agricultural animal industries, researchers will someday devise a way to profitably raise bullfrogs to meet market demand.

CRAYFISH

Crayfish (also known as crawfish or crawdads) are raised commercially in several states. The largest producing state is Louisiana, where over 10 million pounds are produced each year. Other states that produce crayfish include Oregon, California, Washington, Texas, and Mississippi, Figure 9–14. With proper management, over one thousand pounds per acre can be produced.

Crayfish are grown in constructed earthen ponds that are not over two feet in depth. They are often grown along with such crops as rice. Crayfish are omnivorous—they eat both plants and animals—but most of their diet is made up of decomposing plant material. Crops like rice leave large amounts of stubble behind when harvested. As the stubble decays, a large

Figure 9-14. Crayfish are produced in at least six states. *George Lewis, Cooperative Extension Service, The University of Georgia.*

amount of food is created for the crayfish. In addition to decomposing plant material, crayfish also eat worms and insect **larvae**.

The crayfish are put in the ponds in the spring. The water is slowly drained off in the late summer. As the water is lost, the crayfish burrow into the bottom where they reproduce. The adult females lay large numbers of eggs that hatch during the summer and early fall. These young are the crayfish that provide the harvest.

In the summer, a cover crop such as rice is planted and the pond is flooded. The crayfish are harvested in the late fall, winter, and early spring. Harvesting is usually done with a trap made from chicken wire that is closed on the bottom and has an inverted funnel at the top, Figure 9–15. Canned dogfood, cottonseed cake, or chunks of fish are placed in the traps for bait. Trapping is more effective at night because the crayfish are more actively searching for food. Once harvested, the crayfish are packed into porous bags and shipped. As long as the bags are kept cool and wet, the crayfish will survive and reach the processing plant in good condition.

Alligator Farming

At one time, alligators were hunted to the point of extinction because of the high value of their hides. Today, because of extensive conservation efforts, the numbers in the wild have greatly multiplied until they are no longer endangered. As a result of the efforts of conservationists, techniques of growing alligators were perfected. This technology is now used to commercially produce the animals on farms located primarily in Louisiana, Florida, and Georgia.

Alligators can be harvested at about twenty-six months of age when they have reached lengths of five to six feet, Figure 9–16. The hides are sold to make exotic leather goods such as bags, boots, shoes, and belts. The meat is quite tasty and is sold to the restaurant trade. Specialty items such as skulls and teeth are also sold.

Brood alligators are obtained from other producers or can be gotten from the wild. To use wild alligators, producers have to secure a permit from the state game commission and agree to release a predetermined number of alligators a year or more in age back into the wild.

Figure 9-15. Crayfish are harvested by catching them in traps. *George Lewis, Cooperative Extension Service, The University of Georgia.*

Figure 9-16. Alligators can be harvested at about 26 months of age when they reach a length of five to six feet.

The females build nests from vegetation and mud, lay around forty eggs in the nest, and cover them over. Producers remove the eggs from the nest as soon as they are laid because the eggs are a favorite food of several predators. As the eggs are removed, they are marked so that the proper end will be placed in an upright position throughout incubation in order to ensure that the eggs hatch. The eggs are then wrapped in hay and are kept very moist. The wet hay harbors bacteria that helps decompose the shell of the egg. A partially decomposed shell enables the young alligator to break through without much difficulty.

The temperature during incubation is critical in determining the sex of the young animals. Temperatures lower than 86°F produce all-female broods; temperatures above 93°F produce all-male broods. If the temperature is held at about 88°F the brood is mixed in gender.

When trawlers harvest the sea, fish are brought up in the nets that are not desirable for human consumption. These are the fish that producers use to feed alligators, Figure 9–17. Some animals are fed by-products from poultry-processing plants, even though research has shown that this diet is somewhat too high in fat content. Other sources of food include carcasses of animals that are raised and slaughtered for their fur.

Summary

Aquaculture is one of the newest components of animal agriculture. Because of the high efficiency of aquatic animals and the healthy, nutritious, contribution they make to our diet, this area of animal agriculture is likely to grow in the future. The operations are very labor intensive and expensive to operate, but the demand for fish and other products from aquatic animals continues to grow as the oceans of the world cannot keep up with demand. Research will ultimately show us how to produce the animals more efficiently and at a lower cost.

Figure 9-17. Alligators are fed fish that are left from the trawling industry. *James Strawser, Cooperative Extension Service, The University of Georgia.*

◆ Review Exercises

TRUE OR FALSE

1. Aquaculture can be traced back only three hundred years to the mid-1600s. ______.
2. In the United States, fish culture is one of the fastest-growing agricultural enterprises. ______.
3. A steer generally has only about 35–40 percent of body weight in edible meat, whereas fish range from 55–85 percent edible meat. ______.
4. The oxygen needed by fish is dissolved in the water by green aquatic plants that release oxygen in the process of photosynthesis. ______.
5. On a hot night without aeration the entire population of a pond may die. ______.
6. Fish-production operations need little labor and are low-risk. ______.
7. Catfish are cold-water fish that do not produce well when the water temperature is above 70°F. ______.
8. Eggs collected from female catfish are placed in tanks or jars in the hatchery. ______.
9. Tilapia are very hearty fish, but they cannot survive high temperatures, low oxygen levels, or overcrowded conditions. ______.
10. Most of the salmon consumed in the United States comes from ocean fishing or from Norway. There is not much of a future in farming it here. ______.
11. Taxpayers' money is thrown away each year on hatcheries run by the government to produce fish for sport. ______.
12. Crayfish are grown in constructed earthen ponds that are not over two feet in depth. ______.
13. As long as the bags containing crayfish are kept cool and wet, the crayfish will survive and reach the processing plant in good condition. ______.
14. Temperatures lower than 86°F during incubation of alligator eggs will produce all-female broods; temperatures above 93°F produce mostly males. ______.

FILL IN THE BLANKS

1. Aquaculture is the growing of such animals as fresh and ______________ fin fish; crustaceans such as ______________, ______________, and crayfish; mollusks (______________ and ______________); and amphibians such as ______________; and reptiles (______________).
2. Until a few ______________ ago, the demand for fish and ______________ had been easily met by the ______________ fishing industry that harvested ______________ fish from the ______________ and from freshwater ______________.
3. While it takes about ______________ pounds of ______________ feed for a steer to put on a pound of gain, a fish will gain a pound on about ______________ pounds of feed.
4. Oxygen is passed into the water directly from the atmosphere through waves blown by the ______________, ripples produced by a moving ______________, and waterfalls that allow water to ______________ through the ______________.
5. To deliver fish, specially equipped ______________ trucks are used that closely monitor the ______________ for ______________ and ______________ levels.
6. Catfish are hardy fish that produce well in ______________ ponds and survive in ______________ levels of oxygen than most fish.

7. Young catfish, called ________________, are placed in ponds or put in ______________ where they will remain until they reach a weight of ________________________ to ______________ pounds.
8. Raising catfish in cages has several advantages: fish can be ________________ more closely, less ____________ is wasted, there are fewer problems with ______________ such as herons, and the fish are more easily ________________.
9. Trout are highly desirable as a food fish not only for their ________________ quality, but also because such a ______________ percentage of their body is ______________ meat.
10. In the coastal states of Washington and ________________, salmon are stocked in ________________ net cage enclosures that are __________________ in coves and ________________.
11. Among some of the difficulties in raising bullfrogs is the problem of their territorial nature. They claim a certain ____________ and do not allow other ________________ into their area.
12. Crayfish are omnivorous—they eat both ________ and ____________—but most of their diet is made up of ______________ plant ________________.
13. Trapping crayfish is more effective at _____________ because they become more ______________ and are searching for ________________.
14. Alligators can be harvested at about ________ months of age when they have reached lengths of ________________ to __________ feet.
15. Alligators are mainly fed fish caught in ________________ that are not fit for human __________________.

DISCUSSION QUESTIONS

1. For how long have people been engaged in aquaculture? How do we know this?
2. Why are fish more efficient users of feed than are traditional agricultural animals?
3. List at least two important problems associated with the production of fish.
4. Why are catfish so widely produced?
5. What are the characteristics of tilapia that give them such a high potential for production?
6. What are the advantages of raising fish in cages?

   Why are trout and salmon raised in concrete raceways?
7. What are three problems that make bullfrogs difficult to raise?
8. Why are crayfish grown with such crops as rice?
9. What are the commercial uses for alligators?
10. What governmental regulations apply to the raising of alligators?

STUDENT LEARNING ACTIVITIES

1. Visit a large grocery store that markets seafood. Make a list of all the types of food from aquatic animals that the store sells. Determine which of the aquatic animals were caught in the wild and which were produced in aquaculture operations. List the differences in price for each category.
2. Locate and visit an aquaculture operation in your area. Determine the problems and solutions in running the operation. How are the animals marketed?
3. Talk with a local conservation officer about the types of fish that are released in lakes and streams in your area. Find out where the fish come from, the stocking rate, and plans for introducing different types of fish in the future.

Chapter 10

The Small Animal Industry

Student Objectives in Basic Science

As a result of studying this chapter, you should be able to

1. Describe how humans first adopted pets.
2. Explain behavior characteristics that make some animals good companions.
3. Describe how different breeds were developed.
4. Analyze the health benefits of owning pets.
5. List some diseases that may be transmitted from pets to humans.
6. Describe ways of preventing the transmission of diseases from pets to humans.

Student Objectives in Agricultural Science

As a result of studying this chapter, you should be able to

1. Describe the importance of the pet industry to the economy of the United States.
2. Tell how companion animals were and still are used in agriculture.
3. Explain how dogs are classified according to use.
4. Describe how by-products are used in the processing of pet food.
5. Explain the regulations governing the raising and importing of companion animals.

Key Terms

companion animals
exotic pets
service animals
assistance dogs
hippotherapy
zoonoses

A very large and rapidly growing animal industry is the raising of and caring for pets. In the United States, almost two-thirds of households own pets. Americans spend over $20 billion a year on companion animals. Dog and cat food purchases alone account for about $8 billion of the total spent on pets. By comparison, people spend approximately $1 billion on baby food.

Dogs and cats are the most common pets, and, in recent years, cats have outnumbered dogs as the favorite pet in the United States. There are about 60 million cats and 52 million dogs in American households, Figure 10–1. Many cat owners have more than one cat, so, although cats are more numerous, dogs are found in more households. Other pets commonly owned by people are horses, rabbits, hamsters, gerbils, guinea pigs, or fish. There is a trend to own exotic animals such as pot-bellied pigs, reptiles, ferrets, fancy birds, or even tarantulas. This wide range of pets provides companionship for people of all ages.

Figure 10-1. Dogs are in more homes than any other pet. *Valerie Mosley, Wrightsville, Georgia.*

The popularity of dogs and cats as companions is due to a combination of factors. One of these is the ability of dogs and cats to form lasting social bonds with humans. They are able to establish a close relationship with their owners. They can also be house-trained fairly easily. House training is impossible with many **companion animals**; for example, horses. Dogs and cats also have the advantage of being large enough for humans to interact with and play with, and yet small enough to be kept in the house. Many of these pets are bred and raised in commercial operations that supply animals to be used for pets. A gigantic industry has grown around the care for pets or companion animals. Americans spend almost 7 billion dollars each year for pet food and an additional 8 billion for veterinary services, Figure 10–2. As the affluence of the average person is increased, the expenditures for pets and pet care will increase.

The History of Pets

Pets have been around for many thousands of years. Evidences of people keeping pets can be found dating several thousand

Figure 10-2. Americans spend more than 8 billion dollars each year for pet food.

years BC. The first companion animals were probably domesticated from wild animals and served other purposes than strictly as pets. Dogs were used for hunting and herding animals; they also served as watch animals to warn humans of the approach of wild animals or strange humans.

Dogs

Dogs have been associated with humans as far back as the Stone Age. All modern dogs developed from wild dogs that resembled wolves. Dogs are scavengers, and archeologists think that dogs may have adopted humans rather than the humans adopting the dogs. The theory is that dogs began hanging around villages to scavenge the leftover food of the humans. People probably discovered that the dogs could help in the hunts by tracking and herding animals, so they began to raise the animals for hunting, Figure 10–3.

A characteristic of dogs is that there can be quite a mixture of traits in the animals born in the same litter due to genetic variation. Later, as humans began to see traits they liked in certain dogs, they began to select animals with those characteristics and bred the males and females to produce the type animal they wanted. This is the way different breeds of dogs developed. Within the dog family there are breeds that weigh scarcely two pounds and breeds that weigh over 200 pounds. They come in all sizes, shapes, colors, and temperaments. In fact, there are over 400 recognized breeds of dogs in the world.

Breeds of dogs are divided into seven major groups. The *Sporting Group* includes hunting dogs such as the Labrador Retriever, the Irish Setter, and the Brittany Spaniel, Figure 10–4. The *Hound Group* is used for tracking and treeing game; it includes such breeds as Beagles, the Bloodhound, and the Black and Tan Coonhound. The *Terrier Group* are smaller dogs that includes the Fox Terrier, the Welch Terrier, and the Bull Terrier. The *Working Dog Group* was developed to provide service as sled dogs, guard dogs, and messenger dogs.

Figure 10-3. People discovered that dogs could be useful in herding animals. *The Working Border Collie Magazine.*

Figure 10-4. The Brittany Spaniel belongs to the Sporting Group of dogs. *Bryan Herren, Watkinsville, Georgia.*

Examples of this group are the Alaskan Malamute, the Boxer, and the Doberman Pinscher. The *Herding Dogs* were bred to help in the raising of livestock by herding the animals and protecting them from predators. These include the Border Collie, The Old English Sheepdog, and the Australian cattle dog. The *Toy Dog Group* are the smallest of the dogs and include such breeds as the Chihuahua, the Pekinese, and the Pug. The last group is the *Non-sporting Dogs*. This group is composed of a wide variety of dogs that are used primarily as companion animals. In this group are the Bulldog, the Poodle, and the Dalmatian.

Cats

Archaeologists were able to distinguish the remains of the first domestic cats from the wild species in their excavations of ancient Egypt; they say the cat was to some extent bred and worshiped in Egypt (1570–1085 BC). However, the popularity of the cat may have stemmed—more than any other reason—from the protection they gave to granaries by killing mice and other rodents. The ancient Romans also valued cats for their service of ridding homes and grain storage areas from destructive vermin, as well as their use as companion animals. The Romans probably were the first people to take cats into Europe and other parts of the world. Today, cats inhabit almost every country in the world, both as wild and domestic varieties.

In the history of the United States, cats were used to rid homes of mice and other vermin; they could be found on farms and in homes all across the nation. Most times these animals earned their keep or by catching and consuming their own food, and they often lived in the barns or grain storage buildings. Today, cats are very popular even though they are not used as much to catch mice. They are excellent companion animals because they are very clean, quiet, and intelligent. Cats provide enjoyment for millions of people, Figure 10–5.

Figure 10-5. Cats are clean, quiet animals that make good companions. *Valerie Mosley, Wrightsville, Georgia.*

Unlike dogs, cats have only relatively recently been bred selectively, and, they have much less variation among breeds. Consequently there are much fewer breeds of cats than dogs. In fact, most cat breeds can be traced back to the late nineteenth century continuing to the present. Most of the breeds that are named after a region such as the Persian, and the Abyssinian probably did not originate in these areas. Breeds of cats are generally divided into two groups: shorthair and longhair.

Exotic Pets

Many people prefer pets other than the traditional dogs, cats, and fish. A large industry has developed around the raising and/or importing exotic animals. At one time, there was a thriving market for animals captured from the wild and imported for use as pets. Now strict laws are in force that regulate the importation and sale of certain animals; in fact, it is illegal to import and sell most animals captured from the wild. Many endangered species are protected from sale to private individuals. These animals generally make poor pets and removing them from their natural habitat increases the risk of their extinction.

The importation of animals also carries a risk of bringing disease into this country. A classic example is New Castle Disease, a highly contagious viral disease that decimated much of the poultry industry in the United States. The disease is thought to have been brought in by a parrot smuggled into New England from Mexico. All animals brought into this country must go through a quarantine period to check for disease and parasites.

Reptiles

The fastest growing category of pets in the Unites States today is reptiles. There are more than seven million pet reptiles living in homes in the United States, Figure 10–6. Since reptiles are cold blooded, they have traditionally been considered as not making good pets. Through better education on the characteristics of these animals, their use as pets has increased. Snakes such as boas and pythons are common; iguanas are the most popular pet among reptiles. These lizards are the size of a domestic cat when full grown. They are clean, odorless, and can be house-trained. Most are grown in Central and South America. Since they are often grown under very unsanitary conditions, salmonella poisoning is a danger when handling these pets. Owners should always thoroughly wash their hands with antibacterial soap after handling these animals.

Figure 10-6. The fastest growing category of the pet industry is reptiles.

Reptiles require special care since they are exothermic animals. This means that they get their internal body temperature from the environment. They need lights and some source of heat such as a heat rock to keep them comfortable.

Health Benefits

Evidence is increasing that pets are more than just companions, they are good for people's health, Figure 10–7. Scientists are now discovering that living with a pet contributes to both the physical and mental well-being of

Figure 10-7. Pets are more than companion animals; they are good for people's health.

humans. There is a need in humans to have a relationship with something living. Since earliest human history, there has been the need to be around and to have dealings with other human beings. Now, scientists say that, at least in part, the need for companionship can be supplied by companion animals. The theory is that relationships with animal companions appear to be beneficial to humans because they are uncomplicated. Animals are accepting, attentive, and responsive to affection. They are not judgmental; they never talk back;they never criticize or give orders. Pets give people something to be responsible for and make them feel special and needed.

In the right circumstances, companion animals are a good influence on children of all ages because they help children develop a sense of security. They have been used to encourage shy or withdrawn children to open up, Figure 10–8. Children who are normally hyperactive often become much calmer around a companion animal. Caring for animals also helps develop a sense of responsibility that can be useful in all areas of life.

Figure 10-8. Under the right circumstances, companion animals can have a good influence on children. *Valerie Mosley, Wrightsville, Georgia.*

New evidence of the beneficial effects that companion animals have on human health is continually being discovered. We now understand that pets make people feel good, and a sick person who feels good mentally is likely to get better faster. People who have pets report fewer minor health problems like colds and flu. Studies have also shown that petting a dog or cat or watching fish in an aquarium can help lower a person's blood pressure and heart rate.

Companion animals can play an important role in the lives of older individuals. Due to the increasing mobility of today's society, many elderly people no longer have family members living close by. Many of them would feel isolated and alone without the pets that provide them with companionship and a sense of being valued and needed. Older people may actually live longer, healthier lives because of their relationship with their companion animals.

Today about 50 percent of nursing homes use animals in some capacity to aid in the care of the elderly and invalids. Bird aviaries and aquariums as well as cats, rabbits, and guinea pigs are popular among nursing home residents. Some authorities do not recommended that dogs live in nursing homes full-time because they tend to be overfed by the residents, who can't seem to resist feeding them cookies and treats. The dogs can become obese and health problems result. Today many volunteers take their dogs to visit the residents of nursing homes and hospitals in their community.

Service Animals

There are many people in the United States and other parts of the world who are challenged by some form of disability.

Figure 10-9. Dogs can serve as helpers to people who are physically challenged. *Courtesy Guide Dog Foundation for Blind Inc.*

Figure 10-10. Guide dogs must be able to recognize dangers. *Courtesy Guide Dog Foundation for the Blind Inc.*

Companion animals serve a tremendously valuable role in assisting with everyday aspects of life. Dogs serve as the eyes, ears, or legs for thousands of people who need assistance in moving about and tending to daily routines, Figure 10–9. There are a number of agencies in the United States that train **assistance dogs** to give more independence and mobility to people with disabilities. Training usually takes between four and eight months, depending on the difficulty of the tasks that must be learned and the aptitude of the dog. Although training an assistance dog can cost thousands of dollars, many agencies provide them to people who need them at little or no cost.

The best known example of an assistance dog is the guide dog for the blind. The most commonly used breeds are German Shepherds and Labrador Retrievers. To work as a guide dog, the dog has to be exceptional. It has to be able to walk through crowds, climb stairs, ride on elevators, ride on buses, and in cars. Most importantly, the dog has to be able to think for itself, Figure 10–10. It must learn to disobey a command if it could bring harm to its master.

Many of the organizations that train guide dogs have volunteers raise the puppies for the first year. Some of the volunteer puppy raisers belong to the 4-H organization. The volunteers are not expected to train the puppies as guides, but they are required to follow some basic rules. The puppies must be exposed to many of the activities they will need to be able to handle with ease as a guide. The puppy raisers are encouraged to take the puppies to the mall, to the park, to nursing homes, schools, or any other setting that involves the public. The puppies must be kept on a lead in public, and they must sleep next to the bed of the volunteer, just as it will when acting as a guide.

The volunteers are warned not to play ball, tug-of-war, or other games with the puppies because these games could turn into bad habits when they become guide dogs. The puppy raisers must never feed their pets human food because guide dogs must not be tempted by the sight or odor of human food since they may accompany their owners to restaurants. They can't be jumping up on tables begging for food.

Only about half of the puppies raised to be guide dogs successfully complete the training. During training, the dogs learn to work in a harness, and they learn commands like "Forward" and "Find the door." The dogs are trained through repetition and praise; they learn to ignore crowds, noises, squirrels, and cats. The dog must imagine itself to be as wide and as high as a human in order to successfully guide a person through doorways and buildings. It must also be aware that obstructions that it can walk around or under easily, like awnings or branches, would impede its owner. A guide dog wearing a harness is on duty and should never by petted by other people; however, when the dog is out of the harness, it is like any other family pet.

Hearing-ear dogs are trained to listen for those who cannot. Also called *signal dogs*, these animals can respond to more than 30 common household sounds like doorbells, telephones, alarm clocks, and fire alarms. They can even be trained to respond to a crying baby. The dogs alert their deaf or hard-of-hearing owners by walking back and forth from the source of the noise to the owner. Signal dogs are taught not to bark when alerting their master to a sound since the person would be unable to hear the barking. The dogs can also be trained to respond to sign language commands. Because the size of the dog is not important, hearing-ear or signal dogs are usually mixed breeds and are often rescued from local animal shelters.

Service dogs are trained to help people who use a wheelchair or have a spinal injury. They are able to respond to more than 40 different commands. Service dogs can open doors, work light switches, pull emergency cords, and pull wheelchairs. Each dog may be trained a little differently in order to address the needs of the individual who will own the dog. Service dogs need to be large and are often retriever breeds.

Another companion animal that has been beneficial to humans is the horse. Horses are used in programs of physical therapy for people with disabilities. This type of therapy, called **hippotherapy**, can be used with people of all ages, but it is especially helpful for physically challenged children, Figure 10–11. The gait of a horse simulates the motion of humans as we walk. When we walk, our bodies move from side to side and up and down. Riding a horse recreates that sensation in people who are unable to walk unassisted.

Through hippotherapy, physically challenged individuals improve their balance, posture, strength, coordination, and muscle flexibility. In addition to the physical benefits, the riders also gain confidence. Hippotherapy can provide them with a whole new perspective and a sense of freedom. A horse can take its rider where no wheelchair could go.

Figure 10-11. Horses can aid in the recovery of physically challenged people. *Courtesy of Cliff Ricketts.*

Volunteers are used in the care and maintenance of the horses as well as for lessons; therefore, hippotherapy programs offer many opportunities to local agriculture programs and youth organizations.

Pet Food

The pet food industry utilizes many of the by-products and surpluses of the human food industry. The main ingredient in dry dog foods is grain: corn, soybean meal, or wheat millings. The main ingredients in canned pet foods are meat by-products, which may include the waste products of meat that was processed for human consumption. Also, carcasses declared unfit for human consumption may qualify for use as pet food. The composition of pet food is carefully formulated to meet the nutritional needs of the animal. Canned, semimoist, and dry foods are equally nutritious, but the canned varieties generally contain a higher percentage of protein and fat.

Animal Health

Americans spend over $10 billion every year on the health care of their companion animals. There are approximately 40,000 veterinarians in the United States; one-third of them treat small animals exclusively, Figure 10–12. Like all veterinarians, those who specialize in the care of companion animals perform a wide variety of tasks everyday. They treat animal injuries, set broken bones, immunize healthy animals against disease, and perform surgery.

Veterinarians today can successfully perform hip-replacements and kidney transplants on companion animals. They have also performed balloon angioplasty to open clogged arteries, open-heart surgery, and dental surgery. In recent years, medical care for animals has become as highly technical as medical care for humans. In fact, new medical procedures are perfected on animals, so, in some instances,

Figure 10-12. Over one third of veterinarians treat small animals exclusively. *Valerie Mosley, Wrightsville, Georgia.*

animals may get even more advanced care than humans. However, the costs of these health care advances may be more than some pet owners are willing or able to bear.

Veterinarians stress preventive measures when they counsel pet owners. They encourage the pet owner to include vaccination programs and regular dental exams in their pet's health care plan. According to veterinarians, one of the most common problems they see in dogs and cats today is obesity. Half of American dogs and almost one-fourth of cats are obese. The animals suffer from overfeeding and a lack of exercise. Some animals simply do not get enough attention, Figure 10–13. When animals get bored, they have a tendency to eat too much just as people do. A sound diet and daily exercise routine should also be a part of the pet's overall health care plan.

Diseases and Afflictions

Unfortunately, pets can give us more than companionship. Each year pets pass along infectious diseases to thousands of Americans. There are approximately 30 varieties of pet-borne illnesses, which are called *zoonoses.*

Figure 10-13. Many animals have a tendency to overeat. *Valerie Mosley, Wrightsville, Georgia.*

Zoonoses are diseases and infections that can be transmitted from animals to humans. They can be passed on through direct contact with the animals or acquired indirectly through contact with animal feces or other contaminants. However, most diseases passed from animals to humans are easy to avoid and are treatable. Good hygiene and safe handling procedures should always be practiced when working with animals. Cats and dogs are responsible for the majority of zoonoses, but birds, fish, and turtles are also culprits.

Rabies is the best known and the most feared example of a zoonosis. It is contracted through the saliva of rabid animals. Although this disease is rare in pets, rabies is increasing in the wild animal population. Because pets may come into contact with wild animals, they should be vaccinated against the disease.

The common roundworm of dogs is a parasite that can infect humans. The parasite is transmitted through contact with the animal's feces or with contaminated soil. Children playing in areas frequented by dogs are especially at risk. Dog owners should make sure their pets are regularly dewormed, Figure 10–14.

Toxoplasmosis is a parasitic disease that can be caused by contact with cat feces. Toxoplasmosis can be especially dangerous to pregnant women, so most veterinarians recommend pregnant women not clean cat litter boxes.

Psittacosis, or parrot fever, is a disease transmitted by parrots, budgerigars, and other related caged birds. Humans can be infected by contact with the feces of contaminated birds. In handling birds and cages, dust masks or protective face shields should be used.

Ringworm is not a worm; it is a fungus that results in skin aggravation in humans. It is primarily passed to humans by kittens and puppies. The animal appears unaffected

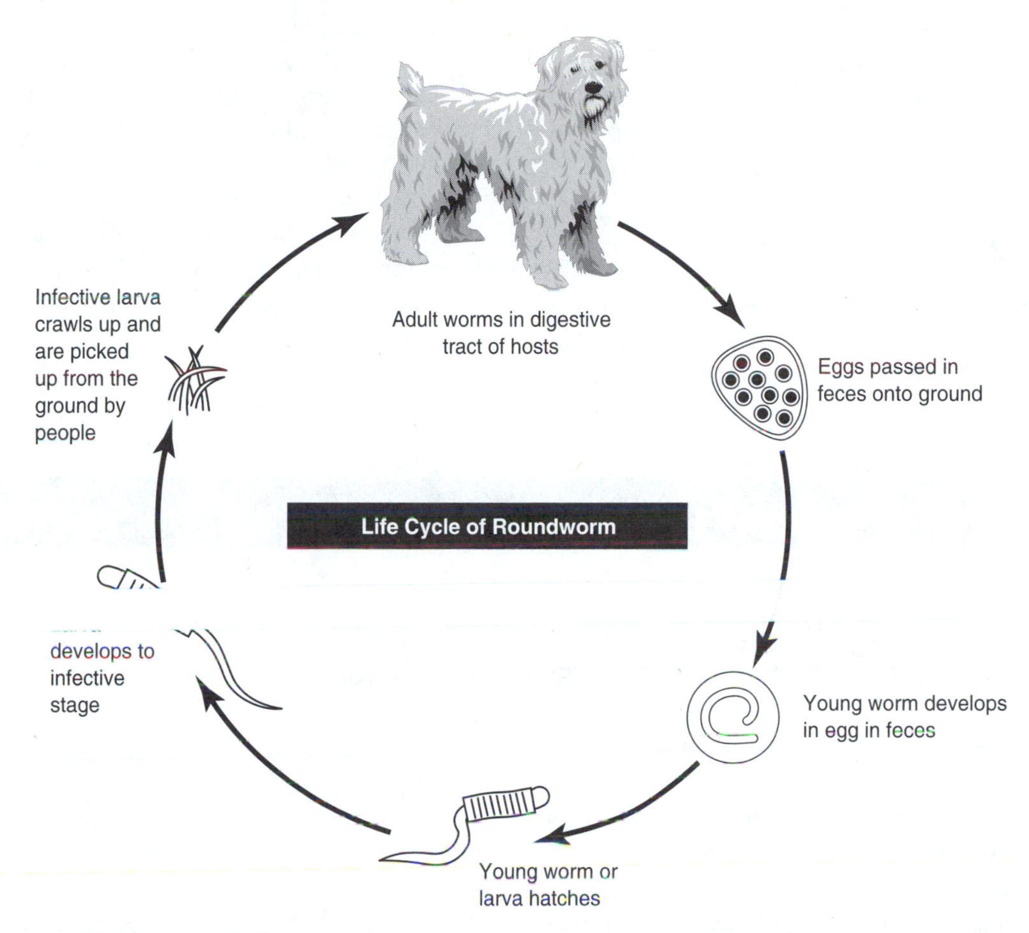

Figure 10-14. The common roundworm can be transmitted by contact with contaminated soil.

because the fungus infects only the animal's fur. It is passed to you when you handle your pet. It is more common in children because adults seem to become more resistant to it with age.

Rocky Mountain spotted fever and Lyme disease are tick-transmitted diseases that can affect both humans and animals. Ticks are found in grassy, wooded areas; they can be brought into the house by dogs and cats that have been outside. Both diseases are treatable with antibiotics. Normal grooming of animals after they have been outside will help locate and eliminate ticks.

Infections due to animal bites and scratches are another concern. The potential for infection varies. Less than five percent of dog bites become infected, but up to 50 percent of cat bites do. Cat scratch fever is associated with cat scratches or bites. This disease is not serious and can be treated with antibiotics. Safe handling techniques are an important measure to prevent bites and scratches as well as to prevent injury to the animal. Prompt and thorough washing of pet bites and scratches with soap and water is always important.

Allergies are probably the most common afflictions that occur when humans have contact with animals. Many people develop allergies to animal hair and their dander, or flaking skin. Allergies cause hay fever-like symptoms that can occur in both children and adults.

SUMMARY

The pet industry in the United State is huge and growing. Americans like pets and are willing to spend large amounts of money to own and care for companion animals. These pets serve many purposes, ranging from service animals to making humans better adjusted and content. Types of animals range widely, and, as the result of a demand for exotic pets, the importation of animals is regulated closely by the government. The business of pets will continue to be a large and growing part of our economy.

◆ Review Exercises

TRUE OR FALSE

1. Americans spend more on pet food than they do on baby food. _______ .
2. There are more dogs than cats used for pets in this country. _______ .
3. Expenditures for pets are influenced by affluence levels of the households who own them. _______ .
4. The first pets were probably wild animals that were tamed. _______ .
5. Dogs were probably first used as hunting dogs. _______ .
6. The most important reason cats were tamed for pets is that they are so affectionate. _______ .
7. Cats are really not very effective in controlling mice. _______ .
8. There are more breeds of dogs than breeds of cats. _______ .
9. In the right circumstances, pets can have a good influence on children. _______ .
10. There is really no scientific evidence that pets help improve a personís health. _______ .
11. Any dog can be trained for use as a service dog. _______ .
12. There are no diseases that can be passed from pets to humans. _______ .
13. Parasites can be transmitted from pets to humans. _______ .
14. Allergies are very uncommon with pet owners. _______ .
15. The size of the pet industry is declining. _______ .

FILL IN THE BLANKS

1. Americans spend over __________ a year on companion animals.
2. Evidence is increasing that pets are more than just companions, they are good for people's ___________.
3. All of the modern breeds of dogs were developed from ______ ________ that resembled __________.
4. Within the dog family, there are breeds that weigh scarcely ________ pounds and breeds that weigh over _________ pounds.
5. Three examples of dogs in the Hound Group are _________, __________, and ___________.
6. Archaeologists were able to distinguish the remains of the first domestic cats in their excavations of _________ _________.

7. Cats are __________, __________, and __________ animals that provide enjoyment for humans.
8. Some authorities recommend that dogs not live in nursing homes because the residents tend to __________ them.
9. The best example of an assistance dog is the __________ dog for the __________.
10. Reptiles need a heat rock because they are __________ animals.

DISCUSSION QUESTIONS

1. List the most popular animals used for pets.
2. Explain why dogs and cats became important to early humans.
3. List the benefits of owning pets.
4. Discuss why companion animals are beneficial to elderly people.
5. Discuss how service animals are trained.
6. What is the concept of hippotherapy as it relates to horses and disabled humans?
7. List the groups that all breeds of dogs belong to.
8. Explain why there are more breeds of dogs than cats.
9. What by-products are used in pet food?
10. What precautions should be observed in handling pets to prevent disease?
11. List some of the common zoonoses.

STUDENT LEARNING ACTIVITIES

1. Choose a pet and list the characteristics of that particular animal that makes it desirable as a pet.
2. Pick a breed of cat or dog and research the origin of the breed. Report to the class.
3. Conduct a survey of the members of your class to determine the number and types of pets owned. Have each person explain why they like that particular type of animal.
4. Take one or two well-behaved pets to a local nursing home and visit with elderly people. Be sure to check with the nursing home administrators before you go. Record the reactions of the people to your pets. Report to the class.

Chapter 11

Alternative Animal Agriculture

Student Objectives in Basic Science

As a result of studying this chapter, you should be able to

1. Explain the meaning of the pH scale.
2. List the animals that are most often used in scientific research.
3. Define *certified laboratory animal.*
4. List the distinguishing characteristics of insects.
5. Discuss why honeybees are essential in the pollination of plants.
6. Discuss the orderly society of honeybees.
7. Discuss the threat of the African honeybee.

Student Objectives in Agricultural Science

As a result of studying this chapter, you should be able to

1. Define *alternative animal agriculture.*
2. List the advantages of raising rabbits as agricultural animals.
3. Discuss the uses of llamas.
4. Explain how fish bait is raised.
5. Explain why the ostrich has potential as a food animal.
6. Explain the importance of the honeybee to agriculture.
7. Explain how bees make honey.
8. Discuss the danger of the African honeybee to American agriculture.

Key Terms

alternative animal agriculture
hutches
USDA
pH
castings
laboratory animals
genetic defects
insect
hives
honeycomb
foundation comb
supers
bee space
propolis
brood chamber
queen excluder
brood cells
pupae
maiden flight
nukes
queen cells
royal jelly
swarm
killer bees

Alternative animal agriculture is the production of animals other than the traditional agricultural animals such as cattle, sheep, horses, and poultry. In modern times, producers have looked for animals other than traditional livestock to raise for a profit. Alternative animal production is usually small in scale and provides a product for a specialty market. Producers may have alternative operations to supplement their traditional operations. People who work full-time in areas outside agriculture may use the production of alternative animals as a hobby or as a way to make a profit in a part-time operation.

Figure 11-1. Domestic rabbits are raised in commercial operations. *American Rabbit Breeders Association.*

Rabbit Production

People have raised rabbits for food for hundreds of years. The Romans produced rabbits as far back as 250 B.C., and they used rabbit meat as a substantial part of their diet. The Phoenicians were great sailors and trading people who are accredited with introducing domesticated rabbits throughout much of the known world as far back as 1100 B.C. In the United States, domesticated rabbits were brought into the country sometime around 1900 and were produced in large rabbitries in southern California. Since that time, the rabbit industry has grown all across the country.

Although the majority of the rabbits grown are produced by part-time growers, there are several large commercial operations in this country, Figure 11–1. So many rabbits are produced in small private rabbitries that it is difficult to determine how many rabbits are produced in this country. Estimates are that seven million to ten million rabbits are produced each year and that Americans consume about 10 million to 13 million pounds of rabbit meat per year. Some of the meat consumed in this country is imported from Europe, where the rabbit industry is larger. France is the largest producer.

The American Rabbit Breeders Association (ARBA) registers and promotes all the breeds of purebred rabbits grown in this country. The association currently has over 36 thousand members.

There are several advantages to raising rabbits over raising other agricultural animals. First, rabbits can easily be raised by anyone under almost any climactic condition. Most are raised indoors in cages called **hutches,** Figure 11–2. Hutches consist of woven wire with boxes for the rabbits to sleep in and to bear their young. The facilities take up little space

Figure 11-2. Rabbits are raised in wire cages called hutches. *Dr. James McNitt, Center for Small Farm Research.*

Figure 11-3. Rabbits are raised in climate-controlled houses. *Bass Equipment Co..*

compared to other agricultural animals such as hogs or cattle. The rabbit houses are usually heated in the winter and cooled in the summer to provide comfort for the animals. However, in areas where the climate is mild, an adequately insulated house may provide the animals with a comfortable environment without artificial heating or cooling. This means that the producer can work with the animals in relative comfort as well, Figure 11–3.

Rabbits gain weight on a relatively small amount of feed. The feed efficiency ratio for properly fed and managed rabbits is about 2.5 to 1. This means that for every two and one half pounds of feed fed to the rabbit, the animal gains one pound in body weight. Rabbits can also be fed on a lower-quality feed than some of the other animals. A rabbit's digestive system allows the animal to make use of roughages such as alfalfa and other fibrous plant material. This is quite an advantage over other agricultural animals because rabbits can potentially be raised at less expense per pound, Figure 11–4.

The demand for rabbits is greater than the supply. As mentioned earlier, rabbit meat is imported into this country, so the potential exists for an expansion of the rabbit industry in the United States. Many restaurants now offer several dishes that are prepared using rabbit meat. The **USDA** points out that rabbit meat is one of the most nutritious meats available. Not only is it high in protein and low in fat and

Figure 11-4. Rabbits have a higher feed conversion ratio and can make use of lower quality feed than most other agricultural animals. *American Rabbit Breeders Association.*

	Calories	Protein (grams)	Fat (grams)	Water (grams)
Rabbit	136	20.05	5.55	72.82
Lamb	267	16.88	21.59	60.70
Veal	144	19.35	6.77	72.84
Beef	291	17.32	24.05	57.26
Pork	398	13.35	37.83	47.86
Chicken	215	18.60	15.06	65.99
Turkey	160	20.42	8.02	70.04

Figure 11-5. Nutritional value of common meats (per 100 grams). Note that rabbit meat is higher in protein and lower in fat than many other meats. *Human Nutrition Service, USDA.*

Figure 11-6. The female rabbit pulls fur from her body to make a nest for her baby rabbits. *Dr. James McNitt, Center for Small Farm Research.*

cholesterol but it is also easily digested and very flavorful, Figure 11–5.

In addition to the meat, rabbits are used for several other purposes. The fur is used in making coats, hats, liners for boots, and for toys. Scientists use the animals in experiments dealing with medical research; manufacturers use them for testing products. In addition, many animals are sold as pets because of their docile nature, clean habits, and cuddly fur.

Rabbits are very prolific breeders. They produce young thirty days after breeding and raise four to five litters per year consisting of up to eight young per litter. Over a period of one year, one pair of rabbits can produce a lot of meat, considering that some breeds of rabbits reach sexual maturity (the ability to have young) at about five months of age. An example of how rapidly rabbits reproduce is that of the wild rabbit in Australia. Rabbits were first introduced there in 1859 when sailors released a pair of wild European rabbits. In only thirty years, over 20 million rabbits inhabited the country. Since then the animals have become a serious pest in both Australia and New Zealand. Harsh measures have been taken to control the wild rabbit population.

In domesticated production, the young are born in small nesting boxes that give the mother security and comfort and offer the young protection from outside stresses. The female will pull fur from her own body in order to make a soft warm bedding for the newborn rabbits, Figure 11–6. After weaning, the young rabbits are put into cages where they are grown out to about eight weeks of age. This is considered the proper age for the animals to be slaughtered for meat.

Rabbits offer potential as a food enterprise, especially in developing countries. Since rabbits can digest roughages such as alfalfa and other plants, the feed can be low-cost compared to feed for other animals, and the initial outlay in beginning production can be less. Research has shown that rabbits can do well on feeds containing very little grain. Since many developing countries have an abundance of roughage that can be eaten by rabbits, the animals can provide a relatively inexpensive food supply of much-needed protein.

Rabbits also have a potential in this country as a relatively inexpensive source of meat.

Figure 11-7. A major drawback of raising rabbits for food production is that they are so cute and cuddly. *American Rabbit Breeders Association.*

Figure 11-8. Llamas are well adapted to thin, cool mountain air. *Charles Hackbarth, Mt. Sopris Llamas.*

Americans, however, do not consume as much rabbit meat as do other peoples of the world. Reluctance on the part of Americans is one major drawback for the industry. While we may accept the production and slaughter of cattle, hogs, and chickens, the thought of slaughtering cute, cuddly animals such as rabbits is repulsive to many, Figure 11–7.

LLAMA PRODUCTION

Llamas are native to South America and belong to the same family as camels. They have been raised in several countries in South America for hundreds of years. In Chile, Peru, and Bolivia, these animals were raised by the ancient Incas and other peoples for work animals. They are well adapted to the cool, thin mountain air of the Andes mountains, but can adapt to most climactic conditions, Figure 11–8. During the past fifteen years, llamas have developed into an animal industry in the United States. It is estimated that there are about 20 thousand llamas in this country, and the number is growing.

Llamas stand three to four feet high at the shoulder, weigh from 250 to 400 pounds when mature, and can carry heavy packs for long distances. Begin related to the camel, llamas can go longer than many animals between drinks of water and can subsist on low-quality forage. They have two types of fiber in their coats: long guard hairs and short fine fiber that helps keep the animals warm. The fiber length may range from three to ten inches.

Because of these characteristics, raising llamas has gained popularity in the United States. Most are produced in the western part of the country where people use them to carry gear on hunting or camping trips into the mountains, Figure 11–9. They are also used to pull carts and are raised as pets. The hair is used for a variety of crafts, such as making rope. A close relative, the alpaca, is raised for its high-quality wool, which is made into fine rugs and blankets.

FISH BAIT PRODUCTION

One of the great outdoor hobbies of Americans is fishing. In our country, there are thousands of streams and lakes that are well stocked with game fish. People take advantage of these waters to catch fish and enjoy a

Figure 11-9. In the United States most llamas are used as pack animals. *Charles Hackbarth, Mt. Sopris Llamas.*

leisurely outing. A very popular way to catch fish is by using natural, live bait. Fish bait is grown all across the country, largely by part-time producers.

Earthworms

Earthworms are grown in beds that have been built up using loose, porous materials, Figure 11–10. These materials might include shredded newspaper, shredded cardboard, garden compost, grass clippings, straw, or well-decayed manure. Usually peat moss is added to the mixture to keep the material loose and to help hold moisture.

Figure 11-10. Earthworms are raised in beds of loose, porous materials. *James Strawser, Cooperative Extension Service, The University of Georgia.*

The **pH** of the bedding is monitored and is kept slightly acid (pH 6.8). pH is the measure used to indicate how acidic or how alkaline a material is. On the pH scale, 7 is neutral (neither acid nor alkaline). A number less than 7 indicates an acid; the lower the number the more acidic the material. A number greater than 7 indicates an alkaline; the higher the number the more alkaline the material. Since the material used for bedding is usually of an acidic nature, limestone is added to help neutralize the acid.

The beds are kept moist, and lights are used to prevent the worms from crawling out of their beds. Worms are sensitive to light and generally come out only at night. As long as the worms see light when they come to the top of the bedding, they will stay near the bottom.

The worms are fed vegetable scraps and cornmeal. They mature at one to two months of age. They are packaged and marketed in small round containers of approximately 100 worms per container for the smaller red wigglers and about 25 to 50 for the larger night crawlers. Mature worms have a broad, raised band encircling the body behind the head, Figure 11–11.

An alternative marketing source is that of selling the worms to gardeners. Earthworms improve the quality of the soil by creating pores as they move through the soil. The pores allow better movement of air and water. Manure from the worms, called **castings**, helps enrich the soil.

Crickets

Crickets are raised in wooden boxes that are covered with screens. The floors of the boxes are covered with sand in which the adults lay their eggs. The sand is covered with

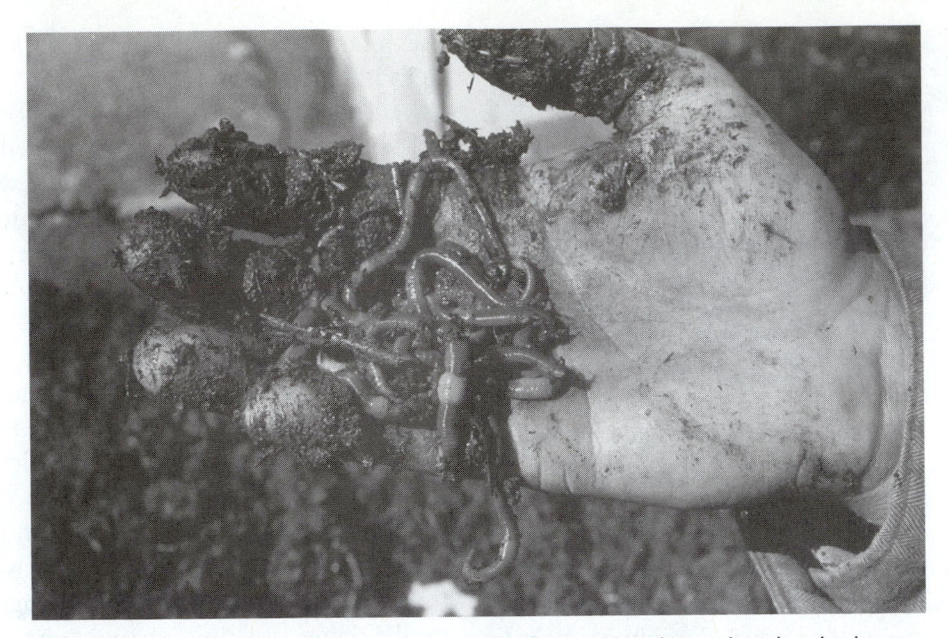

Figure 11-11. Mature earthworms have a broad, raised band encircling their bodies just behind their heads. *James Strawser, Cooperative Extension Service, The University of Georgia.*

fine wood shavings or other shredded material. Heat lamps are used to keep the crickets warm and to keep the sand warm for hatching the eggs. When the young crickets hatch out, they are fed grain mixtures in small trays. Small trays filled with water-saturated cotton provide the crickets with a ready drinking fountain. The crickets are placed in cages and shipped to bait outlets. There they are sold to fishers who put them in their own cricket cages.

Ostrich Production

For many years, ostriches have been grown commercially in South Africa. They are the largest existing bird in the world. Mature males may stand as tall as nine feet and weigh as much as 330 pounds. The sheer size of the birds makes them valuable for meat, leather, and feathers. The leather is of exceptional quality in that it has a very soft, supple texture and it is also durable. The plumage of the male is quite attractive and is used for a variety of purposes in home decorating and clothing.

Although the eggs are currently too valuable to use for any other purpose besides brooding, the potential exists for the eggs to be a food source. One ostrich egg equals the content of 24 chicken eggs. If researchers could develop ostriches that could lay an egg each day, the eggs might someday prove to be an important food source.

Ostriches are raised to a limited extent in the United States. Because of their low numbers, ostriches are quite expensive to buy. This allows people to make a good profit by selling the young ostriches to other people who wish to raise them.

The birds have to be kept within a high fence and require protection from cold weather, Figure 11–12. The birds build nests in the sand, so sand must be provided for the females to lay

Figure 11-12. Ostriches have to be kept within a high fence. *Dr. Jean Sander, Department of Avian Medicine, The University of Georgia.*

in. Since the birds are so valuable, the hatching is done with an incubator, and the young chicks are given close supervision and care. Adult birds are hardy and are not very susceptible to disease, but the chicks need a lot of protection and care.

Ostriches may begin breeding when two-and-a-half to three years of age. With proper growing conditions, they go to market when 10–14 months old. The carcasses yields about 70 pounds of meat and 14 square feet of leather.

One drawback to raising ostriches is that the animals can be quite dangerous. They defend themselves by flailing their legs and kicking. Their toenails are sharp and can severely injure or even kill a person they attack. This means that extra care must be taken in the feeding and daily care of the birds.

Large Game Animals

A growing area of alternative animal agriculture is the commercial production of large game animals. By far the largest component of this industry is the production of elk. The growing of domesticated deer and elk goes back at least 5,000 years in China and other parts of the world. In this country, elk has been grown commercially for the past 100 years. Currently, there are about 1,200 elk farms in North America. These operations have around 70,000 animals in domestication.

These large animals offer several advantages to traditional grazing animals. They convert feed more efficiently than cattle on the range. This means that they gain more weight on less feed than beef animals when raised

Figure 11-13. Americans consume approximately 100 metric tons of deer and elk meat each year. *North American Elk Breeders Association.*

under the same conditions. Elk can also make use of lower quality feed stuff such as browse,

Americans consume approximately 100 metric tons of deer and elk meat per year with most being imported from New Zealand, Figure 11–13. The meat is relatively low in fat content and has a flavor preferred by many people. In addition to the sale of the meat, producers also market the antlers which are used as ornaments, in the making of jewelry, and as an ingredient in herbal medicines.

LABORATORY ANIMAL PRODUCTION

Millions of animals are needed each year for use by scientists conducting research. Almost all materials that come in contact with humans—food, medicines, and cosmetics—have to be tested on animals to prove the effectiveness and safety of the products. Most of the advances in medicines have come about through the use of **laboratory animals**. As pointed out in Chapter 21, there is considerable controversy over the use of animals in experimentation. However, no one can deny the benefits to humans brought about through the use of animals in research.

These animals are produced by commercial and part-time producers. The animals most in demand for research are mice, rats, hamsters, guinea pigs, and rabbits. Other animals, such as primates, may be used for highly specialized research.

Animals that are raised for use in laboratories have to be raised under strict conditions. Measures have to be taken to ensure that the animals have no **genetic defects** and are not harboring disease organisms. Animals of this nature could very well cause an otherwise well-designed research study to turn out wrong. The product being tested or the experiment being conducted could be tainted by a disease organism or a genetic defect that would cause the animal to react differently from a healthy animal. Most producers raise animals that are certified for laboratory use.

THE HONEYBEE INDUSTRY

Honeybees are classified as insects because they fit the characteristics of the **insect** class. They have three distinct body segments: a head, a thorax, and an abdomen. They also have three pairs of legs. Even though the honeybee is an insect, it is still classified as an important agricultural animal. In fact, the argument could be made that bees are the most important of all the agricultural animals.

Although it would be difficult to estimate the value of the services provided by the bee each year, it is safe to say that many crops could not survive without help from bees. Most of the other agricultural animals eat plant-derived feed and therefore rely on bees to pollinate the plants they eat. Bees assist in the pollination of flowering plants by scattering pollen from one flower to the next as they gather nectar and pollen, Figure 11–14. (Refer to Chapter **12** for information on how the bees locate the flowers.)

This is nature's way of ensuring that the flowers are pollinated in order to produce seed. Often the seed and/or the fruit surrounding the seed of a plant is the crop sought by producers.

Honeybees are particularly adept at pollinating. Many insects work flowers, but most go from flowers of one type to flowers of a different type. Bees, on the other hand, work a particular kind of flower for a period of time. For example, honeybees may be working apple blossoms on a particular day. They will work the apple blossoms for several days until the flowers are gone, then work a different type of flower. By doing this, they go from one apple blossom to another apple blossom and spread the pollen. This process ensures that the blossoms are thoroughly pollinated.

Figure 11-14. Honeybees pollinate plants as they move from flower to flower. *James Strawser, Cooperative Extension Service, The University of Georgia.*

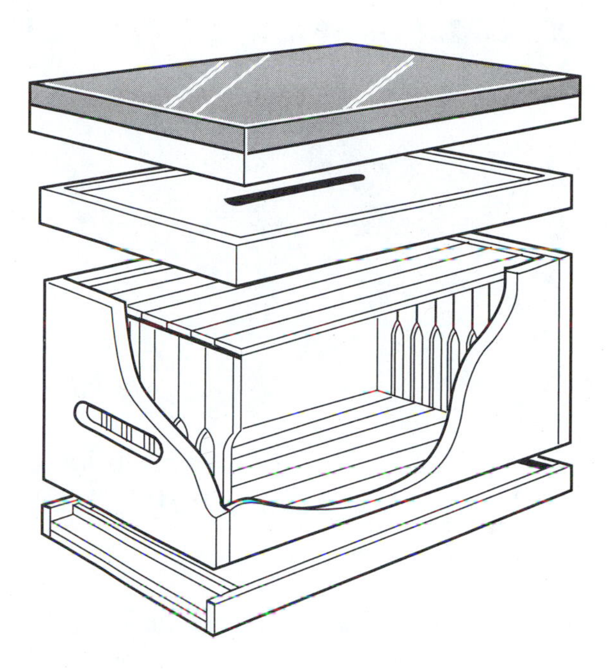

Figure 11-15. Bees are raised in hives. *Brushy Mountain Bee Farm.*

Fruit growers hire beekeepers to bring in truckloads of bees in the spring when the trees are blooming. The bees live in wooden, boxlike structures called **hives,** with a separate colony of bees in each hive, Figure 11–15. The hives are easy to handle and can be loaded on a truck with the bees still in the hive. The beekeeper can move the hives from orchard to orchard for a fee from the fruit or crop producer. In addition, the beekeeper can harvest hundreds of pounds of honey each year to be sold at a profit.

Bees produce and store honey to eat during the winter months. Honey is made from nectar the bees gather from the flowers. Different flowers make honey that varies in

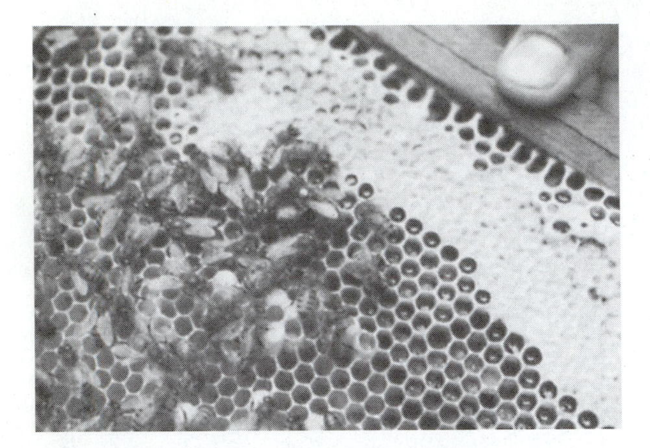

Figure 11-16. Bees fill the cells, concentrate the honey, then cap the cells. Note the light-colored open cells containing pollen. *Dr. Frank Flanders, Agricultural Education, The University of Georgia.*

Figure 11-17. Foundation combs are placed in frames, and the frames are placed in supers. *Dr. Frank Flanders, Agricultural Education, The University of Georgia.*

color and strength of flavor depending on the source of the nectar. The bees store the nectar in six-sided cells that are joined together to create a **honeycomb**. The cells are made from wax that is secreted from glands on the bodies of the bees. The bees work the wax particles loose from their bodies and use them to shape the cells into perfect hexagonal storage containers. Once the bees have deposited the nectar in the cells, the bees begin to concentrate it into honey by using their wings to create air movement that evaporates the excess water in the nectar. Once the cells are full of honey, the bees mold wax across the top of the cells to cap and seal them, Figure 11–16. Some cells are filled with pollen that serves as a source of protein for the bees.

The beekeeper places sheets of comb called **foundation comb** into frames on which the bees build the remainder of the comb to fill with honey. The frames are hung into boxes called **supers**, Figure 11–17. Several supers stacked together compose the hive where as many as 80 thousand bees may live during the peak of the honey season. The space between the frames in the super is critical. If too much space is given, the bees will build across the frames and the frames will be difficult to remove. If too little space is given, the bees will only fill one side of the frame. This critical space is called **bee space** and is about three- eighths of an inch. This is just enough room for two bees to work back-to-back on opposite frames, Figure 11–18.

Figure 11-18. The frames are spaced so that the proper bee space will prevent the bees from building across the frames. *Dr. Frank Flanders, Agricultural Education, The University of Georgia.*

Figure 11-19. The beekeeper uses smoke to calm the bees when honey is being removed. *Dr. Keith Delaplane, Cooperative Extension Service, The University of Georgia.*

When all of the honey in a super is capped, the beekeeper removes the super and replaces it with an empty one. The keeper is always careful to leave enough honey for the bees to live on through the winter. The beekeeper uses smoke from a smoker to make the bees more docile, Figure 11–19. The bees think the hive is on fire and begin to gorge on honey for their departure from the hive. This seems to make the bees more calm and docile. The smoke also serves to disorient the guard bees at the hive entrance before they can give an alarm to the rest of the hive.

The beekeeper must pry the supers apart in order to remove them. The bees use a sticky gum called **propolis** to seal and glue the parts of the hive together. Propolis is made from the sap and gum of trees.

Once the supers are removed, the frames are taken out and the caps cut from the top of the comb using a heated uncapping knife, Figure 11–20. The honey is removed from the comb by using an extractor that consists of a

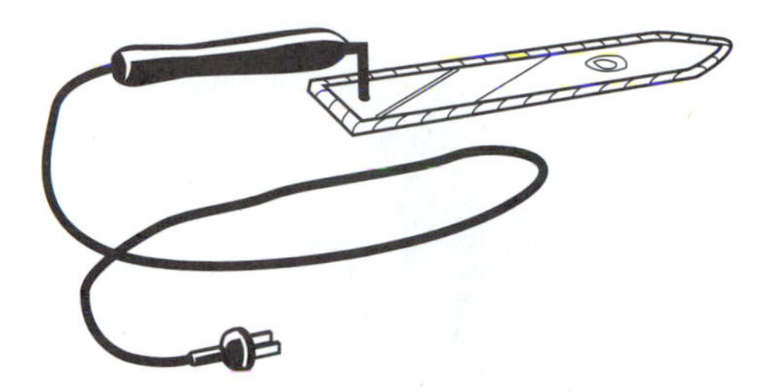

Figure 11-20. The caps are removed from honeycombs by using a hot uncapping knife. *Walter T. Kelly Co.*

metal drum with racks for the frames. The racks containing the frames are revolved in the drum, and centrifugal force slings the honey from the comb without damaging the cells in the comb, Figure 11–21. The empty frames are then put

Figure 11-21. Honey is removed from the comb by using a centrifugal extractor. *Dr. Keith Delaplane, Cooperative Extension Service, The University of Georgia.*

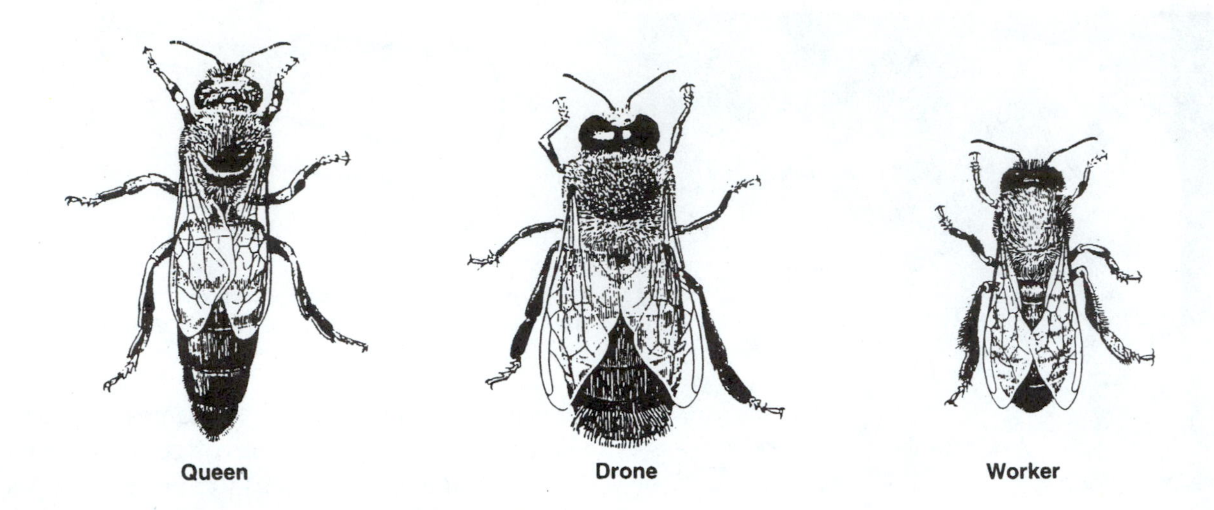

Figure 11-22. There are three types of bees. *USDA.*

back into the super and the super is put back on the hive for the bees to fill again. The honey is then processed by gently heating it in order to prevent the honey from granulating (turning to sugar). The honey is then packaged and sold.

Types of Bees

Bees live in a very ordered society. Each bee seems to have its job, and each works in concert with the rest of the bees in the hive. Within a colony of bees, there are three types of bees: the queen, the drones, and the workers, Figure 11–22.

The Queen

The queen exists to lay eggs for the hive. In her lifetime, she lays thousands of eggs. Other bees feed her and care for her. She is recognizable from the rest of the bees in the hive because she is much larger and is more slender. The queen is kept in the lower part of the hive (called the **brood chamber**) by means of a **queen excluder.** This consists of a screen that has openings large enough for the workers to pass through but too small for the queen to pass through, Figure 11–23. The queen excluder is used to prevent the queen from laying eggs in the comb that the beekeeper will remove in order to extract the honey.

The queen lays eggs in cells called **brood cells** that are slightly larger than those used to store honey, Figure 11–24.

The eggs hatch into larvae that are fed and cared for by worker bees in the hive. The larvae

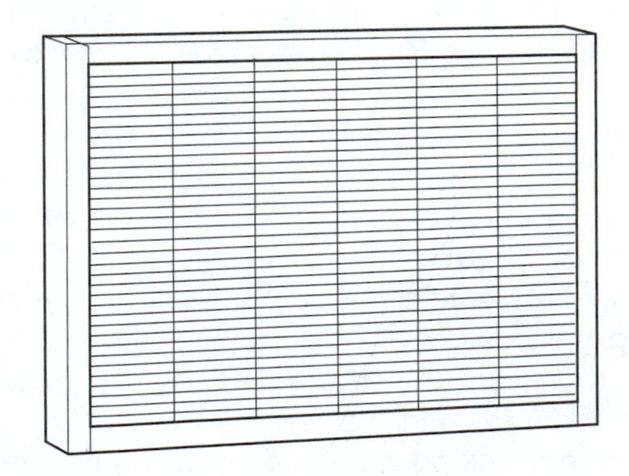

Figure 11-23. Queen excluders keep the queen in the brood chamber. *Walter T. Kelley Co.*

Figure 11-24. The queen bee in the center is laying eggs in the brood cells. The workers are caring for the brood. *Dr. Keith Delaplane, Cooperative Extension Service, The University of Georgia.*

develop into **pupae** and then into adults while remaining in the cell. (See Chapter 19 for a discussion on the metamorphosis of insects.) When the bee has reached the adult stage, it emerges from the cell.

The Drones

The drones are the male bees whose sole purpose is to mate with the queen. When a young queen emerges as the new queen for the hive, she goes on what is called her **maiden flight,** during which she mates with the drones. The queen only mates once during her lifetime. The drones are then removed from the hive by the worker bees and are not allowed to winter in the hive. The next spring, new drones will hatch.

The Workers

The worker bees are all sterile females. Some worker bees bring nectar and pollen to the hive from the field. Some workers care for the queen and brood, some scout the area for nectar, others serve as guard bees at the hive entrance. Worker bees live a short life of only about six weeks. For this reason, the continual producing and growing of brood by the queen and drone bees in the colony is essential. A hive whose queen has stopped laying or has slowed down because of her age is not productive, and the entire colony may die. Beekeepers periodically place new queens into hives to ensure that each hive has a vigorous queen who will lay plenty of eggs. Queens are produced commercially in small hives known as **nukes** and are shipped to the beekeepers in small cages, Figure 11–25. Two or three worker bees are placed in the cage to feed the queen from a sugar cube stuck in the end of the queen cage. Bees are also shipped in two or three pound packages, complete with queen, Figure 11–26.

New Queens

When the hive becomes too crowded, the bees will produce a new queen by drawing

Figure 11-25. Queens are shipped in small cages. *Dr. Frank Flanders, Agricultural Education, The University of Georgia.*

Figure 11-26. Bees are shipped in screened cages of two- or three-pound packages. *Dr. Frank Flanders, Agricultural Education, The University of Georgia.*

special large cells called **queen cells**. Larva in these cells are fed a special substance called **royal jelly** which is secreted from the bees. This food causes the larvae to develop into a queen bee. When the new queen emerges, the old queen will usually leave with a portion of the bees (called a **swarm**) to form a new colony, Figure 11–27.

Scout bees will locate a suitable site for the new colony. In the meantime, the bees will cluster around the queen on a limb, a house, or other object until a new home can be located. Beekeepers often place these swarms into a new hive.

Parasites and Diseases of Bees

Beekeepers have always been plagued by pests of honeybees. As with any animal, bees are susceptible to a number of parasites and diseases that cause beekeepers to spend a lot of money and time trying to maintain the health of their hives. However, within the past 15 years, new parasites have emerged that are having a dramatic effect on the bee industry. Two types of parasitic mites have threatened to wipe out the bee population in this country.

Figure 11-27. When a colony of bees becomes too crowded, part of the colony leaves and forms a swarm. *American Bee Journal.*

In 1984, tracheal mites were first discovered in U.S. beehives. These internal parasites lodge in the air passages of adult bees and suffocate them. In 1987, the presence of varroa mites were found. These mites feed on the blood of pupae and adult bees. Not only is the loss of blood a problem for the bees but also the damage caused by the mites make bees much more susceptible to diseases. Since these two serious pests were first detected, as much as 60 percent of the commercial bees have been lost and 90 percent of the wild bee population have been wiped out. The combined loss of bee colonies from these two mites may have a far-reaching effect on the bee industry as well as on the pollination of many of our crops. Scientists are working hard to find a solution to this devastating problem.

Honey Producers

While there are many large producers of honey in the United States, most of the honey is produced by hobbyist beekeepers, who own less than 25 hives, and the part-time beekeepers, who own 25–299 hives. Commercial producers are those who own 300 or more hives. The value of honey produced in the United States each year well exceeds $100 million.

Killer Bees

In recent years, a real concern for beekeepers has been the threat of the invasion of the African honeybee. These bees have been referred to as **killer bees** because of their extremely aggressive nature, Figure 11–28. Honeybees are not native to the United States; most of the bees in this country are descendants of bees that originated in Europe. They were introduced by European settlers to the new world. The European or yellow bees have a relatively docile nature and are easy to keep.

In 1956, research scientists in Brazil imported bees from Africa to cross with the European bees. Their idea was to develop a new strain of hybrid bees that would be more productive than the pure European bees. Despite efforts to contain the African bees, they managed to escape into the wild. In 1990 these bees were reported as far north as southern Texas.

The media have given the coming of the "killer bees" much attention because of the bees' tendency to attack both humans and animals. Some people have feared that the African bees will eventually destroy the bee industry in the United States by replacing the more docile European bees, Figure 11–29.

The African bees take over by entering a colony of European bees, killing the queen, and replacing her with an African Queen. The African queen then lays eggs that hatch into African bees. Scientists tell us that the African

Figure 11-28. African honeybees are much more aggressive than European bees. *Dr. Keith Delaplane, Cooperative Extension Service, The University of Georgia.*

Trait	African	European
open, exposed nests	common	rare
colony population	smaller	larger
colony honey supplies	smaller	larger
tendency to abandon nest	often	rarely
swarming rate	greater	lesser
stinging behavior	intense	moderate
body size	smaller	larger
development time for young	shorter	longer
adult life span	shorter	longer
commercial honey production	poorer	better
commercial pollination	poorer	better

Figure 11-29. A comparison of the African and European honeybees. *Cooperative Extension Service, The University of Georgia.*

bees are adapted to living in warm tropical areas and do not thrive in temperate climates where the weather turns cold. Also, the effect of the mites mentioned earlier, will probably slow their spread.

SUMMARY

As the industry of animal agriculture grows, many new types of animals are added; each has a unique characteristic or quality that makes it valuable to humans. These animals may now be considered to be animals seen normally in the wild; however, we must realize that at one time all of the animals we now grow were wild animals. As with other aspects of agriculture, research will find new uses for these animals and new and better ways of producing them.

◆ Review Exercises

TRUE OR FALSE

1. Most of the rabbits produced in this country are grown in large commercial rabbitries with between 5 hundred and 10 thousand head. ______.
2. Most rabbits are raised indoors in cages called hutches that consist of woven wire boxes for the rabbits to sleep in and to bear their young. ______.
3. Rabbits are not very prolific breeders, and it is difficult to raise enough to make a profit. ______.
4. Rabbits may be an excellent potential food enterprise in developing countries. ______.
5. Most llamas are produced in the western part of the United States where people use them to carry gear on hunting or camping trips into the mountains. ______.
6. Not only are earthworms raised for bait but they are also sold to gardeners to help improve the movement of air and water through the soil. ______.
7. Crickets are actually collected from the wild, then placed in cages to be shipped to bait outlets. ______.
8. One ostrich egg equals the contents of 24 chicken eggs. ______.
9. Animals raised for research must not have genetic faults or disease organisms. ______.
10. Although honeybees are not animals in the strict sense, their contribution to agronomy as well as the animal industries is invaluable. ______.
11. Beekeepers can move hives from orchard to orchard for a fee from the fruit or crop producer. ______.
12. Different flowers make honey that varies in color and strength of flavor depending on the source of the nectar. ______.
13. The critical space between the frames in the super is called the bee space and is about three inches wide. ______.
14. Most of the honey produced in the United States comes from commercial producers who operate three hundred or more hives. ______.
15. Africanized bees are referred to as "killer bees" because of their extremely aggressive nature. ______.

FILL IN THE BLANKS

1. Alternative animal production is usually ______________ in scale and provides a ______________ for a ______________ market.

2. The feed efficiency ratio for rabbits is ________ to ________, which means that for every ____________ pounds of food fed to the rabbit, the animal gains ____________ in body ____________.

3. Not only is rabbit meat high in ____________ and low in ____________ and ____________, but it is easily ________________ and very ________________.

4. Being related to the ____________, the llama can go longer than many ____________ between drinks of ________________ and can subsist on __________-____________ forage.

5. Earthworms are grown in ____________ that have been built up using ____________, ______________ materials including shredded newspaper, shredded ____________, garden ____________, ____________ clippings, straw, or well-decayed ____________.

6. The sheer ________________ of the ostrich makes it valuable for ________________, ________________, and ________________.

7. An adult ostrich is ______________ and not very susceptible to ______________, but the chicks need a lot of ________________ and ______________.

8. Almost all materials that come into contact with humans—food, ________________, and ________________—have to be tested on ________________ to prove the effectiveness and ______________________ of the ______________.

9. Honeybees live in ______________, boxlike structures called ________________, with a separate ____________ of bees in each.

10. Once the bees have deposited the nectar in the ___________, they begin to concentrate it into ______________ by using their wings to create ________ ___________ that evaporates the excess ______________ in the nectar.

11. When the bees think the hive is on fire, they begin to __________________________ on ______________________ for their departure from the ______________.

12. There are three types of bees in each colony: the ______________, the ______________, and the ______________.

13. When a young queen emerges as the ________________________ queen for the ________________, she goes on what is called her ________________ flight, during which she mates with the ______________.

14. When the hive is too crowded, a new queen is raised and the old queen will usually leave with a ______________ of the bees (called a ______________________) to form a new ______________.

15. Most scientists agree that the African ______________ cannot live in ____________ ____________ where the weather turns ___________.

DISCUSSION QUESTIONS

1. What is meant by an alternative agricultural animal?
2. What are the advantages that rabbits have over other agricultural animals?
3. What is one major drawback in the production of rabbits?
4. What are the uses people have for llamas?
5. What is meant by the pH of a material?
6. Why are lights used in the production of earthworms?
7. What are the uses for the ostrich?
8. Why are elk raised as agricultural animals?
9. Why must laboratory animals be free of disease and genetic defects?
10. Why are honeybees more effective in pollinating than other insects?
11. Name the three types of bees within a hive and give the role of each type.

12. What causes bees to swarm?
13. How were the African bees introduced into the Americas?

STUDENT LEARNING ACTIVITIES

1. Create a list of the alternative animal operations in your area. Choose one of these operations and report to the class on the enterprise. How did the producer get started? What are the markets? What is the potential for expansion?
2. Think of an alternative animal enterprise that was not covered in this chapter. Give a report to the class on your ideas. How would these animals be raised? Where would be the market? Why are there not more of these animals being raised?
3. Search the internet and find information about the effects of the tracheal and varroa mites in your state. Share your findings with the class.

Chapter 12

Animal Behavior

Student Objectives in Basic Science

As a result of studying this chapter, you should be able to

1. Define *ethology.*
2. Discuss the difference between instinctive and learned behavior in animals.
3. Discuss the concept of animal intelligence.
4. Define *conditioning.*
5. Explain how animal behaviors are developed.
6. Describe the types of social behavior in animals.
7. Describe the types of sexual and reproductive behaviors in animals.
8. Describe the types of ingestive behaviors in animals.
9. Explain how certain animals communicate.

Student Objectives in Agricultural Science

As a result of studying this chapter, you should be able to

1. Explain how animal behavior is used to protect sheep.
2. List the social behaviors of agricultural animals.
3. List the sexual and reproductive behaviors of agricultural animals.
4. List the ingestive behaviors of agricultural animals.
5. List the methods used by agricultural animals to communicate.
6. Discuss how the natural behaviors of agricultural animals can be used to provide the animals with a safer, more comfortable environment.

Key Terms

ethology
instinct
imprinting
intelligence
conditioning
docile
indiscriminate breeders
social behavior
stallion
estrus
ingestive behavior
grubs
ruminants
dental pad
colony
scout bees

All animals, whether wild or domesticated, act in certain ways. Scientists have long recognized that animals have patterns in the way they behave. Many of these behaviors are predictable; that is, animals will usually act in a certain way under particular conditions. The study of how animals behave in their natural habitat is called **ethology**. The natural habitat may be the natural area lived in by wild animals or the pastures, pens, or facilities of domesticated animals. The behavior of animals in the wild has been an accepted branch of biological science for many years. Only recently has the science of ethology been used to research methods of producing agricultural animals. Ethology plays an important role in the production scheme of the modern livestock industry because the way animals behave can be better understood, and, in turn, a better growing environment for the animals can be created.

Different types of animals behave in different ways. Most animal behavior can be divided into two categories: instinctive and learned behavior. The most basic is **instinct**. Instinct is the behavior that is set in an animal at birth and causes the animal to respond automatically to an environmental stimulus. This behavior is a result of genetics and is not something the animal learns. A good example is newly born and young animals being able to nurse; most agricultural animals are able to stand and nurse only minutes after they are born, Figure 12–1. This behavior is not taught but is with the animals at birth. Other instinctual behaviors are breeding, eating, and drinking.

Figure 12-1. Most agricultural animals are able to stand and nurse shortly after birth. *James Strawser, Cooperative Extension Service, The University of Georgia.*

Animals can also learn behaviors. One of the most basic types of learning is **imprinting**. Imprinting means that an animal will attach itself to or adopt another animal or object as its companion or parent. This usually occurs shortly after the animal is born or hatched. For instance, if a hen sits on duck eggs, the resulting ducklings will accept the hen as their mother. Goslings (baby geese) have been known to adopt dogs or humans in this manner if the human or dog becomes a companion within the first 36 hours after hatching. Dogs also have been known to adopt other animals as "their own." Some sheep producers make use of this behavior by placing very young pups with a flock of sheep. The dog is raised among the sheep and accepts the sheep, Figure 12–2. The

Figure 12-2. Dogs raised with sheep often act as guardians. *Calvin Alford, Cooperative Extension Service, The University of Georgia.*

dog then acts as guardian to its adoptive family and keeps predators from the flock.

Several species of fish, especially salmon, are imprinted when they hatch as to that location. After several years, the mature salmon return to the place where they were hatched to spawn. This drive in the fish is extremely strong, and they will go over and through severe obstacles on their journey to the place of their hatching.

Different species of animals have different abilities to learn. This is called **intelligence**. Obviously, the most intelligent animals are humans. Primates, such as chimpanzees, are next in order of intelligence, followed by ocean mammals such as dolphins and whales. Among agricultural animals, the pig is considered to be the most capable of learning and therefore the most intelligent. In fact, some scientists rank the pig just under chimpanzees and dolphins in intelligence, Figure 12–3.

Learning in animals comes about through several means. One is **conditioning**. This means that an animal learns by associating a certain response with a certain stimulus. A Russian scientist named Ivan Pavlov was famous for his theories of conditioned reflex. His experiments involved feeding meat to a group of dogs and ringing a bell as the animals were fed. Since the only time the bell was rung was when the animals were fed, they associated the ringing of the bell with eating. After a time, the animals would begin to salivate when the bell was rung even if there was no food in sight. Agricultural animals are conditioned to perform certain reflexes. For example, as cows

Figure 12-3. Some scientists rank pigs just under chimpanzees in intelligence. *Rick Jones, Cooperative Extension Service, The University of Georgia.*

enter the milking parlor, they let down their milk because they associate going into the milking parlor with being milked.

This principle can be used by animal producers to teach animals to respond in a certain way. Animals can also learn on their own by trial and error. For example, a horse may learn to open a gate by tinkering with the latch mechanism until it learns the proper sequence to open the latch and get where it wants to go. A pig learns that by lifting a lid, access may be gained to feed in a self-feeder, or it may be able to get water from a self-waterer by applying pressure in the correct location. Horses are trained by being given positive rewards when the animals respond in a manner desired by the trainer. Animal trainers use the natural abilities and instincts of animals to teach them to do tricks, perform work, or be more productive. Dogs that have a natural instinct to herd animals are trained to herd sheep, hogs, and cattle, Figure 12–4. Scientists tell us that this instinct comes from the wolves who were the ancestors of the herding dogs. Wolves in the wild work together in a pack to circle a group of animals until a vulnerable one is singled out for the kill. Breeds such as the Border Collie retain much of this instinct and can be trained to care for and move animals instead of herding them as prey, Figure 12–5.

Figure 12-4. Some breeds of dogs have an instinct for herding animals. *Calvin Alford, Cooperative Extension Service, The University of Georgia.*

Other breeds, such as the Blue Tick Coonhound, are not as easily trained to herd animals, but they are easily trained to track animals and to tree them. Certain breeds of horses, such as the American Saddlebred, are trained for pleasure riding and for show. Other breeds, such as the Belgian, are trained as draft animals and are used less for pleasure riding.

When humans first started domesticating animals, they discovered that not just any animal was suitable to be tamed for raising. Animals possessing certain natural characteristics made them more desirable for domestication. Animals were chosen that had some use either as food, as workers, or as companions. The animals also had to have certain behaviors that made them conducive to domestication. In order for humans to raise them, the animals had to be **docile** and easy to handle, Figure 12–6. Those animals that were more docile were selected for breeding and eventually developed into animals that could be handled by humans. In addition, the animals had to be

Figure 12-5. Scientists say that the instinct to herd comes from wolf ancestors. *The Working Border Collie Magazine.*

Figure 12-6. In order for humans to raise them, animals had to be docile and easy to handle. *The World Bank.*

indiscriminate breeders. This meant that the animals could not be paired for life, but any male would breed with any female. This allowed for selective breeding and the development of the characteristics deemed desirable by the producers.

There are several types of behavior that livestock producers use to make raising livestock easier, more efficient, safer, and more comfortable for the animals. These behaviors have the natural instincts of the animals as their basis, but they may have been developed or enhanced through years of selective breeding. At any rate, there are patterns of behavior that seem to run in all species of agricultural animals.

Figure 12-7. Most farm animals are gregarious; they tend to herd together. *James Strawser, Cooperative Extension Service, The University of Georgia.*

Social Behavior

Social behavior refers to the manner in which animals interact with each other. Most farm animals are gregarious. This means that they tend to want to herd or flock together, Figure 12–7. Even in the wild, cattle, sheep, and horses tend to want to group together. This is the result of a natural instinct to group together for defense purposes. Young, old, and weak members of the herd or flock can be better protected if the animals remain together in a group.

There appears to be a preference among animals as to the size of the herd or group. In the 1830s, Charles Darwin wrote about large groups of cattle (10 thousand to 15 thousand head) breaking up into smaller groups. Even after being mixed together and stampeded, the cattle would come back into groups of 40 to 100 head. Many cattlemen today divide their herds into this approximate size. Sheep will band together in much larger groups, some reaching several hundred in number. Pigs in the wild tend to band together in groups of about ten. Wild horses usually live in small groups of a **stallion** and his harem of mares, Figure 12–8.

Figure 12-8. Wild horses usually live in small groups.

Other males may live in a small group together with one of the stallions being the leader or dominant male.

Producers have made use of this gregarious behavior by moving, feeding, and caring for the animals as a group. As mentioned earlier, the animals can be moved and controlled more efficiently through the use of herding dogs. Without gregarious behavior on the part of the animals, the use of the working dogs would not be nearly as efficient.

Within each group of animals, there is a hierarchy or order of social dominance. Some animals within the group are recognized by the other animals as having dominance or the ability to exert social influence or pressure over others in the group. In poultry, this is known as the pecking order, Figure 12–9. Certain chickens in the flock are allowed to have priority for space, food, water, etc.

Research has shown that there is a social dominance order in other agricultural animals. For instance, according to some research studies, baby pigs have been competitive about which teat they will nurse. The teats closer to the front seem to be the preferred teats. Each piglet has a particular teat it will go to for nursing, Figure 12–10. Research has indicated that

Figure 12-9. Chickens establish a social order known as the pecking order. *Joe Mauldin, Cooperative Extension Service, The University of Georgia.*

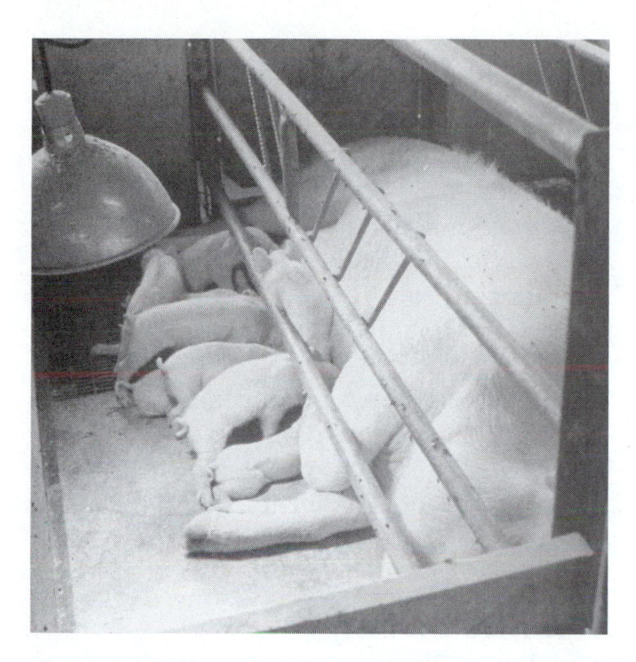

Figure 12-10. Each piglet has a particular teat to which it will go. *Rick Jones, Cooperative Extension Service, The University of Georgia.*

the social dominance established in nursing is carried over when the pigs are weaned. Dominant pigs are allowed to be first to the feed trough, and they establish where they want to sleep in the pen.

Social dominance patterns have also been noted in most of the other agricultural animals. For example, if two or more males are in the same flock or herd, one will be the dominant male who will mate with most of the females. This dominance is usually established by fighting among the males with the strongest and most vigorous emerging as the dominant male, Figure 12–11. This is nature's way of ensuring that the heartiest animals are the ones that breed and raise the next generation of the species.

Producers must consider social dominance as they plan growing their livestock. For example, males must be kept separate to prevent injury. If animals are on a limited ration, they must be separated or the dominant animals will get too much feed and the subordinate animals will get too little.

Figure 12-11. Male animals will often fight to establish dominance. *Calvin Alford, Cooperative Extension Service, The University of Georgia.*

SEXUAL AND REPRODUCTIVE BEHAVIOR

All animals have certain behaviors associated with mating and reproducing. (The process of reproduction is discussed in Chapter 15.) Most female agricultural animals come into **estrus** (heat) in preparation for mating. As this happens, the females engage in behavior that indicates their condition. Cows may bellow, mill around restlessly, allow other cows to mount, or mount other cows. Sows will mount other sows, appear restless, frequently urinate, and grunt loudly. Sheep and other female agricultural animals show signs of estrus in a similar manner.

Males actively seek out females that are in estrus to complete the mating process, Figure 12–12. Often during mating, the males will become more aggressive or belligerent toward other animals and humans. Particular breeds of animals may display a behavior different from other breeds of the species. For example,

Figure 12-12. Males actively seek out females that are in estrus. *Calvin Alford, Cooperative Extension Service, The University of Georgia.*

Brahman cattle usually prefer to breed at night rather than in the daylight.

As the end of the gestation period approaches, females display behavior that indicates the approach of birth. Sows, if in open pasture, will usually try to build a nest out of grass, soil, or other materials they may find. Cows about to give birth generally appear nervous and isolate themselves from the herd. Sometimes they may even hide if there is enough cover from trees or undergrowth in the pasture. A mare will usually bite at her flanks, switch her tail, and lie down and get back up repeatedly.

After the offspring is born, the behavior of the mother changes. She almost always becomes more aggressive and protective of her young. Even females that are normally very docile can become very belligerent after the birth of offspring. This is nature's way of protecting the young from predators.

Most mothers of agricultural animals generally recognize their own offspring and will only allow that individual to nurse. An exception is pigs. A sow will usually accept orphan pigs if she has enough teats for all of the pigs to nurse. It is more difficult to get cattle and sheep to accept young that are not their own. The cows or ewes recognize the scent of their newborns and will accept only that smell. Sometimes producers can fool the mothers into accepting an orphan by changing the way it smells. A cattle producer may rub both the mother's calf and the orphan in a strong-smelling solution. Because the cow cannot distinguish which calf is hers, she may accept both calves. A sheep producer may take the skin from a dead lamb and cover an orphan with it to trick the ewe into thinking the orphan is indeed her lamb. After nursing begins, the ewe will soon accept the orphan.

INGESTIVE BEHAVIOR

Ingestive behavior means the manner in which animals eat and drink. Different animals have different habits or ways in which they take food. Obviously, most of these differences in habits reflect the way the animals are made—that is, the type of digestive system the animals have and the type of food they prefer.

Pigs that run outside in a pasture or in a lot tend to root or dig in the ground for food, Figure 12–13. This is a carryover from the time before they were domesticated when their diet consisted of roots, **grubs**, insects, seeds, and nuts. Even the most modern breeds of pigs will revert to the habit of digging in the ground with their snout. Their digestive system is a simple stomach system, and they are not capable of digesting large amounts of fiber as are **ruminants**. Therefore, the type of food found by their rooting action is proper for their digestive system.

People who are not familiar with agricultural animals think of pigs as animals that overeat. Expressions such as "eat like a pig," "pig out," etc., indicate that pigs are animals

Figure 12-13. Pigs tend to root or dig in the ground for food. *Rick Jones, Cooperative Extension Service, The University of Georgia.*

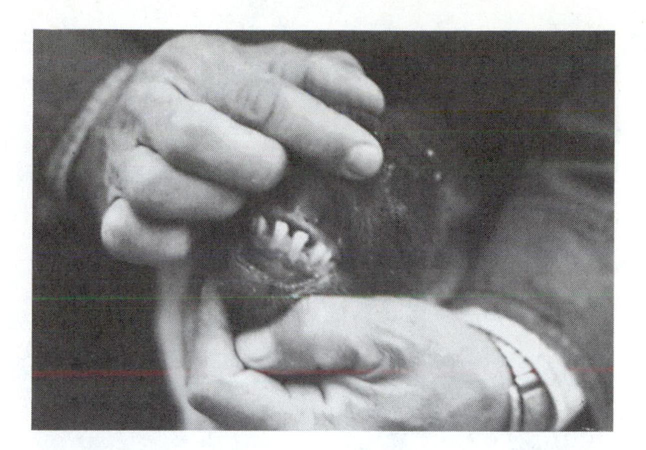

Figure 12-15. Sheep and cattle have a dental pad instead of upper teeth. *Calvin Alford, Cooperative Extension Service, The University of Georgia.*

that eat so much that they make themselves sick. Contrary to this belief, pigs will only eat what they need and never make themselves sick by overeating, Figure 12–14. In fact, given the opportunity, pigs will eat the right amounts of particular feeds and balance their own diet. As mentioned earlier in this chapter, pigs are among the most intelligent of agricultural animals.

Ruminant animals, such as sheep, goats, and cattle, have digestive systems designed to handle large amounts of roughage such as

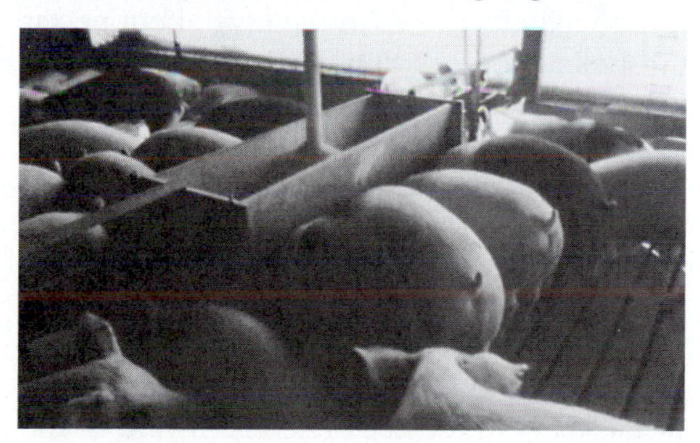

Figure 12-14. Pigs eat only what they need and never make themselves sick by overeating. *Rick Jones, Cooperative Extension Service, The University of Georgia.*

grass or other plants. Even though ruminant agricultural animals eat basically the same type of food, there are differences in the way they gather and ingest food. Cattle, goats, and sheep have no upper front teeth and must rely on a thick **dental pad** in the top of their mouths to tear off plants as they graze, Figure 12–15. Cattle wrap their tongue around the plants and tear them off between their lower teeth and upper dental pad. For this reason, cattle prefer to graze in forages that are at least six inches high. On the other hand, sheep cut off the forage by nipping it with their teeth and dental pad and gathering it into their mouth with their lips. This is why sheep can graze much closer to the ground than cattle can, Figure 12–16.

This behavior of sheep was the basic cause of the range wars of the late 1800s in our western frontier. Sheep that were moved into cattle country grazed the land so close that in some areas the grass would not grow back. This angered the cattle ranchers because they felt that cattle grazing better protected the land. Later research revealed that a system of grazing sheep and cattle on the same ground could be beneficial if managed properly. Cattle prefer to

Figure 12-16. Sheep graze by nipping off grass between the dental pad and teeth. They cut the grass off shorter than do cattle. *James Strawser, Cooperative Extension Service, The University of Georgia.*

eat grasses and sheep prefer to eat plants that are leafy and coarser. If the two species are control-grazed on land that contains both types of plants, both the cattle and sheep benefit.

Cattle tend to graze for a longer period of time per day than do sheep. Cattle usually graze from four to nine hours a day, whereas sheep graze from nine to eleven hours per day, Figure 12–17. As ruminants graze, periods of eating are followed by periods of rest. This allows time for rumination, or the digestion of the plants. During these periods animals regurgitate and chew the plant material they have swallowed.

Figure 12-17. Cattle generally graze four to nine hours per day. *James Strawser, Cooperative Extension Service, The University of Georgia.*

Although horses eat large amounts of plant material, they do not ruminate. Instead, they have a large section in their digestive system called a cecum that processes the roughage. Since horses have both upper and lower front teeth, they bite the plants off as they graze. They usually prefer pasture forages, but will eat brushy plants if no other forages are available.

ANIMAL COMMUNICATION

The ability to communicate means that animals are able to pass information from one to another. We as humans tend to think of communicating as being able to speak and express ourselves within the context of a broad and diverse vocabulary. Obviously animals do not talk as we do, but nonetheless they do pass information from one to another. Their form of communication may be through body motions, through sounds they emit, or through smell.

Perhaps the most dramatic example of animal communication is that used by honeybees to tell other bees in the hive about nectar and pollen sources they have located. Within the bee **colony**, certain bees serve the purpose of looking for food sources. For this reason they are called **scout bees**. Through a series of elaborate moves and dances, the scout bees tell the workers the direction, distance, and amount of the nectar source. The scout bees also bring back samples of the type of nectar that was located. By smelling and tasting the nectar, the worker bees become aware of the smell and taste of the particular nectar. Recent research has shown that the bees also use sound in communicating the distance to the food source. Scientists believe that scout bees also use the sun to orient their flight and communicate this orientation to the worker bees in the hive.

Other agricultural animals communicate as well. Through the use of certain sounds, chickens will call other chickens to feed. A mother hen has a certain cluck that calls her chicks. Different stances depict social standing within the flock. A chicken that stands tall with stiff legs, unruffled feathers, and tail feathers held close together indicates a submissive animal. On the other hand, a chicken that stands in a semicrouch with legs bent, head held high, tail feathers spread wide, and body feathers ruffled is a dominant chicken. The other chickens in the flock recognize this form of communication and act accordingly.

When a sow lies on her side for the piglets to nurse, she will grunt in a particular manner. This distinctive grunt calls the piglets to come and nurse. In addition, specific sounds are emitted by pigs that warn the others of danger. In the wild or in the pasture, pigs may rub trees, stones, or other objects as a way of marking their territory. The rubbing leaves an odor that can be detected by other pigs.

Horses communicate through several different means as well. For example, the direction the ears are pointed can transmit a horse's mood. Anger is expressed by laying the ears straight back toward the neck. On the other hand ears that are pulled forward show that the horse is interested in something, Figure 12–18. Horses also communicate by the sounds they make. A neigh or whinny may indicate that the horse is frightened or concerned. A snort may be used to warn other horses of approaching danger. A squeal may indicate that a horse is angry. Of course, people are able to communicate their commands to horses. Horses are trained to respond to verbal commands of riders, or they may respond to a tug of the halter or a nudge with the knee. A good rider or horse handler will be able to determine the mood or attitude of his or her horse.

Figure 12-18. A horse will point its ears forward when it is interested in something. *James Strawser, Cooperative Extension Service, The University of Georgia.*

Cattle communicate by using their voices. A cow that is in estrus will bellow in order to find a mate. A cow communicates vocally with a calf to call it to her or to warn it of impending danger. Cows may even hide their calves in a wooded area. Body stances are also used by cattle to relay messages. A lowered head with the horns or the top of the head thrust forward indicates that the animal is ready to fight. A bull pawing the ground is a sign of aggression, Figure 12–19. A twisting or slinging of the head issues a warning. A head that is held high with the back swayed and the tail head raised indicates that the animal is about to take flight.

A good livestock producer can recognize animals' communication and treats the animals accordingly. The best producers study the behavior activities and requirements of agricultural animals and use their studies to provide safer and more comfortable environments for the animals. Space requirements are adjusted to

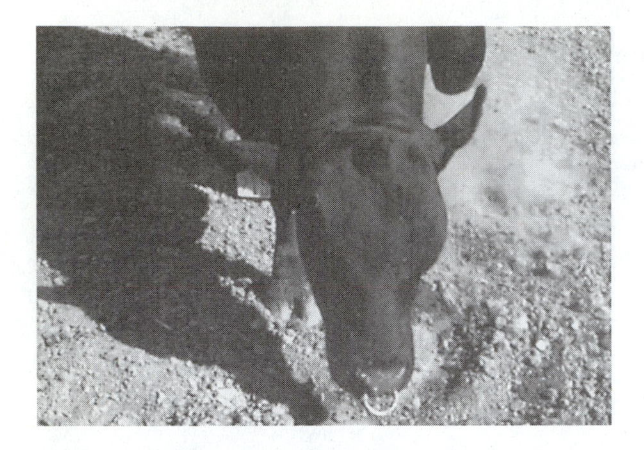

Figure 12-19. A bull pawing the ground is a sign of aggression. *James Strawser, Cooperative Extension Service, The University of Georgia.*

provide the animals with the proper amount of room; for example, research indicates that pigs like to be in contact with each other and that touching other pigs is important to them, Figure 12–20. Pigs raised in isolation do not do as well as those raised with other animals. Chickens are given enough space in cage operations to make them comfortable. The proper space for all animals has been well researched.

Producers can also use the instincts of animals to design facilities. For instance, cattle and sheep urinate and defecate indiscriminately. This means that these animals deposit waste materials anyplace within their living space. On the other hand, horses and pigs will eliminate waste only in a certain part of their

Figure 12-20. Research shows that pigs like to be in contact with each other. *Rick Jones, Cooperative Extension Service, The University of Georgia.*

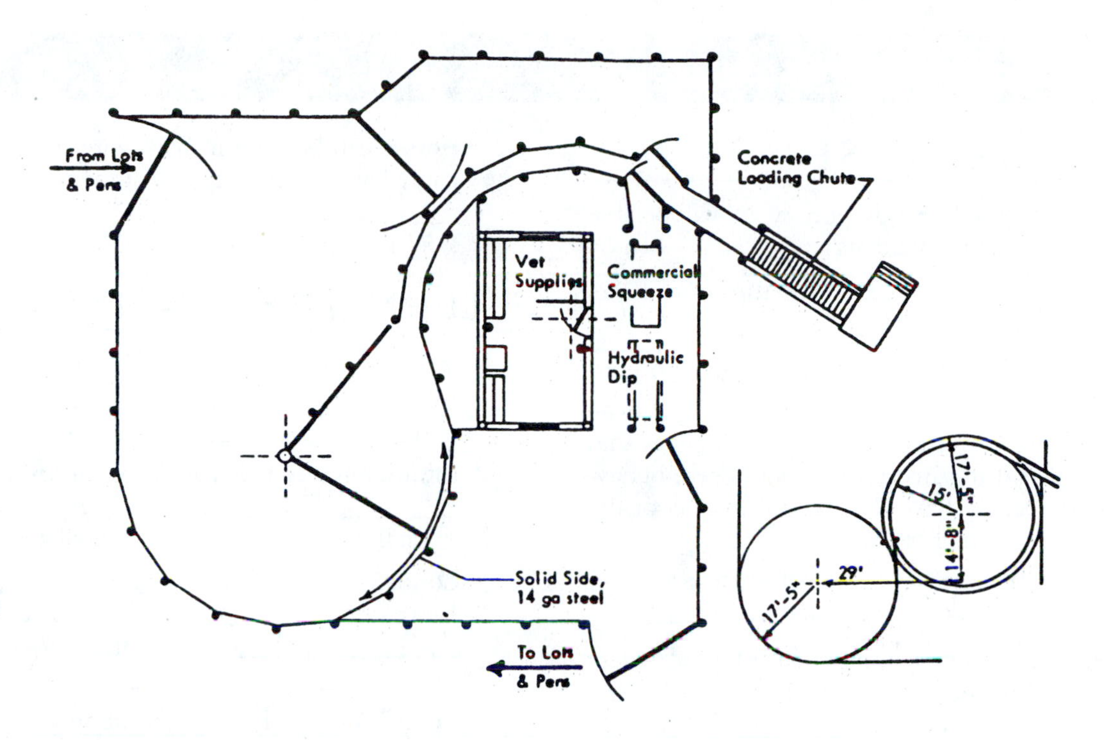

Figure 12-21. Handling facilities are designed to take advantage of cattle's tendency to follow a curved passageway. *Cooperative Extension Service, The University of Georgia.*

space. As facilities for these different animals are designed and constructed, the instincts of the animals have to be taken into consideration. Cattle have a natural tendency to follow a curved passageway and will usually want to circle to the right. Therefore, handling facilities should be designed to accommodate this behavior. Cattle can be more effectively and safely handled in corrals and chutes that are completely opaque (the animals can't see through them) and that circle to the right, Figure 12–21. Horses should not be kept in barb wire fences because they have a tendency to cut themselves on the wire if they become spooked or excited. Cattle are kept very well in barb wire fences and seldom stampede into the fence. By keeping the nature of the animals in mind, modern producers can raise healthier, more productive animals at a more profitable rate.

SUMMARY

To a large degree, how animals behave determines their usefulness as agricultural animals. By studying how animals act in their environment, scientists can better understand how to keep the animals contented and safe. Also, this knowledge helps in designing production systems that can make best use of the animal's nature. Through an understanding of animal behavior, producers are better equipped in providing for and producing animals.

◆ Review Exercises

TRUE OR FALSE

1. Ethology is the study of how animals behave in their natural habitats. ______.
2. Imprinting happens after the animal has reached adulthood. ______.
3. Dogs do not have a natural instinct to herd animals and must be trained to do so. ______.
4. Producers look for animal behaviors that make raising livestock easier. These behaviors usually have the natural instinct of the animal as their basis. ______.
5. Within each group of animals there is a hierarchy or an order of social dominance. ______.
6. Although social dominance plays a part in breeding behavior, food will be divided equally between dominant and subordinate members. ______.
7. After offspring is born, the mother almost always becomes more aggressive and protective. ______.
8. All ruminant animals gather and ingest food in the same manner. ______.
9. Animals communicate with each other through body motions, through sounds they emit, or through smell. ______.
10. Dominant chickens stand in a crouch, while submissive ones stand tall. ______.
11. The direction in which a horse's ears are pointed can communicate its mood. ______.
12. Even the best livestock producer cannot recognize the means of communication given by animals. ______.
13. Pigs need a great amount of space and do not do well if raised in groups. ______.
14. Producers can use the instincts of animals to help design facilities. ______.
15. Horses can be kept in barb wire fences, but cattle have a tendency to cut themselves up. ______.

FILL IN THE BLANKS

1. Most animal behavior can be divided into two ______________: ______________ and ______________ behavior.
2. Intelligence is the ability of an animal to ______________; humans are considered to be the ______________ intelligent.
3. Conditioning means an ______________ learns by ______________ a certain ______________ with a certain ______________.
4. Animals chosen for domestication were used as __________, ______________, or as ______________.
5. Most farm animals are gregarious; that is, they tend to want to ______________ or ______________ together.
6. Often, during mating, the males will become more ______________ or ______________ toward other ______________ and ______________.
7. A cattle producer may rub both the mother's ______________ and the ______________ in a __________-______________ solution to get her to accept them both.
8. Ingestive behavior means the ______________ in which animals ______________ and ______________.
9. During the rest period between grazing times ruminants ______________ and chew the ______________ ______________ they have ______________.
10. In bee colonies, the scout bees tell the __________ the direction, ______________,

and __________ of ____________ source through a series of elaborate ___________ and ____________.

11. Horses are trained to respond to a __________ command of a rider, or they may respond to a __________ of the ___________ or a ________ with the knee.
12. A cow with head raised, back ___________, and tail __________ indicates that it is about to take _____________.
13. The best producers study the ___________ activities and _____________ of agricultural animals and use their studies to provide a _______ and more comfortable ________ for the animals.
14. Cattle can be effectively and ___________ handled in corrals and _____________ that are completely _______________ and that circle to the ______________.
15. By keeping the _____________ of the animals in mind, modern producers can raise ______________, more ______________ animals at a more ________________ rate.

DISCUSSION QUESTIONS

1. Why is the study of animal behavior (ethology) important to producers of agricultural animals?
2. What is the difference between instinctive and learned behavior?
3. What is meant by imprinting?
4. How have producers used imprinting in raising animals?
5. Which of the agricultural animals are considered to be the most intelligent?
6. Give two examples of how conditioning is used in animal agriculture.
7. What were the characteristics of animals that early humans looked for in selecting animals to tame and raise?
8. List some ways in which animals determine social dominance.
9. Briefly discuss at least two sexual and reproductive behaviors of agricultural animals that are important to producers.
10. What aspect of animal ingestive behavior caused range wars during the 1800s?
11. Explain how the differing ingestive behaviors of sheep and cattle make them beneficial to each other.
12. List some ways in which agricultural animals communicate.
13. Explain why cattle chutes should be built in a curving pattern.

STUDENT LEARNING ACTIVITIES

1. Choose a herd or flock of agricultural animals to observe. During a period of several hours ,make a list of all their behaviors. Decide which of the behaviors are learned and which are instinctive. Compare your list with others in the class.
2. Visit a livestock producer and obtain permission to tour the operation. Make a list of all of the aspects of the operation (facilities, work schedules, feeding times) that deal with animal behavior.
3. Choose one animal and observe it for several hours. Record all attempts of the animal to communicate with other animals. This communication might include the animal's attempting to communicate with you!
4. Visit the local animal shelter and interview the workers. Determine what they have learned and observed about animal behavior. List how they use the animals' behavior in their work.

Chapter 13

Animal Genetics

Student Objectives in Basic Science

As a result of studying this chapter, you should be able to

1. Explain the basic function of deoxyribonucleic acid (DNA).
2. Describe the function of ribonucleic acid (RNA).
3. Define *allele.*
4. Describe how traits are passed from parents to offspring through genetic transfer.
5. Explain the concept of dominant genes versus recessive genes.
6. Discuss the concept of codominant genes.
7. Discuss how computers are used to predict genetic differences in animals.
8. Explain how the sex of an animal is determined.

Student Objectives in Agricultural Science

As a result of studying this chapter, you should be able to

1. Explain how producers use the laws of genetics to produce the type of livestock they want.
2. Describe how the concept of heritability is used in the selection of livestock.
3. Explain the difference between phenotypic and genotypic characteristics.
4. Explain how performance data are used in the selection process.
5. Describe how computers are used in the modern selection process.

Key Terms

fertilized egg
phenotype
genotype
chromosomes
deoxyribonucleic acid (DNA)
helix
nucleotides
ribonucleic acids (RNA)
differentiation
allele
homozygous
heterozygous
codominant genes
epistasis
gamete
zygote
heritability
yearling weight
most probable producing ability
growthability
estimated breeding value
siblings
pedigree record
expected progeny difference
thurl
sickle-hocked
recessive gene

Of all the billions of animals in the world, there are no two exactly alike. Even those animals that originate from the same **fertilized egg** are not alike in every aspect. Even though they have the same genetic makeup, one may be slightly taller, a little heavier, or may grow faster. Differences in animals are brought about by two groups of factors: genetic factors and environmental factors.

One set of differences is said to be the animal's **phenotype**. Phenotypes are the physical appearance of the animal such as color, size, shape, and other characteristics. The phenotype can be caused by the environmental conditions under which the animal is raised. For instance, the amount and type of feed an animal receives influence the way it looks. The amount of stress, climactic conditions, exposure to parasites, and diseases can all have an impact on the animal's appearance and performance. Obviously, the producer has a lot of control over the animal's environment.

The other reason for the animal's phenotype is the **genotype** or actual genetic makeup of the animal. Characteristics of individual animals are controlled by the animal's genes that were passed on by its parents, Figure 13–1. By controlling the type of animals used as breeding animals, the producer may (to some degree) also control the genotype. The phenotype is what the producer is able to observe and is used in the selection process.

Figure 13-1. Characteristics of an animal are passed on from parent to offspring.

Through the application of the science of genetics and with the aid of computers, the producer is also able to use the genotype in the process. Whatever methods the producer uses in the selection of animals for breeding, the entire process is built around the concepts of gene transfer.

Gene Transfer

An animal's characteristics are passed on to the animal by its parents. It gets half of its genetic makeup from each of its parents. The information about how the animal will be structured is passed along through the **chromosomes** that are contributed by each parent. Chromosomes are composed of long strands of molecules called **deoxyribonucleic acid (DNA)**. DNA is a very complex substance composed of large molecules that are capable of being put together in an almost unlimited number of ways. Segments of DNA, called *genes*, are connected and arranged on the chromosomes. Each segment or gene is responsible for developing a particular characteristic of the animal.

The code for how the animal is to be formed (all of its characteristics) is contained in the DNA that makes up the genes. The molecules forming the DNA have a spiral shape called a **helix** that resembles a corkscrew, Figure 13–2. If this corkscrew-shaped helix were to be straightened out, it would resemble a ladder with rungs where the segments fit together. The helix is composed of two halves that separate when the cell divides.

At each point on the helix where the two halves are connected, different substances such

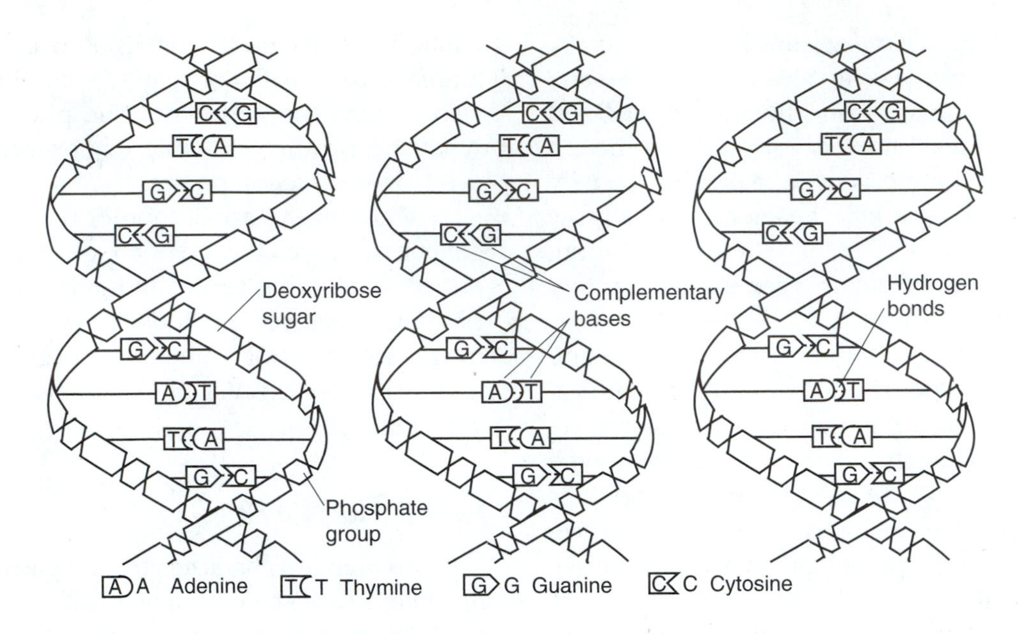

Figure 13-2. Molecules forming DNA are shaped like a spiral.

as adenine (A), thymine (T), guanine (G), and cytosine (C) are attached to each other. These substances, called **nucleotides**, are shaped so that each one can pair only with one particular nucleotide. Adenine (A) can only pair with thymine (T), and cytosine (C) can only pair with guanine (G). When the cell divides, the strands of DNA separate and each half replicates itself so that two strands, exactly alike, are formed. This process is called DNA replication, Figure 13–3.

The messages encoded in the DNA are transferred to the rest of the cell by means of a messenger substance known as **ribonucleic acids (RNA)**. RNA uses the model of the DNA molecule to transfer the pattern or blueprint for how the animal is to be constructed. In genetic engineering, segments of DNA are spliced into existing strands of DNA. This places new genetic information into the cell, and characteristics are produced that are different from what would otherwise be expected.

As the embryo begins to grow and develop, the protein cells begin to differentiate. This means that some of the cells begin to form muscle, some hair, some bone, some skin, and some internal organs. The process by which **differentiation** occurs is not fully understood.

At conception, the chromosome halves from each parent are combined to form fully paired chromosomes, and the chromosomes from each parent are united to form the genetic code for the fertilized egg. There is an almost infinite number of ways that the DNA molecules can be arranged in the gene. The arrangement of the molecules and how the molecules are paired at conception determine the makeup of the new animal.

Every gene that comes from the male is paired with a gene of the same type from the female. For example, the gene that controls the color of the animal's coat is made up of a pair of "coat color" genes—one from the father and one from the mother. A pair of genes that controls a specific characteristic is called an **allele**. If both of the genes are the same—that is, both call for a black coat or both call for a white coat—the genes are said to be **homozygous** and the animal will be the color called for by the genes.

But what happens if the gene from the father calls for a black coat and the gene from the mother calls for a white coat? In this case the genes of the offspring are said to be **heterozygous**. The color of the offspring's coat will be determined by the dominant gene. This means that one gene will override the effect of the other gene. If a white sow is mated to a black boar, the piglets will probably be black because the black gene will be dominant. However, each of the piglets will carry a gene for white color and a gene for black color. If *B* represents the dominant black color and *w* represents the recessive white color, the pairing of the genes for the piglets will be *Bw*.

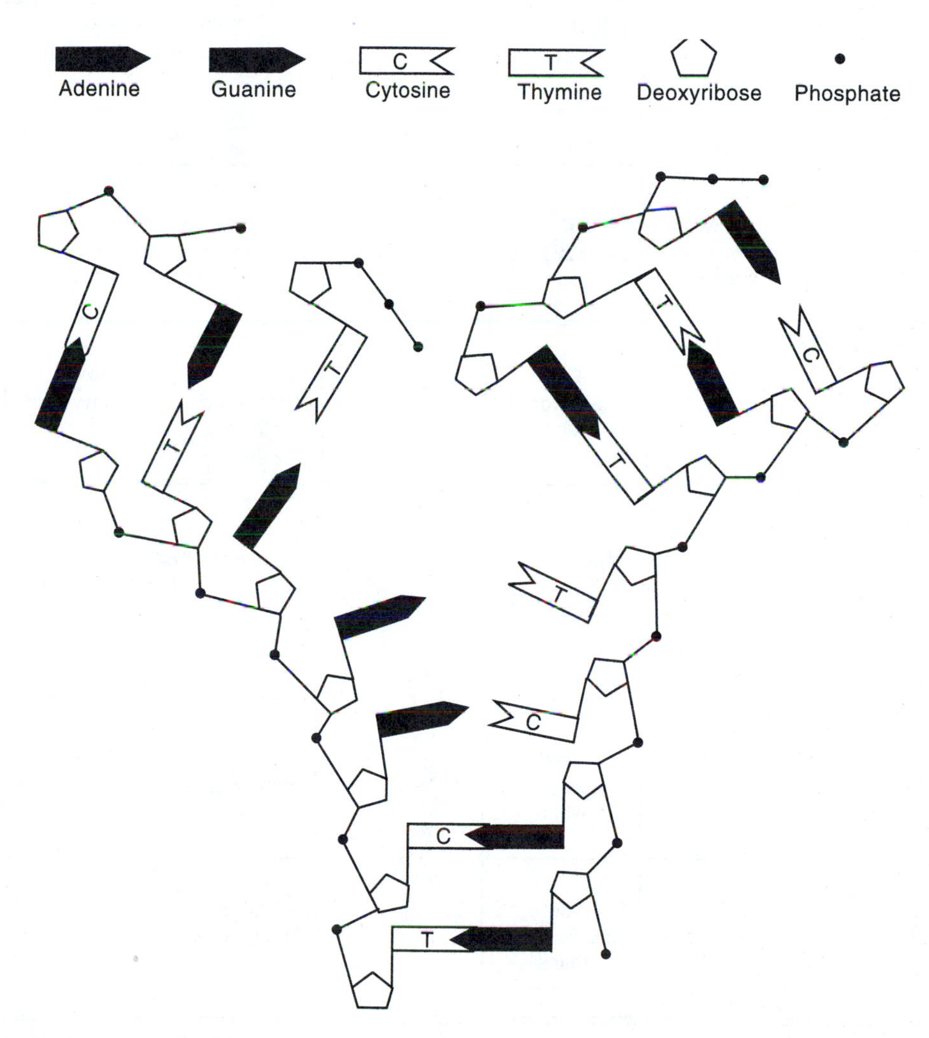

Figure 13-3. When the two sides of the helix are separated, each side replicates itself.

Now suppose that the females from the litter are mated to a purebred black boar (all genes homozygous). Half of the genes from the females that control color will be *B* and half will be *w*. All of the genes from the boar will be *B*. Those genes from the male that match with the females' *B* genes will result in a *BB* genotype. The offspring will be black, and they will possess genes for the black color only. Those genes from the male that match with the *w* female genes will result in a *Bw* genotype. The pigs will also be black, but they will possess genes for both the black and the white color, Figure 13-4.

If the *Bw* females are bred to a purebred white boar, the outcome will be different. Half of the female genes will be *w* and half of the genes will be *B*. Since the male is a purebred, all of his genes for color will be the same (*ww*). The *w* genes of the male that are matched with *w* genes of the female will result in white

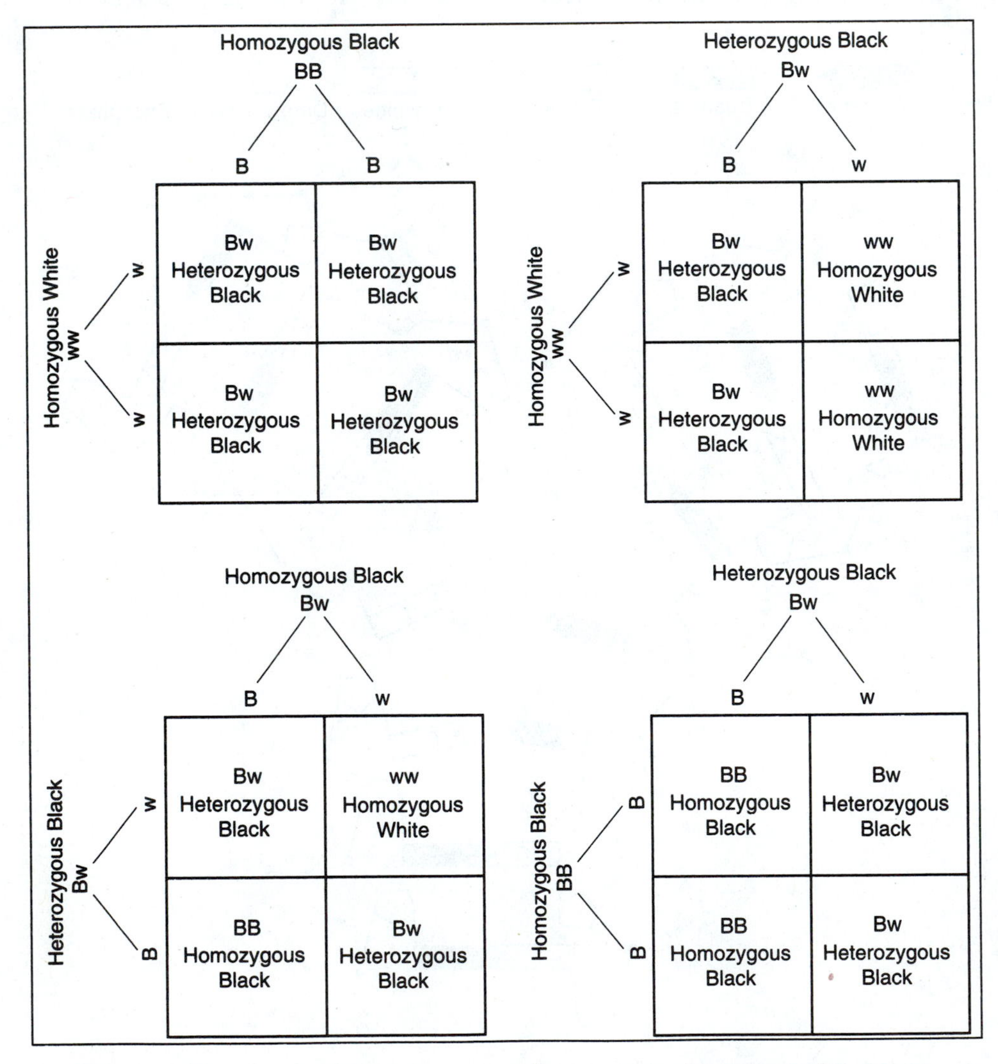

Figure 13-4. Black and white parents can produce a variety of genetic combinations.

piglets. The *B* genes of the female that match with the *w* genes of the male will result in a *Bw* genotype and will produce piglets that are black. This is why both black and white pigs can be born in the same litter.

However, there are exceptions to the rule of dominance. Some pairs of genes may not have a gene that is dominant over the other. They are of equal power and are said to be **codominant genes**.

A very good example of this may be found in Shorthorn cattle. Purebred Shorthorns may be red, white, or roan. Cattle that are completely red carry genes that call for red color only (*RR*); cattle that are completely white carry genes for white color only (*WW*); and cattle that are roan or spotted carry one gene for red and one for white (*RW*). In this case, neither the red (*R*) nor the (*W*) gene is dominant. The color of the animal will be a combination of red and white; the animal will be spotted or roan in color. These cattle can have a variety of color combinations and still be registered as purebred Shorthorn cattle.

Color coding is a good example of how genes transfer traits. The same general principles can be applied to other traits such as horned or polled, tall or short, etc. However, the entire process of defining the characteristics of the animal by the genetic makeup is tremendously more complicated. For instance, genes that are not alleles (matched pairs that control a characteristic) may interact to cause an expression that is different from the coding on the genes. This interaction is called **epistasis**.

Another factor in genetic transfer of characteristics is that of the additive expression of genes. This means that a number of different genes may be added together to produce a certain trait in an animal. For instance, the amount of milk produced by the female is not controlled by a single pair of genes, but by several pairs of genes. The size and body capacity of the female, the ability to produce the proper amounts of hormones, and mammary size and functioning ability are all controlled by different pairs of genes. Yet all these factors contribute to the female's overall ability to produce milk, Figure 13–5. The same thing may be said about the animal's rate of gain or its ability to efficiently reproduce. Several genetically controlled factors, such as body size and structure, can affect the animal's ability to grow rapidly and efficiently. A heifer's pelvic size, shape of the genital tract, and output of sex hormones are all controlled by different genes and are all factors in reproduction efficiency.

Occasionally, an accident will happen within the genetic material, and traits are not passed on as intended. In this case, the animal will take on characteristics unlike the parents; for example, animals are sometimes born with extra legs or two heads. These are referred to as *mutations*. Sometimes mutations can be used to introduce new kinds of animals. A good example is the Polled Hereford breed of cattle. Hereford cattle are naturally horned. Around the turn of the century, an Iowa Hereford

Figure 13-5. A wide variety of genes may control the amount of milk a mother produces. *National Livestock and Meat Board.*

breeder named Warren Gammon noticed that Hereford calves were sometimes born without horns. In these calves, the gene that transmitted the horned characteristic had failed. He began collecting these calves from other breeders, and he used these animals to begin the Polled Hereford breed, Figure 13–6.

THE DETERMINATION OF THE ANIMAL'S SEX

Whether a mammal is male or female is a factor controlled by the matching of chromosomes from the mother and father and is determined at conception. Each body cell contains one pair of chromosomes called the sex chromosomes. Each **gamete** (sex cell from the parent) contains one half of the sex chromosome from the parent.

The female chromosome is usually referred to as *XX*. When the chromosome divides and half goes into the gamete (egg), each of the chromosome halves are the same (*X*), Figure 13–7. The male chromosome is referred to as *XY*, and when divided into the gametes (sperm), contains either *X* or *Y* chromosome halves.

When the two halves (in the sperm and the egg) are united at conception, the **zygote** (fertilized egg) will be either *XX* or *XY* chromosomes. The zygotes containing the *XX* sex chromosome develop into females and the zygotes containing the *XY* sex chromosome develop into males. For this reason, the number of male or female offspring is dependent on the male gamete.

USING GENETICS IN THE SELECTION PROCESS

In Chapter 14, the selection of livestock by physical appearance is discussed. The modern

Figure 13-6. The Polled Hereford breed was developed from mutations. *American Polled Hereford Association.*

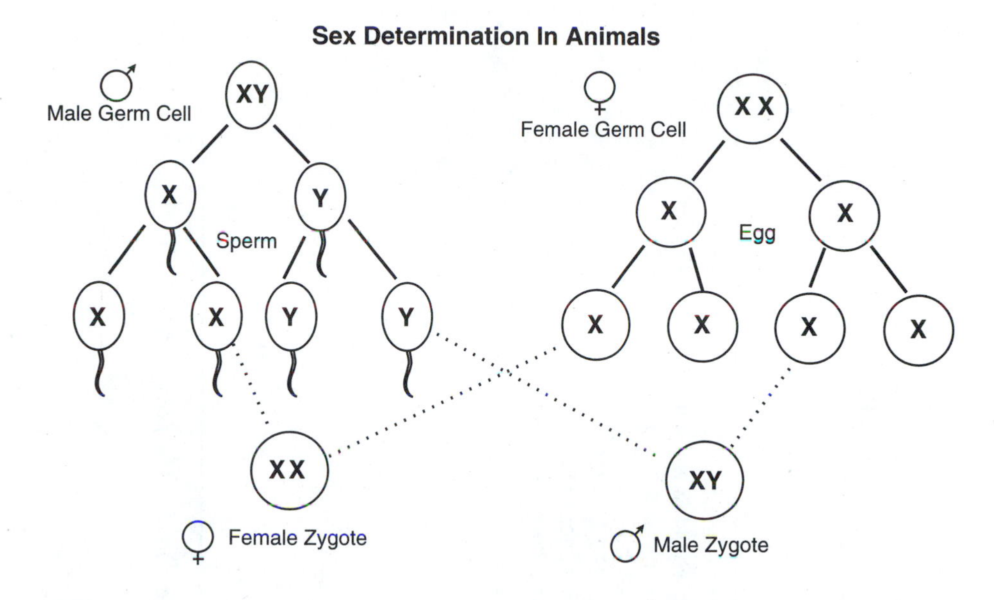

Figure 13-7. The sex of an animal is determined by the combinations of *X* and *Y* chromosomes.

producer has a number of genetically based tools to use in the selection of livestock. One such measure is known as **heritability**. Heritability is the measure of how much of a trait was passed on to the offspring by genes. In other words, how much of a trait is inherited and how much is due to the environment in which the animal lives? Heritability measurement varies from zero to one. The higher the number, the stronger the degree of heritability.

Figure 13–8 lists the heritability estimates for traits in agricultural animals. Notice that the heritability for milking ability in sheep is .25. This means that the variation among different sheep in milking ability is due to factors other than genetics. These factors might include such environmental factors as quantity and quality of feed available to the ewe and climactic conditions. Obviously, those traits that are more highly heritable are the traits that can be used by the producer in selecting breeding stock.

PERFORMANCE DATA

Through the use of data collection and computer analysis, the records of how an animal has performed and the analysis of how the animal's ancestry and progeny have performed can be a valuable tool in determining the animal's use as a breeding animal. The following are some of the measures of performance that can be used in selecting breeding animals.

Indexes

An index is a measure of how well an animal has performed as compared to the animals raised with it. This can be a measure of the genetic differences in the animals because the animals are presumably raised under the same conditions and are treated alike. An index is measured based on a scale of 100, with 100 being the group average. The formula for calculating an index is:

HERITABILITY ESTIMATES FOR BEEF, SWINE, AND SHEEP

BEEF

Trait	Heritability Estimates
Birth weight	.40
Weaning weight	.25
Yearling weight	.40
Feedlot gain	.45
Efficiency of gain	.40
Fat thickness	.45
Rib eye area	.70

SHEEP

Trait	Heritability Estimates
Number born	.13
Weaning weight (90 days)	.30
Yearling weight	.40
Fleece weight	.40
Milking ability	.25
Rib eye area	.53

SWINE

Trait	Heritability Estimates
Litter size (weaned)	.12
Weaning weight (3 weeks)	.15
Number of nipples	.60
Age at 220 lb.	.30
Length at 220 lb.	.60
Backfat thickness at 220 lb.	.50
Percentage carcass muscle	.45

Figure 13-8. Heritability estimates indicate how likely it is that a particular trait of a parent will be passed on to its offspring.

$$\text{Index} = \frac{\text{Individual Animal Weight}}{\text{Average Group Weight}} \times 100$$

For instance, if a group of calves are weaned when they are 205 days old and the average weight of the group is 583 pounds, an index of 100 is equal to 583 pounds. If one of the calves weighs 614 pounds, the calf is said to have a 205-day weaning weight index of 105.3. This means that the animal has outperformed its peers. If a calf has a 205-day weaning index of 82, this would mean that particular calf has performed only 82 percent as well as the other calves with which it was raised. Other indexes are used for comparing animals on **yearling weight**, birth weight, and other measures. Other factors, such as the age of the mother, can also be figured into the formula in order to give a more accurate comparison of the animal's performance. Remember though that an index is only good for comparing animals within their own groups. Comparing indexes of animals that were raised in separate groups is of little value because the animals almost assuredly were raised under dissimilar conditions.

Measurements of Mothering Ability

One of the most valuable traits that a female agricultural animal can possess is being able to produce enough milk to feed her young. Mothers that produce a sufficient quantity of milk will wean young that are larger and faster growing than those from mothers that milk poorly. This is true for all agricultural animals whether it be sheep, cattle, or swine.

The **most probable producing ability** (MPPA) is a measure of a cow's ability to wean a superior calf. It is figured using the 205-day weaning weight index and the number of calves the cow has produced. MPPA is an indication not only of the cow's ability to pass **growthability** on to her calf but also her ability to produce enough milk and to care for her calf well enough to wean a large calf. A similar measurement for pigs is used. This measure is called the *Sow Productivity Index* and takes into account such factors as the number of piglets born alive, the average for the other sows in the herd, and the 21-day litter weight (for both the sow and the herd). The term for this measurement in sheep is the Ewe Index, which takes into account the ewe's ability to wean above-average lambs. The most probable producing ability (MPPA) for cows, the Sow Index for pigs, and the Ewe Index for sheep are all ways of putting the mothering ability of individual animals into numbers so they can be compared.

Estimated Breeding Values

Cattle breed associations make use of a measure referred to as the **estimated breeding value**, which is an estimate of the breeding value of an animal compared to other animals. It is computed by using a very complicated formula using the animal's own performance data as well as that of the animal's half brothers and half sisters. Through the use of artificial insemination, a particular animal can have a large number of **siblings** (brothers and sisters). Data on a large number of siblings can be very useful in determining how well the desired characteristics are passed on by the animal's ancestors.

The larger the number of siblings and the larger the amount of data on both siblings and ancestors, the more accurate the estimated breeding value will be. The accuracy is also used in the determination of how much emphasis a producer will place on an animal's estimated breeding value in the selection process. Measures of performance are summarized on the animal's **pedigree record**, which is a part of the animal's registration papers. Notice that all the types of data that have been discussed are on the record for the producer to use in evaluating the animal.

Expected Progeny Difference (EPD)

The **expected progeny difference** (EPD) is used to predict the differences that can be expected in the offspring of a particular sire over those of other bulls used as a reference. The data for calculating EPDs are obtained from the performance data of the progeny from the bulls. This is another example of how artificial insemination is useful in the determination of the desirability of a sire because of the tremendous number of offspring made possible by using artificial insemination, Figure 13–9. This data can be tremendously helpful to a producer who wishes to increase or improve certain traits in the calves produced. For example, if a bull has an EPD of +18, his offspring can be expected to have a weaning weight of 18 pounds more than the average bull of the same breed. Of course, the bull may have calves that are below average weight, but it should be expected for most of his offspring to have higher weaning weights than average for the breed.

Linear Classification

A modern tool used by dairy producers for selecting replacement animals is a process known as linear classification. This process combines the use of visual and genetic selection. Cows are evaluated and classified by assigning a value between 1 and 50 for certain traits that are considered to be of importance to the animal's production ability. The Holstein-Friesian Association uses 17 functional traits for evaluation and assigns such a number to each trait, Figure 13–10. For instance, a wide pelvis (**thurl**) is related to ease of calving, so a wide pelvis is desirable in an animal that is to be used as a replacement animal. An animal that is appraised as having an extremely narrow pelvis might be given a point score of one to five points, depending on how severely narrow the thurl is. A cow with an extremely wide pelvis (thurl) may be assigned 45–50 points.

However, a score of 50 does not necessarily mean that the animal has the most desirable form of the particular trait. To be comfortable in walking and standing, an animal must have strong legs. The rear legs must have the proper curve or set to the legs in order to provide the most flex and cushion as the animal walks. However, a cow with too much set to the rear legs is not desirable because too much stress will be placed on the muscles and tendons of the legs. A post-legged cow (with straight "posty" rear legs) will be given one to five points; a cow with an intermediate set to the rear legs will be given 25 points; and a cow with an extreme amount of set (**sickle-hocked**) will be given 45–50 points. In this case, the most desirable condition is given the midpoint score of 25.

The animals are scored periodically by a professional evaluator who has been trained by the breed association. These evaluators come to the dairy producer's farm and evaluate each animal for a set fee. Information from the animals is then compiled, and the producer receives a data sheet on the linear classification of the animals. This information can then be used by the producer to select bulls for use in artificially inseminating the cows. For instance, if the cows in the herd have problems with the structure of their feet and legs, a bull is selected that is known for siring daughters that have strong and structurally correct legs. The offspring from the cows should then have stronger feet and legs than their mothers and can be used as replacement heifers.

The Holstein Association lists the following advantages of using the linear classification system:

- Provides unbiased and accurate evaluations of cows

TRI-STATE BREEDERS – 1991 Beef Sire EPDs

SIMMENTAL		C.E.H EPD	C.E.H ACC.	C.Ec EPD	C.Ec ACC.	Birth Wt. EPD	Birth Wt. ACC.	Wean. Wt. EPD	Wean. Wt. ACC.	Yearl. Wt. EPD	Yearl. Wt. ACC.	Mat. C.E.H EPD	Mat. C.E.H ACC.	Mat. C.E.c EPD	Mat. C.E.c ACC.	Mat. Wean. Wt. EPD	Mat. Wean. Wt. ACC.	Mat. Milk EPD	Mat. Milk ACC
14SM340	AF Redlands 35Y	−6	.07	−2	.07	+ 3	.16	+18	.15	+31	.14	+ 1	.06	+ 0	.06	+11	.08	+ 2	.08
14SM341	ASR Polled Pacesetter 413Z	−1.0	.16	+0	.16	− 1	.21	+16	.20	+32	.18	+ 0	.15	+ 0	.15	+25	.17	+17	.17
21SM270	Bold Leader	−4.7	.68	−1.3	.68	+ 2.8	.92	+26.6*	.91	+47.4*	.88	− 2.1	.68	− .5	.68	+24.5*	.78	+11.2*	.77
14SM334	F N Stamina	+4.5	.09	+1.2	.09	− .4	.18	+ 6.7	.17	+16.8	.17	+ 1.2	.09	+ .4	.09	+ 1.0	.11	− 2.3	.10
14SM338	Hancocks Pineview Regal	−7	.18	−2	.18	+2	.35	+13	.32	+28	.19	+ 0	.18	+ 0	.18	+12	.18	+ 6	.18
14SM326	HCC Prophet	+6.3	.28	+1.6	.28	− .5	.73	+ .3	.65	+ 4.5	.58	+ 3.6	.19	+ 1.0	.19	+ 1.5	.24	+ 1.4	.23
36SM145	HF Phantom	−6.1	.37	−1.8	.37	+ .8	.70	+18.6*	.66	+32.3*	.61	+ 5.0	.37	+ 1.3	.37	+ 4.7	.45	− 4.6	.43
14SM327	HMF Gold Bar 304W	+3.4	.16	+1.0	.16	+ .7	.53	+ 1.4	.46	+ 9.1	.40	+ 1.2	.14	+ .4	.14	+ 8.5	.18	+ 7.8	.17
14SM330	HMF Polled Siegfried 230U	−3.2	.17	− .8	.17	+ .9	.55	+ 9.0	.46	+25.0	.40	+ 2.3	.15	+ .7	.15	+ 6.5	.20	+ 2.0	.19
14SM336	Keystone	−10.8	.53	−3.4	.53	+ 5.9	.66	+31.3*	.62	+50.5*	.61	+ 6.9	.53	+ 1.7	.53	+11.1	.55	− 4.5	.55
14SM339	LCHM Black Baron 235X	+2.2	.17	+ .7	.17	− .9	.45	+11.0	.40	+25.3	.35	+ 4.3	.17	+ 1.1	.17	+ .8	.19	− 4.7	.19
14SM337	Mr GF Train	−7	.17	−2	.17	+ 2	.20	+22	.20	+35	.18	− 2	.17	− 1	.17	+15	.18	+ 5	.17
14SM320	Paymaster	−1.5	.19	− .3	.19	+ .1	.75	+ 2.9	.68	+21.4	.66	− 5.9	.21	− 1.7	.21	+ 3.8	.41	+ 2.4	.39
14SM011	Pineview Apache	−7	.12	−2	.12	+ 1	.35	+32	.32	+57	.18	+ 1	.12	+ 0	.12	+23	.14	+ 7	.13
14SM142	Pineview Jazz	−5.5	.81	−1.5	.81	+ .9	.93	+16.1*	.92	+38.8*	.91	− 2.2	.80	− .6	.80	+12.7*	.86	+ 4.7	.85
36SM154	Pineview Presley	−4.8	.34	−1.3	.34	− .6	.77	− 7.6	.72	− 6.2	.65	+ 3.1	.23	+ .8	.23	− .5	.32	+ 3.3	.30
14SM342	R&R Magician Z504	−2	.11	+0	.11	− 2	.34	+14	.31	+31	.17	+ 0	.10	+ 0	.10	+14	.12	+ 7	.12
14SM332	R&R The Wizard 504X	+4.1	.14	+1.1	.14	− 3.6	.42	+15.3	.34	+28.6	.30	− .9	.11	− .2	.11	+10.6	.14	+ 3.0	.14
14SM318	Royal Can Am	+ .4	.16	+ .2	.16	+ 2.2	.46	+ 9.1	.41	+10.9	.36	+ 2.6	.15	+ .7	.15	+ 9.2	.19	+ 4.6	.18
36SM165	S&S Eclipse	− 3.3	.27	− .9	.27	+ .9	.67	+20.1	.59	+37.3	.57	− 2.4	.20	− .6	.20	+ 9.8	.26	− .3	.24
14SM321	Sunny K Blackjack	+5.4	.18	+1.4	.18	− .4	.57	+ 2.6	.50	+ 9.4	.45	+ .6	.15	+ .2	.15	+ 1.1	.19	− .2	.18
36SM184	Super Light 19U	+ .9	.18	+ .3	.18	− .1	.37	+ 5.1	.32	+ 9.6	.30	+ 3.5	.17	+ .9	.17	+ 5.0	.19	+ 2.5	.18
14SM324	The Greek	−8.7	.21	−2.7	.21	+ 6.0	.67	+36.7	.55	+58.9	.46	+ 7.7	.17	+ 1.9	.17	+ 3.0	.21	−15.4	.20
14SM357	TNT Mr T	−10.8	.60	−3.4	.60	+ 2.5	.90	+20.9*	.87	+41.6*	.85	− 1.7	.55	− .4	.55	− 7.9	.66	−18.3	.65
14SM325	Triumph	−13.8	.65	−4.6	.65	+ 5.6	.85	+19.9*	.81	+36.3*	.78	+ 3.7	.64	+ 1.0	.64	+ .2	.68	− 9.8	.67
14SM322	WRS Alien	−21.4	.15	−8.3	.15	+ 7.7	.70	+25.8*	.63	+51.9	.53	+ 6.9	.10	+ 1.7	.10	+10.5	.16	− 2.4	.15
36SM156	WRS Enterprise	−2.2	.17	− .5	.17	− .2	.34	− .7	.31	+ 6.9	.29	− .3	.17	+ .0	.17	+ 2.4	.20	+ 2.8	.19
14SM331	Y1 Yardleys P&B R248	+1.8	.25	+ .6	.25	+ 1.9	.73	+ 7.1	.66	+24.9	.58	− 1.4	.27	− .3	.27	− 5.3	.46	− 8.9	.45

*Trait Leader

Figure 13-9. Expected progeny differences (EPDs) are used in determining which sire to use in an artificial insemination program. *Tri-State Breeders.*

Page No. 1

Owner Dean Dairyman

Address 1500 Farm Lane

Holstein WI 53535

	ANIMAL NO.	PEDS	MATE	ANIMAL BARN ID / DATE OF BIRTH	LACT. NO.	Stage Code	DATE OF CALVING MO.	DAY	YR.	FORM: STATURE	STRENGTH	BODY DEPTH	DAIRY FORM	RUMP: RUMP ANGLE	THURL WIDTH	LEGS/FEET: REAR LEG SIDE VIEW	FOOT ANGLE	UDDER: FORE UDDER ATTACHMENT	REAR UDDER HEIGHT	REAR UDDER WIDTH	UDDER CLEFT	UDDER DEPTH	TEATS: FRONT TEAT PLACEMENT	TEAT LENGTH	RESEARCH TRAITS: REAR LEG REAR VIEW	UDDER TILT	(NOT DEFINED)	(NOT DEFINED)	(NOT DEFINED)
1	10342984	4	0	Regina 10-05	7		12	30	89	34	33	34	41	25	36	25	30	26	32	37	45	28	18	37	40	28			
2	10922314	4	0	Veronica 10-04	7		1	22	89	42	40	37	42	18	34	22	40	10	36	36	34	15	14	36	37	32			
3	11600072		0	Margie 6-10	5		2	02	90	checked cow																			
4	11717850			Bon Bon 6-08	4		5	24	89	32	31	32	36	12	27	32	28	32	25	17	36	25	37	26	18	27			
5	11990933	3		Rebecca 5-10	4		1	15	90	42	38	40	42	25	37	25	36	34	36	38	27	28	32	22	32	30			
6	11990935	3		Lilac 5-10	4		4	07	89	36	38	37	37	18	36	30	12	42	48	40	40	36	34	26	22	29			
7	12052652	4		Suzanne 5-09	3		7	26	89	50	50	43	40	22	45	27	36	32	28	25	20	15	26	37	18	22			
8	12313697	3		Vera 5-00	3		12	26	89	40	37	40	36	20	38	23	37	22	26	18	32	28	28	32	36	28			
9	12313698	3		Margaret 5-00	3		8	28	89	37	30	34	37	22	30	34	28	40	34	37	36	34	36	23	27	30			
10	12495037			Regal 4-07	2		8	18	89	36	36	24	22	27	34	22	32	34	34	36	33	36	32	30	32	30			
11	12615110	3		Bonnita 4-06	2		4	24	89	34	29	34	44	15	28	36	17	43	26	30	37	40	36	14	18	28			
12	12616706		0	Carmen 4-04	2		7	20	89	24	21	24	34	27	15	21	24	17	34	24	25	19	17	15	35	18			
13	12798352	3		Vanessa 3-09	2		2	26	90	39	22	25	41	34	18	32	27	28	18	20	30	32	28	26	20	21			
14	12857939	3		Margo 3-08	2		12	29	89	34	19	25	40	35	18	28	16	32	34	34	36	34	30	23	25	30			
15	12932831	3		Romance 3-05	1		4	25	89	35	32	32	18	33	27	27	32	36	30	30	38	34	35	22	18	32			
16	12932834			Bobbie 3-05	1		5	05	89	27	20	25	28	31	23	28	17	17	27	17	16	28	25	27	16	40			
17	13225840	3		Sunshine 2-09	1		9	04	89	40	27	28	20	28	27	27	26	27	18	17	32	38	25	28	27	30			
18	13225841			Carmel 2-09	1		10	29	89	32	30	32	31	32	26	18	26	26	18	18	26	39	25	27	26	40			
19	13225857	3		Ramona 2-04	1		2	13	90	30	22	21	24	25	27	15	30	31	26	32	38	38	27	25	32	35			
20	13288385			Mermaid 2-03	1		3	21	90	33	31	31	28	26	27	22	18	12	23	22	25	27	25	22	21	27			
21	13387133		0	Suzette 2-01	1		5	28	90	not in condition																			
22																													
23																													
24																													
25																													

	ANIMAL NO.	PREVIOUS CLASSIFICATION DATE / SCORES	PERMANENT	MULTI E Date Eligible #E	GENERAL	DAIRY	BODY	MAMMARY	FINAL	NEW E
1	10342984	4-12-89 EVVV 88	P		E	E	V	E	90	
2	10922314	4-12-89 +EVG 83	P		V	E	E	G	85	
3	11600072	4-12-89 +EVG 83	P						✓	
4	11717850	4-12-89 FVVG 78	P		G	V	V	+	81	
5	11990933	4-12-89 VVV+ 86			E	E	E	+	88	
6	11990935	4-12-89 +++V 84			+	V	V	E	86	
7	12052652	4-12-89 EEEV 90		09-90 2	E	E	E	V	90	
8	12313697	4-12-89 +VV+ 82			V	V	E	+	85	
9	12313698	4-12-89 ++V+ 83			V	V	V	E	87	
10	12495037	4-12-89 +++G 80			+	+	V	V	83	
11	12615110	4-12-89 +VVV 85			+	E	V	E	86	
12	12616706	4-12-89 +++G 80			G	+	+	G	78	
13	12798352	4-12-89 ++++ 82			+	E	+	G	82	
14	12857939	4-12-89 G+V+ 82			G	V	V	G	81	
15	12932831				G	G	V	V	81	
16	12932834				G	+	+	F	75	
17	13225840				V	G	+	G	81	
18	13225841				G	+	+	G	79	
19	13225857				G	+	+	+	80	
20	13288385				+	+	+	F	78	
21	13387133								N/C	
22										
23										
24										
25										

Figure 13-10. The Holstein-Friesian Association considers seventeen functional traits in evaluating animals. *Holstein-Friesian Association.*

- Defines types of trait trends from one generation to another
- Gives the producer a clear understanding of each animal's strengths and weaknesses
- Compares the producer's herd type pattern to breed averages
- Adds trait appraisals to official pedigrees
- Provides a basis for mating services

The selection process used by modern producers is truly based on science. Principles of genetics are used as well as the findings of research studies that have discovered physical characteristics of animals that indicate the potential of the animal. The use of computers has greatly enhanced the producer's ability to predict how well an animal will perform as well as the performance of the animal's progeny.

SUMMARY

Through an understanding of genetics, producers can better choose animals to use in a breeding program. Traits that are desirable are passed from generation to generation through gene transfer. By knowing which of these traits are dominant and which are recessive, producers can develop a program that will produce the type animals that will pass characteristics on to their offspring. Although we know a lot about how genes are transferred, we are still a long way from understanding how the process works. When we unlock the mechanisms of gene transfer, the way we select and produce animals will be revolutionized.

◆ Review Exercises

TRUE OR FALSE

1. Genotypical differences in animals are controlled by the animal's genes that were passed on by its parents. ______.
2. Chromosomes are composed of long strands of molecules called deoxyribonucleic acid (DNA), which is composed of large molecules that are capable of an almost unlimited number of ways in which they can be put together. ______.
3. If both genes are the same from the parent animals, they are called heterozygous. If they are different, the genes of the offspring are said to be homozygous. ______.
4. Genetic dominance means one gene will override the effects of the other gene.
5. Some pairs of genes may not have one that is dominant over the other. They are of equal power and are called codominant. ______.
6. All animal characteristics, such as milk production, are controlled by only a single pair of genes. ______.
7. Whether the animal is male or female is controlled by the matching of chromosomes from the mother and father and is determined three to five days after conception. ______.
8. In heritability measurements the lower the number (from zero to one), the stronger the degree of heritability. ______.
9. Indexes are used to compare common characteristics of animals in different regions. ______.
10. Mothering ability has to do with the mother's ability to protect her young, not with such physical features as milking ability. ______.
11. The Most Probable Producing Ability (MPPA) for cows, the Sow Index for pigs, and the Ewe Index for sheep are all ways of putting the mothering ability of individual animals into numbers so they can be compared. ______.

12. Expected progeny difference records are used to evaluate bulls used for breeding, but not to evaluate cows. ______.

13. Linear classification combines the use of visual and genetic selection. ______.

14. In linear classification a score of 50 is the best. ______.

15. Linear classifications are based entirely on the evaluation of an individual. Computers are never used. ______.

FILL IN THE BLANKS

1. Phenotypes are the ____________ appearance of the animal such as color, _________, shape, and other ___________ and can be caused by the ____________ conditions under which the animal is ____________.

2. DNA replication happens when the cell divides and the ________________ of DNA separate and each half ____________ itself so that two strands, exactly_______ ___________, are ____________.

3. At conception, the _____________ halves from each _____________ are combined to form fully ____________ chromosomes, and the chromosomes from each ____________ are _________ to form the ____________ code for the fertilized egg.

4. In pig coat color, the *B* genes of the female that match with the *w* genes of the male will result in a __________ gene and will produce pigs that are _____________.

5. Epistasis involves genes that are not ______________ (matched ___________ that control a ______________) that interact to cause an expression that is different from the ___________ on the _________.

6. The zygotes containing the *XX* sex ___________ develop into ___________ and the zygotes containing the *XY* sex __________ develop into ___________.

7. Heritability is the ___________ of how much of a ____________ was ____________ on to the _____________ by ____________.

8. Through the use of data collection and ____________ analysis, the records of how an animal has performed and the ____________ of how the animal's ancestry and _____________ have performed can be a valuable ___________ in determining the animal's use as a _______________ animal.

9. The Sow Productivity Index takes into account such factors as the ____________ of piglets born _____________, the ____________ for the other sows in the herd, and the ____________________-day litter _______________ (for both the ________ and the _________).

10. The estimated breeding value for cattle is computed by using a very complicated formula using the animal's own _____________ ______________data as well as that of the animal's half _________________ and half ____________.

11. Measures of performance are ___________ on the animal's ______________ record, which is a part of the animal's ___________ ____________.

12. EPDs are another example of how artificial ______________ is useful in the determination of the ____________of a ___________ because of the tremendous number of ______________ made possible by using ____________ ____________.

13. In linear classification, cows are _________ and ____________ by assigning a value between 1 and _____________ for certain traits that are considered to be of importance to the animal's ___________ ___________.

14. The producer can use the information from linear classification to select ____________ for use in ______________ ______________ the cows.

15. Principles of ________________ are used in selection processes as well as the findings of ______________ ___________ that have discovered ______________ characteristics of animals that indicate the ________________ of the animal.

DISCUSSION QUESTIONS

1. Explain the difference between an animal's genotype and its phenotype.
2. What is DNA? What purpose does it serve?
3. What is a helix?
4. What part does RNA play in the passing of traits from parent to offspring?
5. Explain what is meant by a dominant and a recessive gene.
6. Explain how a red Shorthorn bull mated with a white Shorthorn cow will produce a spotted or roan-colored calf.
7. If a male rabbit that is black in color (pure gene for the dominant black color *BB*) is bred to a white female rabbit that has the pure **recessive gene** *WW* for the white color, what percent of the young would you expect to be black?
8. If a black male rabbit with the gene *Bw* is bred to a white female rabbit (*ww*), what percent of the young would you expect to be black? What percent should be white?
9. How is the sex of an animal determined?
10. What does heritability measure?
11. If a calf has a weaning weight of 478 pounds and the other calves that were raised with it have an average weaning weight of 634 pounds, what will be the weaning weight index for the calf?
12. What measure of mothering ability is used for cows? for sheep? for swine?
13. What do the letters EBV stand for? How is this measure used?
14. What is meant by the expected progeny difference?
15. What is meant by linear evaluation?
16. Explain how the use of artificial insemination and the use of computers have greatly aided in the selection process.

STUDENT LEARNING ACTIVITIES

1. Choose a species of agricultural animal such as sheep, cattle, swine, or horses. List the physical characteristics you feel are of economic importance to that type of animal. Decide whether or not these characteristics are influenced more by genetics or environment.
2. Obtain copies of the pedigree papers of several animals of the same breed. Compare the animals based on their pedigrees and performance records.
3. Write to the American Breeders Society or to Select Sires to obtain a copy of their dairy cattle sire catalog. From a dairy producer, obtain copies of the linear classification data from the dairy's herd. Using the data on the females from the records and the data on the different sires in the catalogs, choose the most desirable sires for the cows in that particular herd.

Chapter 14

The Scientific Selection of Agricultural Animals

A part of this chapter was contributed by William Vernadore

Student Objectives in Basic Science

As a result of studying this chapter, you should be able to

1. Explain the concept of natural selection.
2. Explain how humans have influenced the development of animals.
3. Illustrate how scientific research has influenced the development of animals by humans.
4. Cite examples of how problems have developed in animals because of the selection process controlled by humans.

Student Objectives in Agricultural Science

As a result of studying this chapter, you should be able to

1. Justify the selection of different animal traits by the producer.
2. Trace the stages in the development of modern swine.
3. Describe problems associated with overly muscled pigs.
4. Interpret the reasoning behind the selection for sex character.
5. Rationalize the selection of animals for structural soundness.
6. Describe the physical characteristics associated with growth in animals.
7. Describe the modern beef and swine animal.

Key Terms

natural selection
reproductive efficiency
fertile
efficiency
performance data
Porcine Stress Syndrome (PSS)
PSE pork
loin eye
pin nipples
blind nipples
inverted nipples
gilts
infantile vulva
vulva
testicles
viable
sheaths
ligaments
pasterns
splayfooted
pigeon-toed
cow-hocked
cannon bone
backfat
rib eye
frame size
hip height
retail cuts
double muscling
concentrate
pelvic capacity
sex character
brisket
cutability
type
balance
carcass merit
style
soundness
smoothness
muscling
yearling
ram
conformation
vacuum packaging

Humans have always selected the type of animals they want to produce. As was discussed in Chapter 7, breeds were developed because humans chose to select animals with certain characteristics for use in breeding. As breeds developed and animals bred true for the characteristics of that breed, animals were selected for desirable traits within that breed. Animal producers throughout history have selected animals based on what they thought were the traits that would improve the next generation of animals and would be more profitable for the producer. Only within recent history has there been a truly scientific basis for the selection of animals. In the wild, animals developed traits that would help them survive in their environment. Animals having traits that aided them in survival stood a better chance of living and reproducing than did animals without the traits. For example, wild pigs that had the longest tusks, the thickest hide, and the fiercest nature had a better chance at survival than did those with small tusks, a thin hide, and a docile nature. Wild cattle that could run fast for long distances and had long horns to defend themselves stood a better chance of evading or fighting off predators. Only the strongest animals survive in the wild, and they are the ones that breed and pass their characteristics on to the next generation. This process is known as **natural selection**. In domestication, animals no longer need many of the characteristics that increase their chance of survival in the wild. On the contrary, many of the traits that were essential for survival in the wild are a great disadvantage to animals in domestication. For instance, the fierce nature of a wild pig is far less desirable than the docile nature of the domesticated pig. Cattle no longer need to be able to run swiftly or to possess long horns for defense. These traits have been bred out of domesticated agricultural animals.

Through selective breeding, humans have attempted to produce those animals that do well in a domesticated state. Obviously the conditions under which the animals are to be raised (the environment) dictate the characteristics that an animal needs to thrive.

For example, during the nineteenth century, the longhorn breed of cattle was developed in Texas. This breed needed to retain some of the characteristics of wild animals like long horns for defense and the toughness to survive under harsh conditions, Figure 14–1. During the

Figure 14-1. Nineteenth century longhorns retained some of their wild animal characteristics in order to survive under harsh conditions. *Soil Conservation Service.*

time when rangeland was almost limitless and labor was in very short supply, this type of cattle very well met the requirements of producers. These animals could tolerate living on an open range and foraging for food with very little help from people. Indeed, longhorns still play a part in some cattle operations, but most situations in modern animal production call for a completely different operation. To meet the demands of the modern livestock industry, animals are selected based on what is desired by the people who produce them and by those who buy them.

Of the animals that are raised for food, there are basically two categories of animals. The first are those that are produced for slaughter; the second are those animals that produce offspring that are to be raised for slaughter. There are many considerations in the selection of the type of animals for these categories. Consumers have to be pleased with the type of product they find at the meat counter. In order to make a profit, the meat packer has to approve of the carcasses that are sent to the meat packing plant, Figure 14–2.

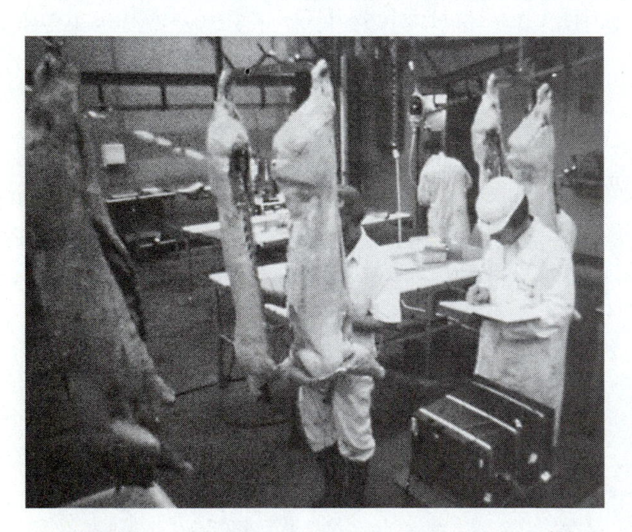

Figure 14-2 Packers want carcasses that will make a profit. *Rick Jones, Cooperative Extension Service, The University of Georgia.*

The buyers want animals that will remain healthy until they reach the slaughterhouse. The growers want animals that can gain weight quickly at an acceptable cost and with a minimum of care. The breeders want an animal that can reproduce efficiently. All of these criteria make the selection of the modern animal a complicated process. There are three basic traits that are desirable in the modern agricultural animal: **reproductive efficiency**, growthability, and efficiency.

Reproductive efficiency means that breeding animals must be selected that produce offspring at a regular rate. If animals are producing young at a steady rate, producers are more likely to make a profit than they would be if the animals produced fewer young, Figure 14–3. This means that the males must be **fertile** (produce sufficient numbers of healthy sperm); they must be healthy, aggressive breeders; they must live a long, productive life. Females must be able to regularly come into estrus, conceive readily, produce an adequate number of healthy offspring, and produce enough milk to ensure that the young are weaned at an adequate size and weight.

Growthability refers to an animal's ability to grow rapidly. The faster an animal grows, the

Figure 14-3 Animals must produce and raise young at a steady rate. *American Salers Association.*

more likely the producer is to make a profit from growing the animal. This trait is inherited from the parents and is greatly influenced by the type of care offered by the producer.

Efficiency is the ability of the animal to gain on the least amount of feed and other necessities. The producer sees that the animals are well cared for and fed. Those animals that gain the most on the least amount of feed are the most desirable. If one steer can gain a pound for every nine pounds of feed it consumes, and if another steer gains a pound of body weight for every eight and a half pounds of feed it consumes, the steer that required less feed per pound of gain is said to be more efficient.

These characteristics have always been the important traits that producers have wanted. In years past, producers had a much more difficult time predicting which animals would possess these traits. Now as a result of modern research, producers are able to predict with much more accuracy which animals will possess the desired traits. For example, research has shown that there are certain physical characteristics of animals that will predict the reproductive capability of an animal. Breed associations have developed a bank of data on the performance of the offspring of a particular sire or dam. This has been brought about through the use of artificial insemination and embryo transfer. As was discussed in the previous chapter, **performance data** have been compiled into values that indicate an animal's usefulness as a breeding animal.

The Selection of Swine

Prior to the 1950s, swine were raised primarily for lard. People used the lard not only for frying and cooking food but also in the manufacture of cosmetics and lubricants. With the advent of cooking oils made from vegetable oils and cosmetics and lubricants made from petroleum-based synthetics, the demand for lard was dramatically reduced. Efforts were then put into developing a hog that produced meat instead of lard. Pigs that were used for breeding were especially selected for the degree of muscling they possessed. The idea was to produce an animal with the maximum amount of meat and the minimum amount of fat. These efforts culminated in the late 1960s and early 1970s with what has been referred to as the "super pig," Figure 14–4. Pigs were selected for huge, round, bulging hams and overall thickness of muscling. Although producers were successful in producing a very lean pig that had a lot of muscle, some problems arose with the highly muscled, extremely lean pigs. This type pig failed to be the "ideal" type for three reasons:

1. **Porcine Stress Syndrome (PSS)**. Extremely heavy muscled pigs are associated with a condition know as Porcine Stress Syndrome (PSS). Apparently this condition is genetic and is passed on to the offspring by the parents. Pigs suffering from PSS have very little tolerance to stress associated with hot weather, moving about, and some management practices. When put under such stress, animals with this condition have muscle tremors and twitching, red splotches develop on their underside, and they suffer sudden death. Obviously this condition in hogs is not in the best interests of the pigs or the producer.
2. **PSE Pork**. Pigs with extremely heavy muscle tend to produce lower quality pork. Although they may have a large quantity of muscling, the meat has little or no intermuscle fat (marbling), is very pale in color, and is soft and watery (exudative). These

Figure 14-4 Selective breeding has brought many changes to the form of the swine. Bottom: 1960s classic outline; Middle: In the 1970s, a different form was developed; Top: The modern swine is designed to be rugged and capacious.

characteristics account for the name of the condition known as Pale, Soft, and Exudative (PSE) Pork. Consumers reject this type of pork because the pale color is not appealing, and when cooked, the meat is dry and lacking in taste.

3. Reproductive efficiency is lessened. Heavily muscled, tightly wound boars have problems moving about and mounting females that are in heat. In addition their sperm count is often very low. These conditions make them less desirable as herd sires. Females that are too heavily muscled are less fertile and have problems conceiving. Those that do become pregnant often have problems farrowing because the birth canal is so tightly bound with muscling that it cannot expand properly. Also the number of piglets born is often fewer than those from less-muscular females.

In an effort to correct these problems, during the 1970s producers developed long, tall, flat-muscled pigs. The idea was to produce animals that could move freely and reproduce efficiently. Market emphasis was on pigs that were extremely long, tall, and could move freely. Pigs with extreme bulge and flare to their muscle pattern were highly discriminated against. The tall, flat-muscled pigs had greater resistance to stress and could reproduce more efficiently. Boars were more efficient breeders and sows had larger litters. Although this type pig improved reproductive efficiency, problems were encountered with carcass desirability. The amount of muscle on these animals did not suit the demands of the packers. **Loin eye** areas (an indication of the overall amount of muscle in the carcass) began to be unacceptably small. In addition, the growth rate and feed efficiency of these pigs was lower than the producers wanted. Market conditions also influenced the type of pig that was needed. Added production costs demanded that producers raise a more efficient pig.

In modern swine operations, the three aspects of the swine production industry that make money for the producer are 1) reproductive efficiency, 2) growthability, and 3) carcass quality. Carcass quality is third and least important. With this information in mind, it is easy to understand why the shift is toward the modern type hog.

Beginning in the 1980s, the emphasis on selection in pigs was on what was termed the high-volume pig: very wide down the top, especially at the shoulders. Producers wanted pigs that were very deep and widely sprung at the ribs and deep in the flank and belly, Figure 14–5. The reason for this was that this type of pig had a lot of room in the body cavity for the internal organs such as the heart and lungs. An animal with larger internal organs seems to grow faster and remain healthier. Also, in a radical change from years past, pigs were selected that had large, loose bellies that were capable of holding large amounts of feed. These animals are more efficient in their intake of feed

Figure 14-5 Modern boars have a level top and high capacity. *American Yorkshire Club.*

and in the conversion of feed to body weight. In addition, pigs with larger bellies will produce more bacon. A wide-topped, deep-sided animal has more capacity and internal volume for the internal organs such as heart, lung, and digestive tract. This type animal is a "better doing animal" in that it should have a higher rate of gain than a narrow-topped, shallow-sided animal. The modern market hog has many traits similar to those of the 1980s pig. However, the modern pig is trimmer and has more muscling. The pork industry has shifted to a leaner, lower fat product thai is demanded by the consumer. While emphasis remains on reproductive capability and growthability, carcass merit is also a very strong consideration. The desirable pig still must have high capacity, be free moving, and structurally sound. In addition, it must be heavily muscled and trim.

SELECTING BREEDING HOGS

As was mentioned earlier, the factors in swine that add profitability to the raising of pigs are reproductive efficiency, growthiness, and carcass merit. Reproductive efficiency and growthiness are by far the two most important. With these factors in mind, breeding hogs are evaluated for characteristics that best combine these factors.

To be reproductively efficient, a female must be feminine—that is, she must look like a female. The same substances (hormones) that control the reproductive cycle also account for the development of sex characteristics. The sex characteristics are therefore an indication that the female is producing a large enough quantity of hormones to cause the female to conceive efficiently.

There are several indicators of femininity that a producer can use to select females that are reproductively efficient. The underline should be well defined—that is, the teats should be large and easily seen, Figure 14–6. There should be at least six pairs of prominent and evenly spaced teats. If the teats are too close together, there may not be enough surrounding mammary tissue for good milk production. Pin, blind, or inverted nipples should be avoided. **Pin nipples** are very tiny nipples that are much smaller than the other nipples on the underline. These may not function well enough to feed the young pigs. **Blind nipples** are nipples that fail to mature and have no opening. Obviously these have little use as they are nonfunctional. **Inverted nipples** appear to have a crater in the center and are not functional, Figure 14–7. Usually a female that is producing enough female hormones will have the proper underline to efficiently feed the young she bears.

Figure 14-6 Underlines should be prominent and well defined. *Rick Jones, Cooperative Extension Service, The University of Georgia.*

Gilts and sows should look like females—that is, they should have a head that is shaped like a female's and not like a boar's. Although all pigs selected for breeding should have a large, broad head, gilts should not have the massive head that is characteristic of the boar.

Another defect that adversely affects reproduction is an **infantile vulva**. This condition is

Figure 14-7 Unsound mammaries, such as inverted nipples, should be avoided. *Rick Jones, Cooperative Extension Service, The University of Georgia.*

characterized by a very tiny **vulva** in a breeding-age gilt. This makes breeding very difficult and conception rates are usually very poor. A gilt with an infantile vulva should be culled from the herd, Figure 14–8. Gilts should be selected that have large, normally formed vulvas.

Boars should appear massive, rugged, and masculine. The **testicles** should be large and well-developed inside a scrotum that is well attached. Research has shown that the larger the male's testicles, the more **viable** sperm he will produce and the more aggressive he will be in breeding. Large, pendulous, or swollen **sheaths** should be avoided as these characteristics can lead to breeding problems.

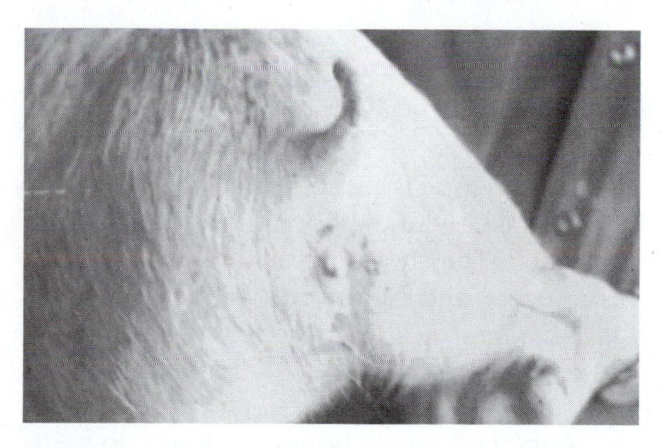

Figure 14-8. Gilts with infantile vulvas should not be kept for breeding. *Rick Jones, Cooperative Extension Service, The University of Georgia.*

Structural Soundness

Structural soundness refers to the skeletal system and how well the bones support the animal's body. Bone growth, size, and shape can have quite an effect on the well-being of the animal. Animals that are structured well are more comfortable as they move about or stand in one place. Structural soundness may affect reproductive efficiency. Boars that have problems moving freely are less likely to be interested in breeding than boars who move freely and are comfortable. Also boars that stand too straight on their legs will have problems mounting females in the mating process.

The vast majority of today's hogs are raised on concrete and should be selectively bred to be comfortable living on the hard surface. Concrete floors are much easier to keep clean and can be kept more sanitary than wood or dirt floors. Structural defects are amplified by the effect of standing and walking on concrete. Soreness, stiffness, and pain in moving greatly reduce reproductive ability. In addition, pigs that have problems standing and moving on concrete generally do not live as long and are not as productive as structurally correct pigs that are more comfortable on concrete.

A skeletal structure that allows a pig to be comfortable on concrete will have a top line that is almost level. At one time, pigs were selected for a uniform arch down the top line; however, pigs with a strongly arched back usually have steep rumps and straight shoulders. As Figure 14–9A shows, the scapula and humerus (shoulder and front leg bone) are more vertical and provide less flex and cushion than the level-topped, more structurally correct pig in Figure 14–9B. Note the vertical position of the

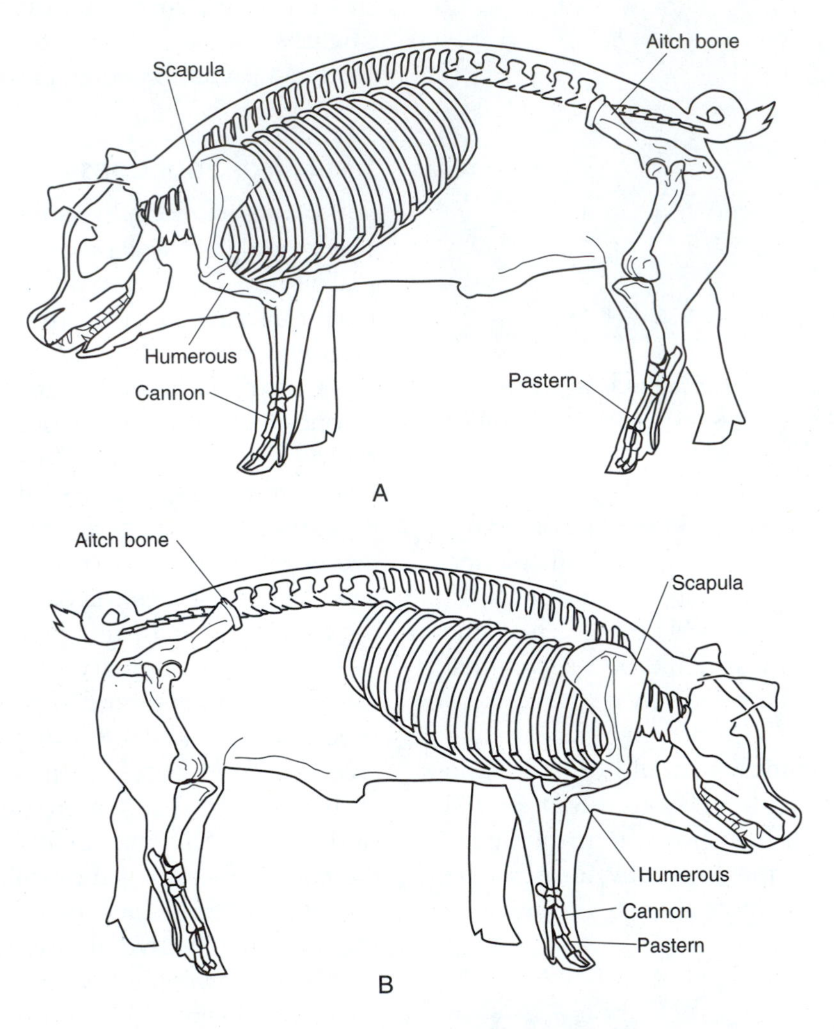

Figure 14-9. The skeletal structure of a high-arched pig (*A*) is more vertical and provides less flex and cushion than the level-top pig (*B*).

aitch bone and femur (hip and rear leg bones) of the pig in Figure 14-9A, as opposed to the same bones that are more nearly parallel to the ground in the pig in Figure 14–9B. Again, bones parallel to the ground add more cushion and flex as the animal walks. If these bones are closer to being vertical to the floor, the ends of the bones will jar when the animal walks. This will eventually cause discomfort to the animal. On the other hand, if these bones are closer to being parallel with the floor, they will act more as a hinge. The shock of walking will be more absorbed by the **ligaments** of the joints and will have less of a jarring effect to the bones. These animals will be more comfortable walking on concrete.

Pasterns (ankle bones) that are too vertical cause too much of a jarring effect to the skeletal system as the animal walks. On the other hand, the pasterns should not be so sloping that they are weak. Bones should be large in diameter and not small and refined. Larger diameter bones are stronger, and research has shown that animals with larger diameter bones will tend to grow faster.

The toes of the animal's feet should be of approximately the same size. Toes that are uneven (most commonly a small inside toe) indicate structural unsoundness. This inherited defect causes misalignment of the feet, weakens the pasterns, and causes an abnormal amount of weight to be placed on the outside toes. This understandably can cause a great deal of discomfort to the animal.

Legs both front and rear should be squarely placed under the animal. Front feet that are turned out (**splayfooted**) or turned in (**pigeon-toed**) should be avoided as well as rear feet that are splayed out (**cow-hocked**) or turned in. The pig should move out with a long easy stride. As the animal walks, the rear foot should be placed about where the front feet were placed. Pigs that take short, choppy steps (goose-stepping) are either structurally unsound, muscle-bound, or both.

Growthiness

In selecting the modern animal, a lot of emphasis is placed on capacity. Preference should be given to those animals that are wide down the top and deep in the side. The animal should be wide between the front legs and have a wide chest floor. The reasoning is that those animals that have greater dimensions in the side, down the top, and through the chest have more room for the vital organs such as the heart and lungs. In addition, pigs with a large, loose belly have a greater capacity for holding feed and thus gain more rapidly.

The ribs should be long and well arched. The rib cage should be rectangular—that is, the fore rib should be about as long as the rear rib. Again, this is an indication that the body cavity has adequate room to house the vital organs.

Other growth indicators are length of **cannon bone** and length of neck. Research studies have shown that animals having longer cannon bones are later maturing and grow more rapidly. Also the longer necked animals have been shown to have more growth potential. Big-headed, broad-skulled pigs that have longer distances between their ears and between their eyes have also been found to be more efficient gainers.

Carcass

As was pointed out earlier, in years past, an overemphasis was placed on thickness and degree of muscling in hogs. Muscling in the modern pig should be long, smooth, and loose. The greatest indicator of overall muscling is the amount of muscle in the hams. Remember, hams are three dimensional; i.e., they have length and width as well as thickness, Figure 14–10. Pigs with short, thick, bulging hams should be avoided. Preference should be given to pigs whose hams get volume in length from tail head to hock and width across the rump. The longer, smoother muscled pigs will move freer and easier. In addition, gilts with this type of muscling will farrow easier than will females that are so muscle-bound that the pelvis cannot give enough to allow pigs to pass easily through the birth canal.

Selecting Market Beef Animals

Consumers play an important role in determining what type of beef animal is raised for slaughter. They usually want beef that is tender, flavorful, and affordable. To produce this

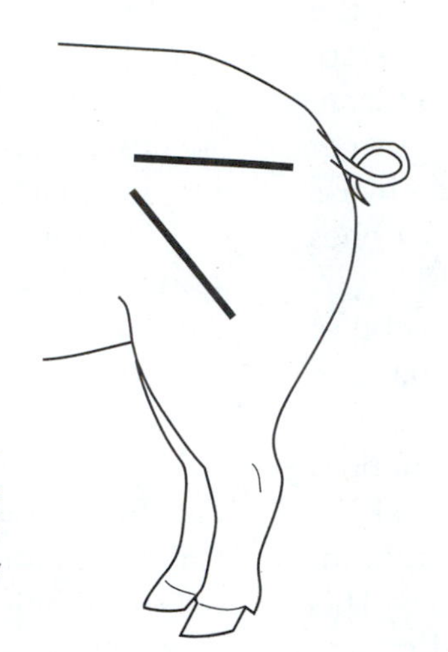
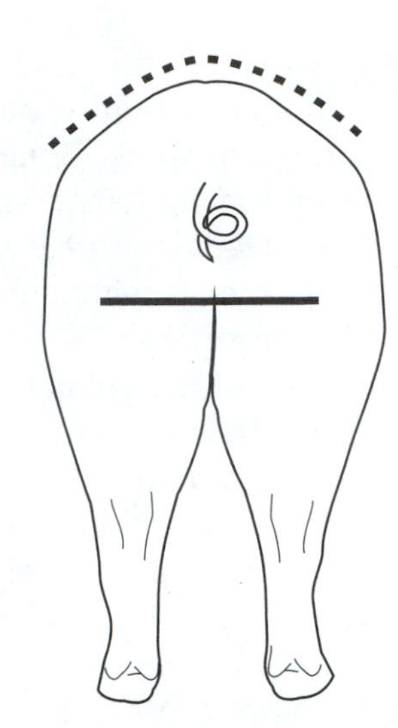

Figure 14-10. The greatest indicator of overall muscling is the amount of muscle in the hams. Total volume of muscle is indicated by length, width, and depth.

type of product, the right type of animal must be produced. Tenderness and taste are both related to the age of the animal. Most of the beef sold in the supermarket as retail cuts come from a choice grade of beef. This means that the animal has reached a degree of maturity where it begins to deposit fat in the muscle.

As animals grow and mature, fat is deposited differently. In young animals most of the energy from feed is put into the growth of bone and muscle so the animal can grow larger. When the animal matures, growth of bones and muscles ceases, and the animal begins to utilize energy from feed to deposit fat. Fat is first deposited in the body cavity of the animal. This serves not only to provide energy storage but to help cushion the internal organs. As the body cavity reaches its peak in terms of fat deposit, the animal begins to deposit fat under the skin. When a certain level of **backfat** is reached, fat is deposited between the muscles and finally inside the muscles. These intermuscular fat deposits, called marbling, are what gives the meat its flavor.

Carcasses are graded to a large degree on the amount of fat that is deposited in the muscles. Cattle, if fed properly, usually reach the proper stage of fatness or finish when they are about two years of age. At this age, the animals are generally still young enough to be tender. The consumer does not want cuts of meat that have a lot of excess fat, so the idea is to produce animals that will put marbling into the meat with a minimum amount of backfat on the carcass.

In addition, consumers are choosy about the size of the meat cuts, Figure 14–11. If the animal is too large when slaughtered, the car-

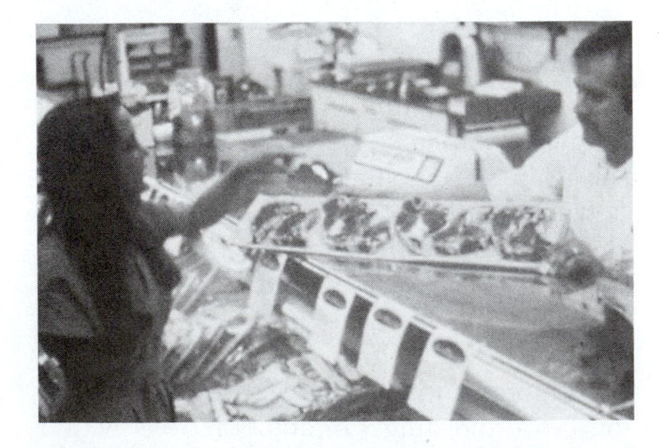

Figure 14-11. Consumers are choosy about the size of meat cuts. *Calvin Alford, Cooperative Extension Service, The University of Georgia.*

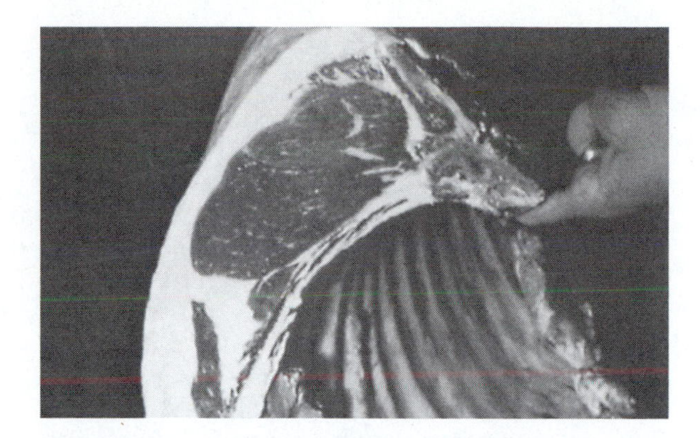

Figure 14-12. Rib eyes larger than fifteen square inches may be too large for some consumers. *Calvin Alford, Cooperative Extension Service, The University of Georgia.*

cass may yield cuts that are too large. Consumers seem to want steaks that one person can consume in one meal. This means that carcasses that have a **rib eye** larger than 15 square inches may be too large for the average consumer, Figure 14–12.

In order to be of the proper size at slaughter, animals have to be the right size when they begin to mature and lay down fat within the muscles. The size of the animal at maturity depends on the **frame size** of the animal. Frame size refers to the skeletal size of an animal at a given age. Small-framed animals mature earlier than do large-framed animals, and they deposit fat in the muscle at a smaller size. A small-framed animal (frame score 1) will probably grade choice around 750–850 pounds, while a large-framed animal (frame score 7) will usually have to weigh 1,350 pounds or more to grade choice. The numerical score for frame size is determined by measuring the height of an animal at the hip at a certain age, Figure 14–13.

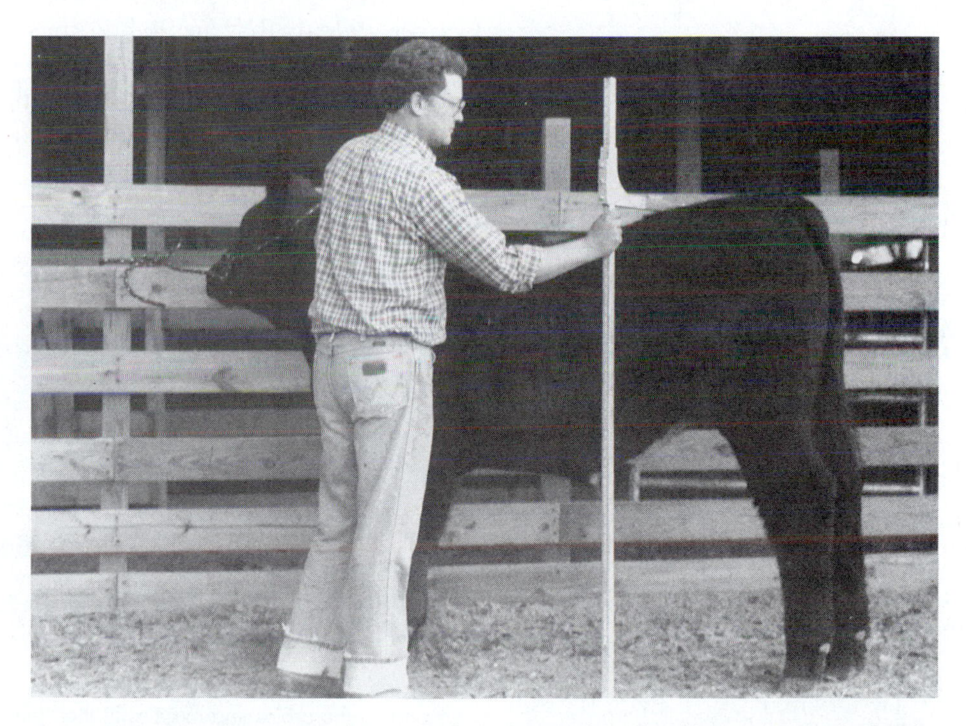

Figure 14-13. The numerical score for frame size is determined by measuring the height of an animal at the hip at a certain age. *Dr. Ronnie Silcox, Cooperative Extension Service, The University of Georgia.*

Figure 14–14 is a table giving the classification of frame scores by **hip height** and age. The base point is 45 inches hip height at 12 months of age for a frame score of 3. Allow 2 inches for each frame score at the same age. Allow 1 inch per month from 5 to 12 months of age, 0.50 inch per month from 12 to 18 months and 0.25 inch up to 2 years.

Commercial packers usually want a carcass that weighs between 600 and 700 pounds. Carcasses within this weight range are more easily managed in the cutting room, and they provide the size of cuts that the consumer wants. Frame size should be large enough for the animal to grade choice at about 1,050–1,200 pounds in order to obtain the desired carcass weight and grade. This means that in selecting slaughter steers, preference should be given to the medium-framed steers that finish at 1,050–1,200 pounds.

The purpose in producing market beef animals is to obtain muscles that can be cut into **retail cuts** of beef for the consumer. It would seem that the more muscle there is on the animal, the more desirable that animal becomes. This is true only up to a point. Just as an animal can have too little muscling, it can have too much. Selection for extreme muscling leads to the development of cattle with a condition known as **double muscling**. Double muscling is undesirable for the following reasons:

1. These animals are difficult to produce. If the goal of the producer is to select animals with double muscling, then breeding stock of the same type must be selected. Fertility in these animals is very poor and calving is much more difficult.
2. The meat tends to be coarse and void of intermuscular fat (marbling). Even though the animal may have a sufficient cover of fat, the marbling tends to be less than adequate. Thus a large percentage of these animals grade standard.
3. Double-muscled calves are difficult to raise because they are more susceptible to disease.
4. Double-muscled feeder animals must be fed a higher proportion of **concentrate** in the ration in order to obtain enough marbling to grade choice.

These cattle can be recognized by their physical appearance. The rump is protruding and round with definite grooves or creases between the thigh muscles. The tail is short and attached high and far forward on the rump, Figure 14–15. The head of the animal is small and long and is carried lower than the top of the body.

Animals that possess an adequate degree of long, smooth muscling should be selected over animals that are lightly muscled or have extremely tight-wound, excessive muscling. Smooth muscling—i.e., muscling that does not bulge too much—allows the animal to move freely and smoothly.

Figure 14-15. Double-muscled animals can be recognized by the protruding muscles in their hind quarters and their short tail. *Dr. Estes Reynolds, Cooperative Extension Service, The University of Georgia.*

FRAME SCORES
MALES

Frame Scores Based on Height (in inches) Measured at Hips

Age in Months	Frame Score 2	Frame Score 3	Frame Score 4	Frame Score 5	Frame Score 6	Frame Score 7
5	36.00	38.00	40.00	42.00	44.00	46.00
6	37.00	39.00	41.00	43.00	45.00	47.00
7	38.00	40.00	42.00	44.00	46.00	48.00
8	39.00	41.00	43.00	45.00	47.00	49.00
9	40.00	42.00	44.00	46.00	48.00	50.00
10	41.00	43.00	45.00	47.00	49.00	51.00
11	42.00	44.00	46.00	48.00	50.00	52.00
12	43.00	45.00	47.00	49.00	51.00	53.00
13	43.50	45.50	47.50	49.50	51.50	53.50
14	44.00	46.00	48.00	50.00	52.00	54.00
15	44.50	46.50	48.50	50.50	52.50	54.50
16	45.00	47.00	49.00	51.00	53.00	55.00
17	45.50	47.50	49.50	51.50	53.50	55.50
18	46.00	48.00	50.00	52.00	54.00	56.00
19	46.25	48.25	50.25	52.25	54.25	56.25
20	46.50	48.50	50.50	52.50	54.50	56.50
21	46.75	48.75	50.75	52.75	54.75	56.75
22	47.00	49.00	51.00	53.00	55.00	57.00
23	47.25	49.25	51.25	53.25	55.25	57.25
24	47.50	49.50	51.50	53.50	55.50	57.50

FEMALES

Frame Scores Based on Height (in inches) Measured at Hips

Age in Months	Frame Score 2	Frame Score 3	Frame Score 4	Frame Score 5	Frame Score 6	Frame Score 7
5	35.75	37.75	39.75	41.75	43.75	45.75
6	36.50	38.50	40.50	42.50	44.50	46.50
7	37.25	39.25	41.25	43.25	45.25	47.25
8	38.00	40.00	42.00	44.00	46.00	48.00
9	38.75	40.75	42.75	44.75	46.75	48.75
10	39.50	41.50	43.50	45.50	47.50	49.50
11	40.25	42.25	44.25	46.25	48.25	50.25
12	41.00	43.00	45.00	47.00	49.00	51.00
13	41.75	43.75	45.75	47.75	49.75	51.75
14	42.25	44.25	46.25	48.25	50.25	52.25
15	42.75	44.75	46.75	48.75	50.75	52.75
16	43.25	45.25	47.25	49.25	51.25	53.25
17	43.75	45.75	47.75	49.75	51.75	53.75
18	44.25	46.25	48.25	50.25	52.25	54.25
19	44.50	46.50	48.50	50.50	52.50	54.50
20	44.75	46.75	48.75	50.75	52.75	54.75
21	45.00	47.00	49.00	51.00	53.00	55.00
22	45.00	47.00	49.00	51.00	53.00	55.00
23	45.25	47.25	49.25	51.25	53.25	55.25
24	45.25	47.25	49.25	51.25	53.25	55.25

The height under inches shown under each frame size is the minimum height for that frame size

Figure 14-14. Frame size table.

Selecting Breeding Cattle

As mentioned earlier, frame size refers to the overall height of the animal at maturity; tall animals are larger framed than short animals. The frame size can be determined when the animal is a young calf. The leading indicator of frame size is the length of the cannon bone (the bone between the ankle and the knee of the front leg). Research has shown that an animal with a longer length of cannon bone will be a taller animal at maturity than an animal with a shorter cannon bone. In fact, some breed associations request that the producer record the length of cannon at birth of any animals they wish to register.

In heifers, consideration should be given to those animals having a longer length between the hooks and pins and those who are wider apart at both hooks and pins. This is an indication of a greater **pelvic capacity**. When a female gives birth, the pelvis must open enough to allow the passage of the calf through the birth canal. If a heifer has a small pelvis, she will probably have problems delivering a calf, Figure 14–16.

Obviously cattle that grow faster (have a greater average daily gain) are more desirable. Research has shown that there are other physical characteristics of cattle that are associated with growthiness. The length of an animal's face (distance from eye to muzzle) and the length of the neck are both indicators of growth. The longer the neck, the faster the animal is likely to grow.

Breeding cattle should have adequate body depth and width to provide adequate room for the internal organs. The larger the internal organs, such as the heart and lungs, the better the animal should do in terms of viability and growth. Indicators of capacity are width through the chest floor; long, well-arched ribs; and depth in the side, Figure 14–17.

Sex character simply means that a bull looks like a male and a heifer looks like a female. Since sex hormones control both the physical appearance of animals and their ability to reproduce, it stands to reason that an animal

Figure 14-16. The heifer on the left has a greater pelvic capacity than the other two.

Figure 14-17. Breeding cattle should have adequate depth and width to provide room for the internal organs. *American Simmental Association.*

with more sex character should be more reproductively efficient.

Sex character in a bull is determined by a broad, massive, bull-like head, Figure 14–18. The shoulders should be bold and well muscled, but care should be taken that bulls with coarse, excessively thick shoulders are not selected as a herd bull. Since this characteristic is passed on to the offspring, calving difficulties can be encountered.

Figure 14-18. Bulls should be bold and masculine appearing. *American Hereford Association.*

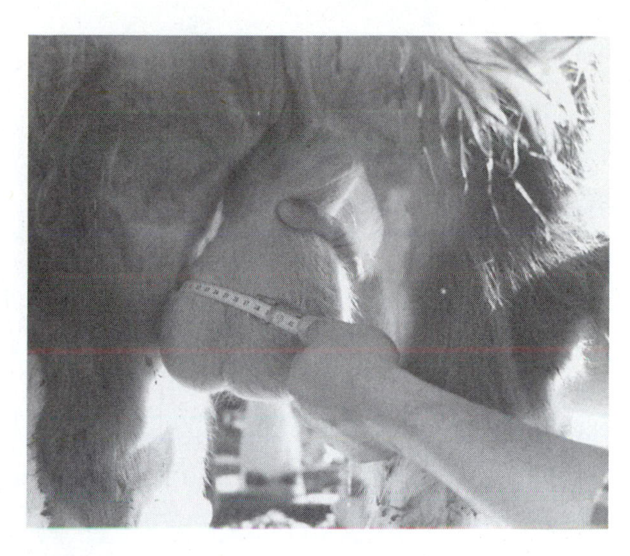

Figure 14-19. A two-year-old bull should have a scrotal circumference of at least 34 cm. *Dr Ronnie Silcox, Cooperative Extension Service, The University of Georgia.*

One of the most important physical traits is that of testicle shape and size. A two-year-old bull should have a scrotal circumference of at least 34 cm when measured at the largest part, Figure 14–19. Research has shown that the larger the testicles, the larger the number of valuable sperm produced. In addition, the scrotum should extend to about hock level and have a definite neck. If the testicles are held too close to the body, the temperature will be too high for ideal sperm production. Bull *1* in Figure 14–20 has a straight-sided scrotum often associated with testicles of only moderate size, and the testicles are held too close to the body. Bull *2* shows the ideal testicle size and scrotal shape, Figure 14–20. Bull *3* has a tapered or pointed scrotum usually associated with undersized testicles that are too close to the body.

The fertile female should be well-balanced and present a graceful feminine appearance. She should be long and clean in the face and throat. Her neck should be long and blend smoothly into smooth, sharp shoulders. She

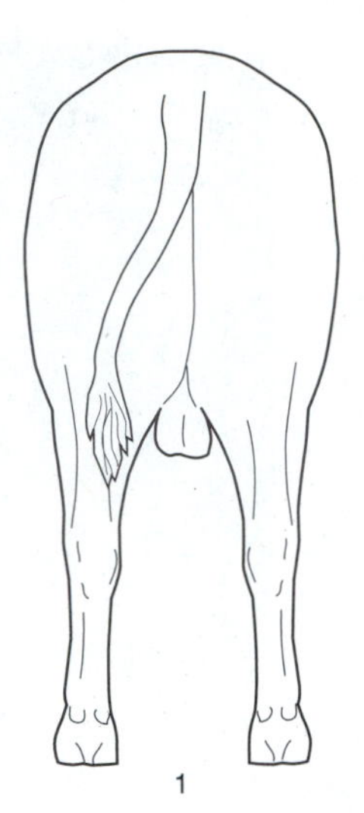

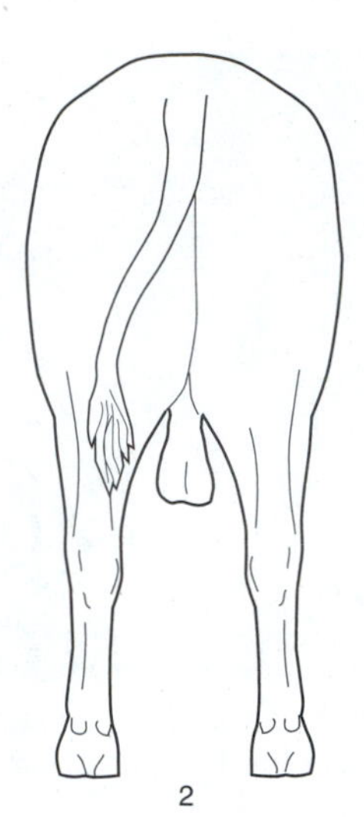

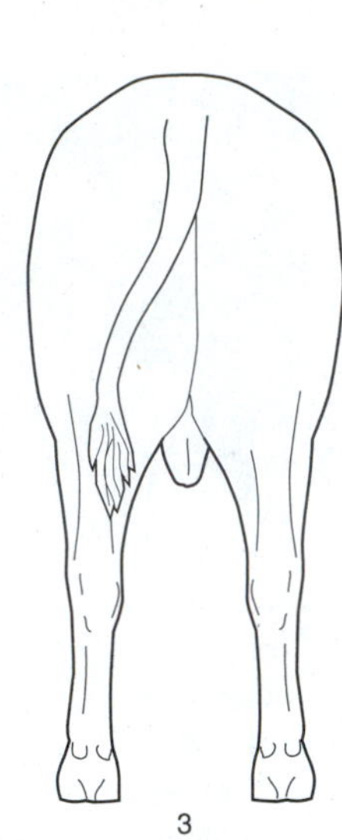

Figure 14-20. Testicles should be carried away from the body and have a definite neck. Bull *2* is the best choice.

should be clean and trim through the **brisket** and middle. The pelvic area should be large and wide for easy calving, Figure 14–21. Distances should be wide from hook to hook, pin to pin, and long from hooks to pins.

Structural soundness refers to the correctness of the feet and legs of an animal. The legs should fit squarely on all four corners of the animal. The correct "set" to the back legs of a bull is shown in Figure 14–22A. If a plumb line is dropped from the pins to the ground, the line will intersect with the hock. The condition known as post-legged is shown in Figure 14–22B. The rear legs are too straight and do not provide enough cushion and flex as the animal walks. In bulls, this condition causes problems in mounting cows. The opposite condition, known as sickle hocked (Figure 14–22C), also causes problems in mating. As the animal mounts, undue stress is placed on the stifle muscle, causing the animal to become stifled, i.e., the ligament attaching the stifle muscle tears. This results in the animal being worthless as a herd bull. When viewed from the rear, the back legs should be straight. Figure 14–23 depicts animals that have structural problems as viewed from the rear. Figure 14–24 depicts animals that are structurally incorrect on their front legs. All cattle are *slightly* splayed in the front, but the front feet should not turn out very much. All animals should move out with a free, easy stride. The rear foot should be placed about where the front foot was picked up. Cattle that take short, choppy steps are either too tightly wound in their muscle pattern or have problems in their skeletal makeup. Either condition is objection-

Figure 14-21. A good breeding female should be clean and trim through the brisket and middle, such as this Charolais. *American International Charolais Association.*

SET OF LEGS AND FEET

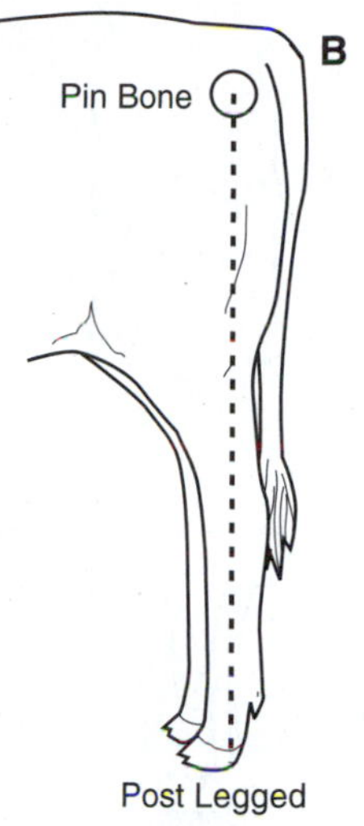

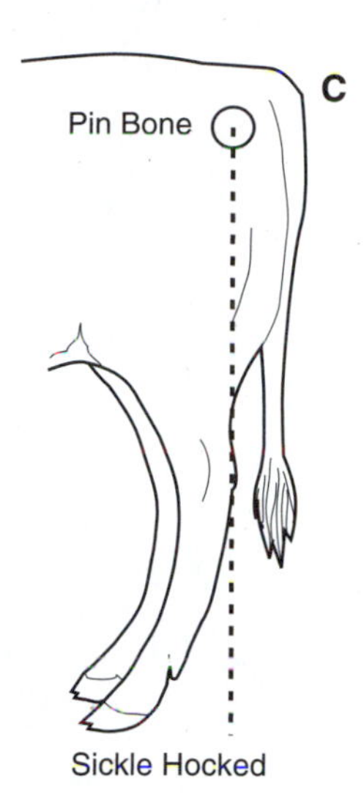

Figure 14-22. *A.* In a structurally sound animal the legs fit squarely on all four corners of the animal; *B:* Common defect known as post-legged; *C:* Common defect known as sickle hocked. *Vocational Materials Service, Texas A & M University.*

able. To feel their best and to grow and do their best, animals must be structurally correct.

The Selection of Sheep

The selection of sheep has always been more complicated than the selection of either beef or hogs because two traits have been traditionally considered—meat and wool. Milk is sometimes considered a third trait. Shepherds had been selecting more productive sheep long before the principles of inheritance were outlined by Gregory Mendel. Prior to and during the 1950s, as much emphasis was placed on the

COMPARISON OF CORRECT WITH DEFECTIVE HIND LEGS

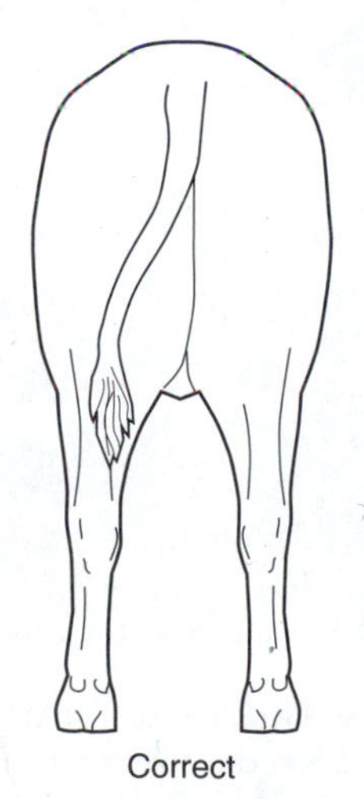

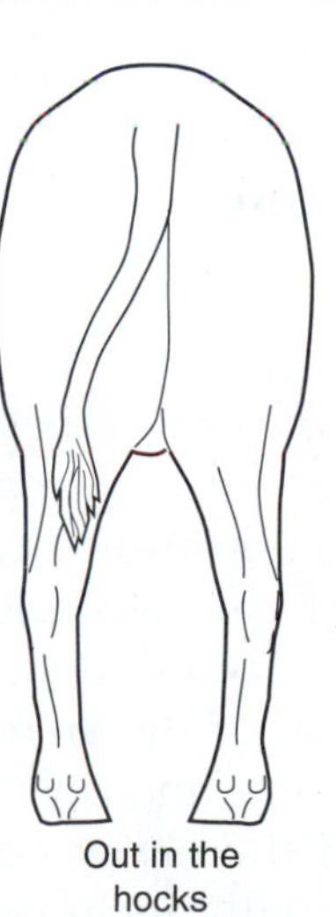

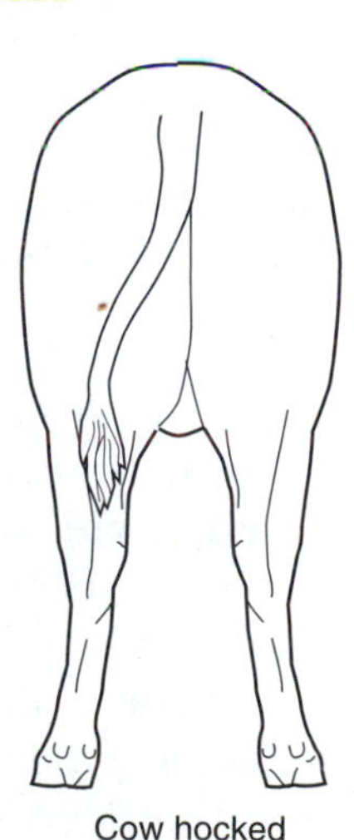

Figure 14-23. Comparison of correct with defective hind legs. *Vocational Materials Service, Texas A & M University.*

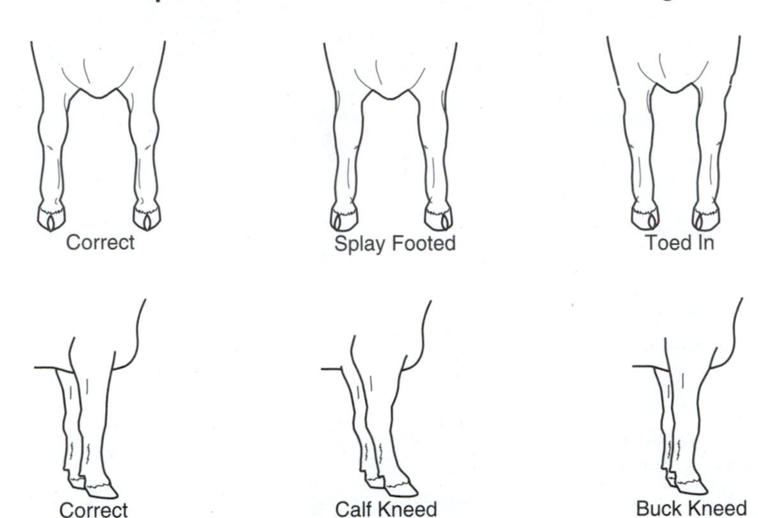

Figure 14-24. Comparison of correct with defective front legs. *Vocational Materials Service, Texas A & M University.*

production of wool as on the production of meat, Figure 14–25.

When synthetic materials began to become more popular, the emphasis was shifted from wool production to meat production within some breeds. During the 1960s, interest and research was concentrated on **cutability**, or the percent of lean cuts a lamb carcass would produce. Research studies put more emphasis on the muscling and leanness of lamb carcasses than on the production of wool.

Figure 14-25. Unlike the other major agricultural animals, sheep are selected on fleece quality as well as conformation. *The Wool Bureau.*

Selection of Commercial or Western Ewes

Sheep are judged similarly to other species of livestock. Selection is made based on **type**, **balance**, muscling, finish, **carcass merit**, **style**, **soundness**, and **smoothness**. However, sheep have another trait to consider besides meat production—wool production.

Type is the general build of the animal. Desirable type changes over time and is greatly influenced by trends in the show ring. However, desirable type usually includes a thick, moderately deep-bodied animal that is smooth, has straight lines, and exhibits good balance.

Balance refers to the general proportions in the structure of the animal. An animal should appear to fit together well. This means that the front end should be in proportion to the rear end.

Muscling is the natural flesh of the animal, not including the fat. Modern lambs should have thick muscling. Meat is composed of muscling, and the more muscling on the carcass the more value to the packer and the producer. A good portion of meat from the carcass is taken from the leg or round and loin. These areas are considered to be the high value portion of the lamb.

Finish is the amount of fat cover on the carcass. A smooth, uniform, thin layer of fat is desirable. The most important thing a livestock evaluator (especially a grader) has to be able to do is evaluate fat. The evaluator who can predict fat can generally predict muscling because the thickness of the animal is composed of either fat, muscle, or a combination of both. Thickness that is not due to fat is made up of muscle.

Carcass Merit is determined by carcass quality (quality grade is a prediction of palatability or eating quality) and yield (yield grade predicts the percent of boneless retail cuts from the leg or round, the loin, the rib or rack, and the chuck or shoulder). A carcass of good merit will yield a large portion of the valuable cuts that are of sufficient quality while maintaining as much leanness as practical.

Soundness refers to the structural soundness or skeletal system supporting the body so that the animal is comfortable and maintains reproductive efficiency. A more detailed discussion of structural soundness is given in another part of this chapter. The same structural correctness desired in other agricultural animals is needed in sheep as well.

Smoothness refers to the lack of awkward bone structure and a smooth even finish along the top and sides. Fat should be distributed evenly along the body of the animal. An uneven finish can indicate an animal that has been off its feed or an animal that is older.

SELECTING BREEDING EWES

Generally, breeding sheep should be sound, healthy, and productive. Ewes should be vigorous, with normal teeth, feet, legs, eyes, and udders. Ewe lambs or **yearling** ewes should be selected for breeding soundness and should be culled for reproductive problems. However, younger ewes will require more attention during their first lambing.

Age is determined by the teeth located on the lower jaw.

The correct ewe type should include the following traits:

- Purebred ewes should show breed traits.
- Feminine appearance.
- Depth in the fore and rear flanks.
- Good chest capacity.
- Proportional length, depth, and width of body.
- Width and thickness throughout the loin.
- A strong, wide, long, and level rump with a wide dock.
- Straight legs with long, full muscling that provides width between the legs.
- Strong pasterns and feet with medium-size toes.
- Strong, wide back and crop with a tight shoulder that blends into the body.
- Balance and smoothness, with all parts blending well together.

- Wool should be indicative of the breed in fiber diameter, length, and quality.

SELECTING RAMS

Great care should be given in selecting the **ram** since half the growth and wool-producing ability of every lamb is contributed by the ram. Generally the same traits listed for consideration in selecting ewes are used in selecting rams. However, rams should be selected with certain goals in mind. Consideration should be given to the strengths and weaknesses of the ewe flock. If the ewes are lacking in a quality, then a ram should be selected that will bring the desired traits into the flock. Different rams would be used as terminal sires than those chosen for purebred flocks. Rams should be selected for genetic capability—whether it is growth rate, wool production, or increased lambing.

Production records should aid in the selection of rams. Rams should have the traits of muscling and structural correctness that were listed for ewes. In addition, they should be rugged, muscular, and masculine, Figure 14–26. Rams should exhibit superior growth, structural soundness, and quality of fleshing. Testicles should be well developed and pliable. A semen test should be obtained or the ram guaranteed as reproductively sound. Finally the ram's semen should be retested just prior to breeding season.

JUDGING MARKET LAMBS

Emphasis for judging market lambs and purebred sheep at livestock shows has changed and affected the industry over the years. In the 1950s and 1960s the ideal lamb was short, fat, and extremely thick. These produced carcasses that were small but thick due to combined muscle and fat. Many of these carcasses had prime **conformation**. Since carcasses were shipped in carcass form, the fat helped reduce shrinkage and allowed the

Figure 14-26. Modern rams are large and growthy. *American Hampshire Sheep Registry.*

retailer to trim the carcass in order to produce good looking—but fat—retail cuts.

During the late 1960s and 1970s, the wastefulness of the excess fat was taken into account and the trend was for producers to select a medium-size lamb that was much leaner. This was due in part to the introduction of **vacuum packaging** and the boxing of lamb carcasses and carcass parts. However, by the late 1970s and until the mid-1980s, show sheep were being selected for height, length, and stylishness with little emphasis on carcass, except for finish (0.1 inch of backfat was considered ideal). These animals were selected for the hind-saddle (leg, loin, rack) muscling without regard for chest capacity and natural fleshing needed by ewes for livability and reproduction.

From the mid- to late 1980s to the present, selection of show sheep has moderated. The following description reflects the modern type of show sheep, Figure 14–27.

Size, muscle, structure, style, and balance are all factors that should be considered when selecting a market lamb at a livestock show. Champion lambs should be heavily muscled, nicely balanced, and correctly finished. Carcass merit should be kept in mind as the animals are evaluated. The hind-saddle (leg, loin, and rack or rib area) makes up about 70 percent of the carcass value. However, capacity is also important, especially in the breeding animal, for livability and reproductive volume.

Different breeds will have a different frame size related to the correct market weight for each lamb. This makes judging large classes made up of different breeds more difficult. Muscling is shown by the thickness and width of the loin, thickness of muscling in the leg, and the prominence of the stifle region. A long hind saddle is desirable, but it must have width and depth as well as length to be of greater total volume. The lamb should have a wide, level rump that blends into a muscular leg. A heavily muscled lamb will stand wide—both front and rear—and exhibit a heavily muscled forearm. Muscling should be firm and smooth, but not too extreme or bulging.

Finish on the carcass is measured one and one half inches from the middle of the backbone between the 12th and 13th ribs. Lamb

Figure 14-27. The ideal modern market lamb is lean and muscular. *National Suffolk Sheep Association.*

carcasses with less than .1 inch fat tend to dry out; and shelf life is reduced. Lamb carcasses with over .25 inch backfat simply require too much trimming to obtain acceptable retail cuts. The ideal amount of finish is about .15 inch or less. Live lamb finish is determined by handling the lamb over the ribs and backbone. This handling requires experience to accurately feel the amount of finish. A good learning practice is to evaluate live lambs and then evaluate the carcass. Videos are especially useful for learning the practice.

Knowing the amount of finish lets the evaluator know the amount of muscling (Flesh − fat = muscling). Other indicators of fat on lambs are that they are heavy fronted (breast), have fat deposits in the cod or udder region, have fat around the dock, and lack trimness in the middle and flanks. A correctly finished lamb should feel firm and have a trim underline. The judge must be able to compensate for the pelt (skin and fleece) in determining finish. This is especially important in lambs with thick hides or in lambs that are not slick shorn. Most market lamb shows require that the lambs be slick shorn.

Bone is important because it relates to frame size and structural correctness. Muscles attach to bones, so sufficient muscle requires a sufficient diameter and length of bone. Structural correctness allows the animal to move in comfort and maintain productivity. It is not as important for market animals to be structurally correct as it is for breeding stock. However, structural correctness is a consideration in determining the overall style and balance of the market animal.

SUMMARY

Since the beginning of the animal industry, selective breeding has been a cornerstone. Choosing the right animals to mate is essential in order to produce the type offspring desired. Although over the years the style and type animal producers want has changed, some of the basics remain the same. The advent of embryo transplant and artificial insemination have greatly aided the selection process. The computer has also had an impact, and we are perhaps just on the brink of realizing how the computer can revolutionize the process.

◆ Review Exercises

TRUE OR FALSE

1. Natural selection refers to the process of only the strongest and best-equipped animals in the wild breeding. ______.
2. Growthability refers to an animal's ability to grow rapidly. ______.
3. Through the years the basic physical types of agricultural animals (swine, beef cattle) have stayed the same. ______.
4. A pig with a strongly arched back has a more vertical scapula and humerus that provide less flex and cushion than the level-topped, more structurally correct pig. ______.
5. The carcass from double-muscled animals is the highest quality. ______.
6. The testicles of a breeding bull should be pointed and close to the body. ______.

7. Tenderness and taste are both related to the breed of animal. ______.
8. Intermuscular fat is referred to as the leanness of the meat. ______.
9. Small-framed animals mature earlier than do large-framed animals. ______.
10. Sex character simply means that a bull looks like a female, and a heifer looks like a male. ______.
11. Traits in sheep production that are considered most are meat and wool production, with milk sometimes being a third trait. ______.
12. When synthetic materials began to become more popular, the emphasis was shifted from wool production to meat production within some breeds. ______.
13. Most of the sheep in the United States are produced in the plains of the Midwest. ______.
14. Vacuum packaging had little effect on the trend towards leaner meat. ______.
15. Judging different breeds of lambs is difficult due to different frame sizes. ______.

FILL IN THE BLANKS

1. To meet the demands of the modern livestock industry, animals are ____________ based on what is desired by the people who ______________ them and by those who ______________ them.
2. The three basic traits that are desirable in the modern agricultural animal are ________________ ________________, ________________, and efficiency.
3. In swine, front feet that are turned out (________________ ________________) or turned in (________________-________________) should be avoided as well as rear feet that are splayed out (________________-________________) or turned in.
4. Frame size refers to the ____________ of an animal at ____________; ______-______ animals are larger framed than ______-______ animals.
5. An animal with a long cannon bone will be a ____________________ animal at ________________ than an animal with a ________________ cannon bone.
6. Consumers usually want beef that is ____________, tastes __________, and is ________________.
7. Cattle that take short, ______________ steps are either too ______________ ______________ in their ______________ pattern or have problems in their ________________ makeup.
8. The purpose in producing market beef animals is to obtain ______________ that can be cut into ______________ cuts of beef for the ________________.
9. Animals that possess an adequate degree of ____________, ____________ muscling should be selected over animals that are ________________ muscled or have extremely ____________________, ________________ muscling.
10. The length of an animal's ____________ (distance from ____________________ to ______________) and the length of the ____________ are both indicators of ________________.
11. Carcass merit in sheep is determined by carcass ________________________ and ____________.
12. Structural soundness refers to the ___________ ___________system support-

ing the body so that the animal is ______________ and maintains ______________ efficiency.

13. In the 1950s and 1960s the ideal lamb was ____________, ____________, and extremely ____________.
14. When selecting a market lamb at a livestock show, ____________, ____________, ____________, ____________ and balance are all factors that should be considered.
15. A ram should be selected for ____________ ____________ capability—whether it is ____________ rate, ____________ production, or ____________ ____________.

DISCUSSION QUESTIONS

1. What is the process of natural selection?
2. Why did animals develop characteristics such as thick hides, long horns, or the ability to run fast?
3. Why are these type of characteristics not desirable in modern agricultural animals?
4. What is meant by selective breeding?
5. What are two basic categories of agricultural animals that undergo the selection process by the producer?
6. In the selection of market animals, what does the consumer want? the packer? the producer?
7. Describe the type of hogs raised in the following eras: the period prior to the 1950s; the 1960s; the 1970s.
8. Describe the modern market hog and tell why these characteristics are important.
9. What is Porcine Stress Syndrome?
10. What is meant by sex character and why is it so important?
11. Why is it so important that animals be correct on their feet and legs?
12. Why should a hog have a level back rather than an arched back?
13. Define *capacity*. Why is it important?
14. How are beef carcasses graded?
15. In what sequence is fat deposited in a beef animal's body?
16. Of what use is an animal's frame size in the selection process?
17. What are double-muscled cattle? Explain why this type cattle is undesirable.
18. Describe at least two traits that are desirable in breeding heifers.
19. Why is testicle size and scrotal shape important in selecting bulls?

STUDENT LEARNING ACTIVITIES

1. Choose a breed of animal (swine, beef, or sheep). By using a set of pictures or by viewing the animals live, make a list of all of the animal's physical traits. From the list, decide which of the characteristics are a result of natural selection and which are the result of selective breeding.
2. Go to the library and research the development of a breed of livestock. Include in your report the place where the animals originated, the traits for which they were selected, and the changes in the breed over the years.
3. Visit with a producer in your area and determine the traits the producer selects for when he/she selects replacement animals.
4. Visit a packing plant and determine the traits that the packer likes in the animals he/she buys for slaughter.

Chapter 15

The Reproduction Process

Student Objectives in Basic Science

As a result of studying this chapter, you should be able to

1. Distinguish between asexual and sexual reproduction.
2. Explain the process by which gametes are produced in both the male and female.
3. Explain the steps involved in meiosis.
4. List and describe the parts and function of the male reproductive system.
5. List and describe the parts and function of the female reproductive system.
6. Describe the functions of the hormones that control reproduction.
7. Describe the phases of the female reproductive cycle.
8. Explain the process by which fertilization takes place.
9. Define and explain the process of cloning.

Student Objectives in Agricultural Science

As a result of studying this chapter, you should be able to

1. List the reasons why artificial insemination is valuable to livestock producers.
2. Explain the procedures used in artificial insemination.
3. Explain the importance of embryo transfer.
4. List and explain the steps used in embryo transfer.
5. Describe the advantages of estrus synchronization.
6. Explain the process of estrus synchronization.
7. Describe new scientific technology that will be of benefit to livestock producers.

Key Terms

asexual
sexual
zygote
mitosis
sterile
meiosis
spermatogenesis
spermatogonia
spermatozoa
chromotids
synapsis
oogenesis
polar bodies
cytoplasm
nucleus
testosterone
llibido
epididymis
vas deferens
seminal vesicles
urethra
cowper's gland
prostate gland
penis
prepuce
estrogen
progesterone
fallopian tubes
cervix
vagina
clitoris
endocrine system
ovulation
copulation
corpus luteum
conception

Key Terms *(continued)*

estrus cycle
ejaculation
motile
fertilization membrane
cleavage
protectant
quarantine
artificial vagina
extenders
straws
estrus
hormones
progeny testing
genetic base
donor cows
recipient cows
superovulation
prostaglandin
catheter
clone
micromanipulator
paraffin

Reproduction in Animals

In order for living things to remain on the earth, they must reproduce. If an organism does not reproduce, that type of organism would disappear as soon as death occurs. By reproducing, animals ensure that their type of animal will continue to exist. There are two basic means of reproducing. One-celled organisms and some plants reproduce by means of **asexual** reproduction—they produce another organism from only one parent. All higher order animals reproduce by means of **sexual** reproduction. This means that animals come from two parents, a male and a female.

Reproduction in mammals is achieved by each of two parents contributing genetic material to the young. Half of the characteristics of the young come from the father and half come from the mother. Each of the two parents create reproductive cells called *gametes*. The male gamete is known as a sperm cell and the female gamete is known as an egg cell. The uniting of the two cells results in the beginning of a new animal similar to the parents. The new cell produced by the uniting of the sperm and the egg is called a **zygote**. The zygote divides by a process called **mitosis**. Both the egg and the sperm contain the material that dictates what the young animal will look like. Even though the gametes are so small that they cannot be seen without the aid of a microscope, they contain all of the material necessary to determine all of the characteristics of the new animal. This material is called deoxyribonucleic acid or DNA. DNA is the matter that carries the code that determines exactly how the animal will develop and how it will look. Specific segments or units of DNA that are grouped together are called *genes*. Each gene has a unique structure that controls a trait of an animal. For example, a gene may determine that Angus cattle are hornless or they may determine that Watusi cattle may have horns that grow to a length of six feet, Figure 15–1.

Groups of genes go together to form threadlike structures called chromosomes. The name comes from chromo, which means "colored," and soma, which means "body." Each cell in an animal's body contains a number of chromosomes. In body cells, the chromosomes

Figure 15-1. Watusi cattle have specific genes that determine that their horns will be long.

CHARACTERISTIC NUMBERS OF CHROMOSOMES IN SELECTED ANIMALS	
Animals	Chromosome Number (2n)
Donkey	62
Horse	64
Mule	63
Swine	38
Sheep	54
Cattle	60
Human	46
Dog	78
Domestic cat	38
Chicken	78

Figure 15-2. Different animals have different numbers of chromosomes.

always come in pairs. Different species have different numbers of pairs. Figure 15–2 provides a listing of the number of pairs of chromosomes for several domestic animals. Each gamete (the sperm and the egg) contains one of each pair or half of the chromosomes. For example, each horse cell contains 32 pairs, or 64 chromosomes. The sperm or egg from a horse would each contain 32 chromosomes. When the egg and the sperm unite at the time of conception, they each contribute 32 chromosomes that go together to form a full set of 64 chromosomes. Each chromosome from the father is matched with a chromosome from the mother. Notice that the donkey has 31 pairs or 62 chromosomes. Horses and donkeys can be successfully mated to produce an offspring known as a mule. However mules cannot reproduce because the donkey's 31 pairs of chromosomes combined with the horse's 32 chromosomes will not divide into an even pairing of chromosomes. As a result, the gamete produced by a male or female mule will not successfully unite to form a zygote. For this reason, mules are almost always **sterile**, which means that mules are not generally capable of reproducing.

Production of Gametes

The formation of the sperm cell takes place in the testicles of the male; the formation of the egg takes place in the ovaries of the female. Within these organs, a process known as **meiosis** takes place. This process differs from mitosis in that meiosis results in a cell that contains only half the number of chromosomes of the original cell, Figure 15–3. In mitosis, the dividing cells contain the same number of chromosomes as the parent cells. Meiosis is necessary in order to allow the contribution of half of the chromosomes by each parent.

The production of the male gamete or sperm is called **spermatogenesis**. In the testes of the male, cells are produced that are called **spermatogonia**. Through a four-step process these cells develop into **spermatozoa**. During the first step, the chromosomes replicate (make an exact copy of themselves) and remain attached. The replicated chromosomes are called **chromotids**. In the next step, the chromotids come together and are matched up in pairs, this process is called **synapsis**. In the third step, the cell divides, the chromosomes are separated, and each cell receives one of each chromosome from each pair. However, remember that each chromosome replicated itself (chromotid) and is still attached to another chromosome. In the final step, the cells separate again and the chromotids separate and become chromosomes. Remember that these cells (the new sperm cells) each contain only half of the chromosomes that the original cell contained. The end result of this process is that four new sperm cells are produced from the original cell. When the sperm is united with the egg at conception, the original number of chromosomes is restored because the sperm furnishes half of the chromosomes and the egg furnishes half.

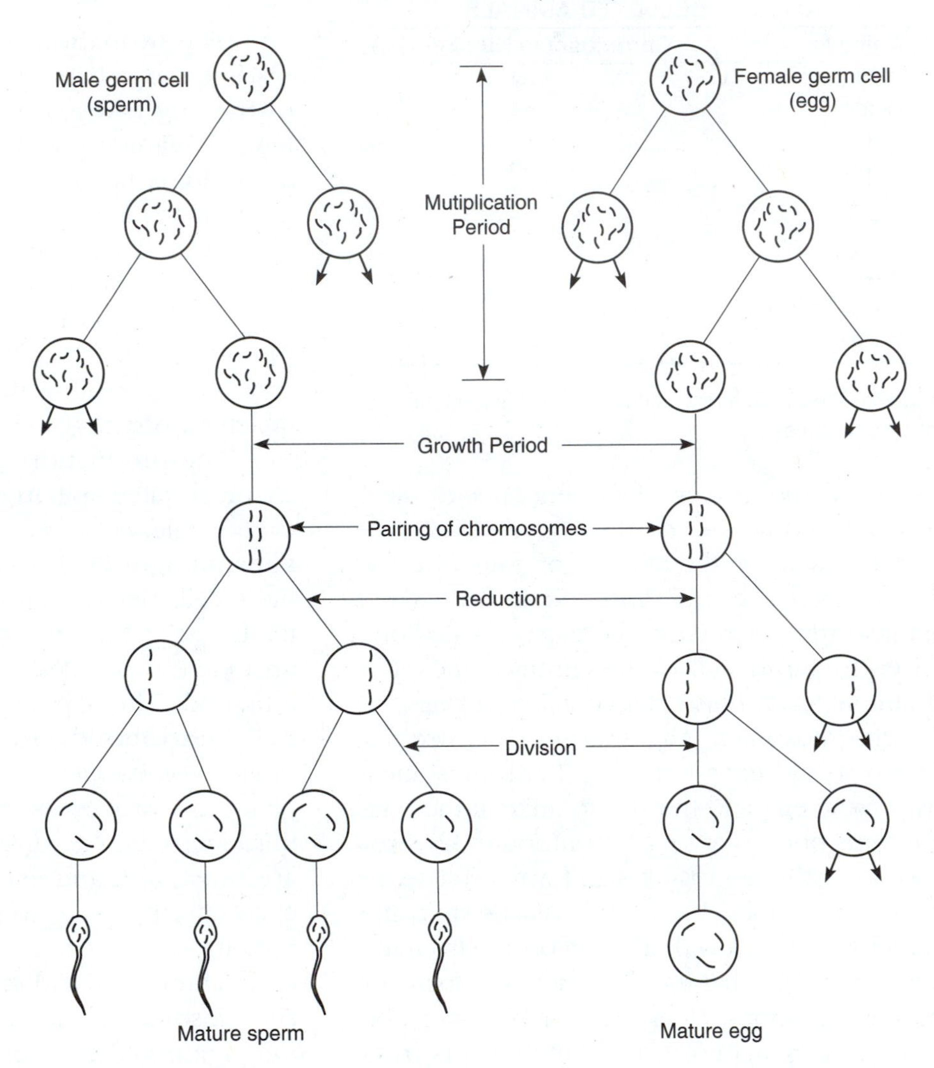

Figure 15-3. Meiosis is the process by which mature sperm and egg cells are developed.

Gamete production in the female is known as **oogenesis**. The stages of egg production are very similar to that of sperm production. The same stages occur, with one important exception. In sperm production, four new sperm cells are produced from the original cell. In egg production, only one egg cell is produced. Instead of producing four new eggs, three of the newly divided cells become what are known as **polar bodies** and only one cell becomes a viable egg. Polar bodies are produced as a result of most of the **cytoplasm** (cell material outside the **nucleus**) from the cells going to the one cell that will become the egg. The function of the polar bodies is to provide sustenance for the egg until conception. The sperm cell is much smaller than the egg cell and needs less to subsist.

The Male Reproductive System

In mammals, both the male and female have specialized systems that function to provide a means of producing and uniting the sperm and egg. In the male, the gamete is produced by the testicles. Ordinarily, a male has two testicles that are suspended away from the body. An exception to this is poultry; the testes are on the inside of the rooster's body. In mammals, the testicles are enclosed in a saclike structure called the *scrotum*. The scrotum functions not only to encase the testicles but also to act as a means of regulating the temperature of the testicles. For sperm production to occur, the testicles must have a temperature lower than the animal's body. This is the reason the testicles are suspended away from the body. The skin of the scrotum is thin, relatively hairless, and contains no subcutaneous (under the skin) fat. This greatly aids in the dissipation of heat in the summer. In the winter, the scrotum is retracted by means of small muscles that draw the testicles toward the body in order to keep them warm. (The reproductive tract of a bull is shown in Figure 15–4.)

In addition to producing sperm, the testicles serve another important function. They produce the hormone **testosterone**. This hormone controls the animal's **libido** (sex drive) and stimulates the development of sex characteristics. For example, the development of large forequarters in a bull and the growth of large tusks and the unpleasant odor in boars are induced by testosterone.

Along the outside of the testicles is a tuberous structure called the **epididymis** that provides a place for the storage and maturation of the sperm produced by the testicles. A long tube called the **vas deferens** leads from the epididymis to the **seminal vesicles** located at the upper end of the **urethra** (the tube through which urine is passed from the bladder). The vas deferens serves as a transportation route for the sperm.

In the male reproductive tract, there are three accessory glands. The first, the seminal vesicles, function to secrete a fluid that is mixed

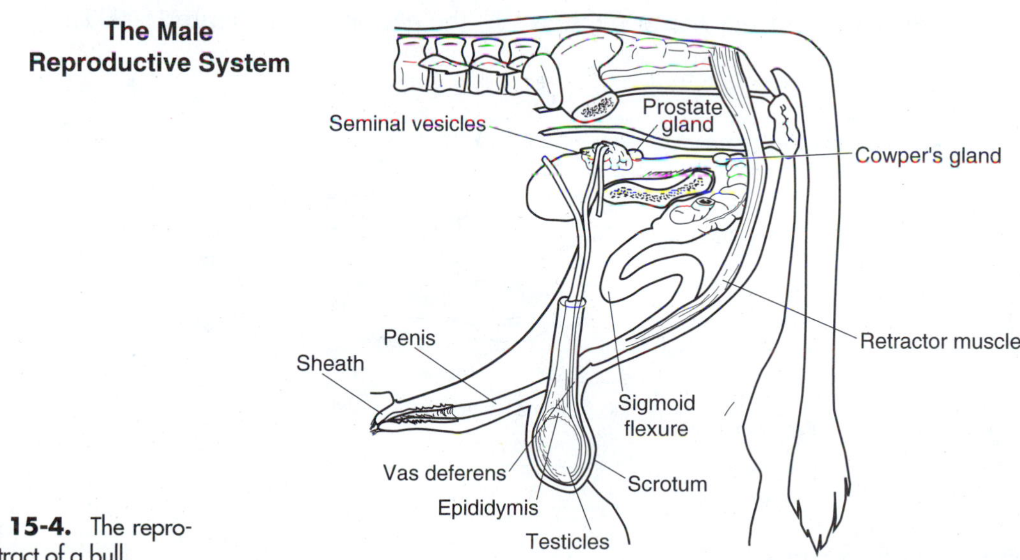

Figure 15-4. The reproductive tract of a bull.

with the sperm to protect the sperm and provide a mechanism by which the sperm can be transported. The two other glands also secrete fluid. The mixture of fluids is referred to as semen. The seminal vesicles also act as a holding place for the sperm. The **Cowper's gland** secretes fluid that helps cleanse the urethra before the sperm is passed along the tube. This secretion also acts to coagulate or to thicken the semen. The **prostate gland** also secretes fluid that is added to the semen mixture. Its purpose is to provide nutrients for the sperm and to expel the semen during the mating process.

The male organ that deposits sperm in the female tract is the **penis**. This organ also serves as the means of expelling urine from the body. The penis of the boar, bull, and ram is composed of a high concentration of connective tissue. The upper end of the penis is S-shaped and flexes to extend the penis outward during mating. The penis of the stallion is made up of a high concentration of vascular tissue that allows the organ to become engorged with blood. This causes the penis to become extended until it is said to be erect. This allows penetration of the female. The external covering of the penis is called the sheath or **prepuce**, its purpose is to protect the penis from injury or infection.

THE FEMALE REPRODUCTIVE SYSTEM

The reproductive system of the female is much more complex than that of the male. Sperm production in the male is constant, but production in the female comes about only in carefully controlled cycles. The cycle produces the egg, places the egg in the proper place, causes the female to accept the male for mating (called *estrus* or *heat*), and ensures that the fertilized egg remains in place throughout the gestation period. This cycle from egg production to fertilization occurs at different intervals in different animals.

The female reproductive system consists of several organs that make a contribution to the process. The *ovaries* are two small organs that are supported in the abdominal cavity by strong ligaments. Inside these ligaments are the arteries and vessels that supply blood to them. The main function of the ovaries is to produce the *egg* or *ovum,* Figure 15–5. This is where the

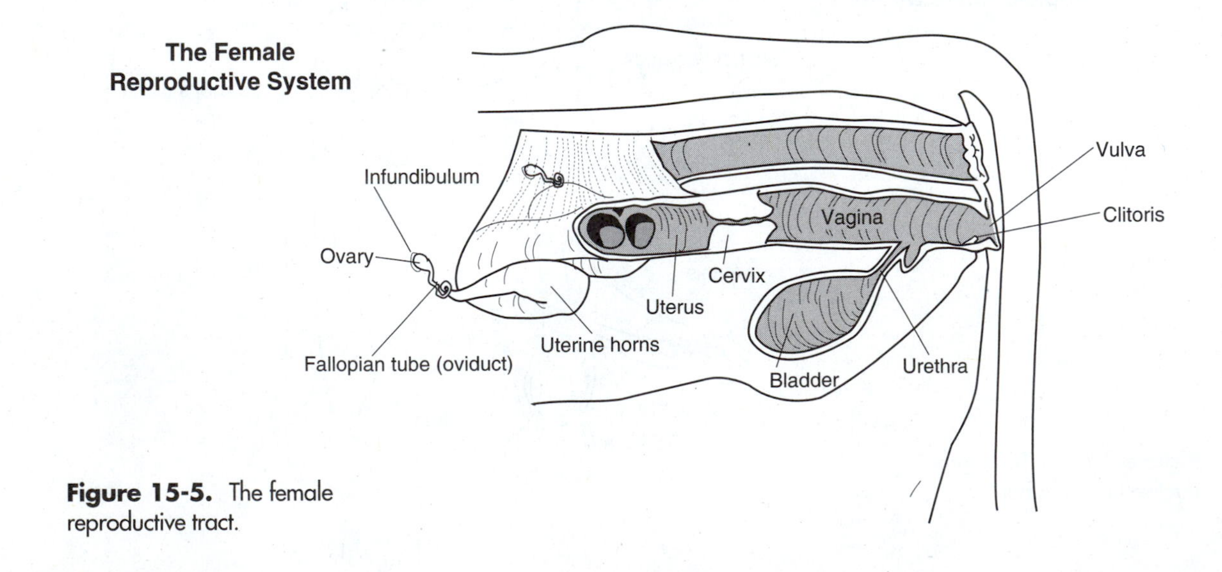

Figure 15-5. The female reproductive tract.

process of oogenesis (the production of the egg) takes place.

Another important role of the ovaries is to produce the hormones **estrogen** and **progesterone**. (These two hormones play essential roles in the reproductive cycle of the female.) Leading from the ovaries are two tubes known as the **fallopian tubes** which serve to transport the egg from the ovaries to the uterus. It is within the fallopian tubes that the egg is united with the sperm and conception takes place. The fallopian tubes open into a muscular saclike organ known as the uterus (sometimes called the womb). The uterus serves as the chamber in which the fertilized egg (zygote) develops into an embryo, then into a fetus, and finally expels the newborn animal. The uterus is sealed by a thick group of circular-shaped muscles called the **cervix**. The cervix acts as a valve that keeps foreign matter from entering the uterus. It contains glands that secrete a waxlike material that serves as a seal to the uterus. When the animal comes into estrus (heat), the cervix opens to allow passage of the sperm. The cervix opens into the **vagina**, which is a sheathlike organ that accepts the male's penis during mating. The semen is deposited here. When the fetus has matured, the vagina serves as the birth canal through which the young animal leaves the uterus. The exterior part of the female reproductive system is the vulva. The vulva provides a closing for the vagina and serves as the end of the urinary tract that expels the urine. Within the vulva is a small sensitive organ called the **clitoris** that provides stimulation during the mating process.

As mentioned earlier, the entire process of the female reproductive cycle is controlled by hormones, Figure 15–6. Hormones are produced by the **endocrine system** and serve to stimulate or inhibit the development or operation of body functions such as reproduction. This system includes the pituitary gland, which is located near the base of the brain in mammals and acts as a type of master control for most of the other glands of the endocrine system. The reproductive cycle of the female begins with a hormone that is secreted from the pituitary gland and stimulates the ovary to produce a blisterlike structure called a *follicle*, Figure 15–7.

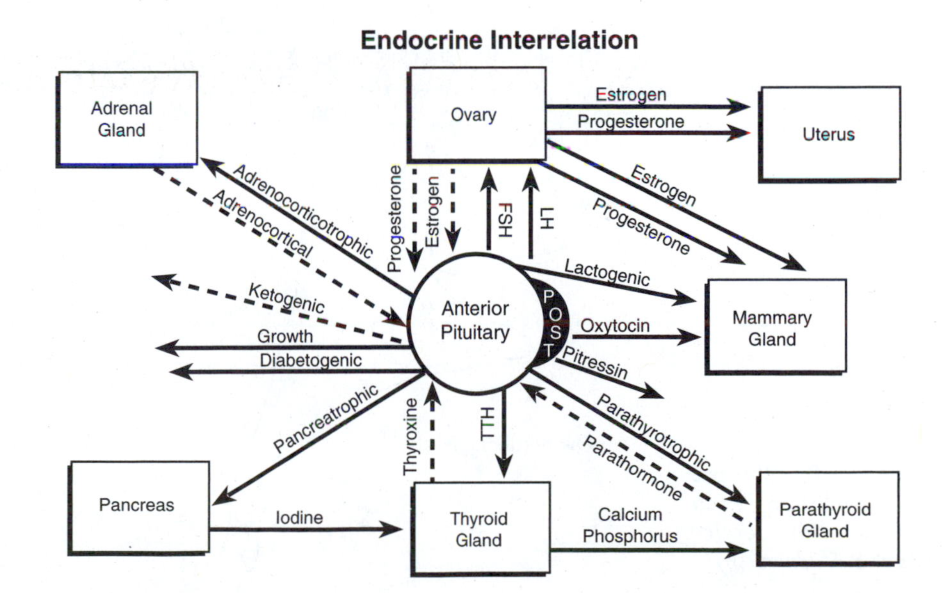

Figure 15-6. Hormone and chemical interaction.

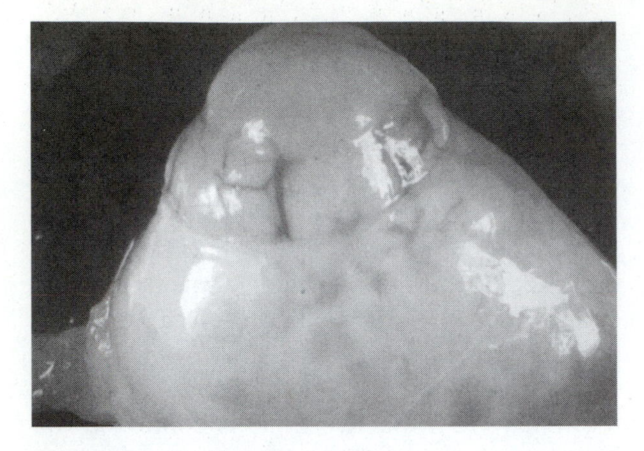

Figure 15-7. The follicle, which appears as a clear blister on the surface of the ovary, secretes a hormone called estrogen. *North Dakota Department of Education.*

The hormone is therefore called the *follicle-stimulating hormone* (FSH). The follicle, which appears as a clear blister on the surface of the ovary, secretes the hormone called *estrogen.* Estrogen acts as a messenger that stimulates the rest of the reproductive system to prepare for the reception of the egg. The follicle continues to produce estrogen and provides a place for the growing and maturing ovum or egg. It is in the follicle that the process of oogenesis occurs. When the gamete (the egg or ovum) is matured, the follicle becomes soft and expels the egg into the fallopian tube. (Some animals such as pigs have more than one young. The follicles of these animals release several eggs instead of one. Occasionally, even cattle may release two eggs instead of only one. This results in the birth of twins.) This process is known as **ovulation**, Figure 15–8. As ovulation occurs, the estrogen produced by the follicle causes the animal to go into the condition known as estrus or heat.

During this time, which may last from a few hours to two or more days, the female allows the male to mate with her. During this process, called **copulation**, the male's sperm are deposited into the female's vaginal tract. The expulsion of the egg from the follicle leaves a rupture that is filled with yellow cells that develop into a body called the **corpus luteum**. The development of the corpus luteum is caused by a hormone from the pituitary gland known as the luteinizing hormone (LH). The corpus luteum that develops where the follicle ruptured

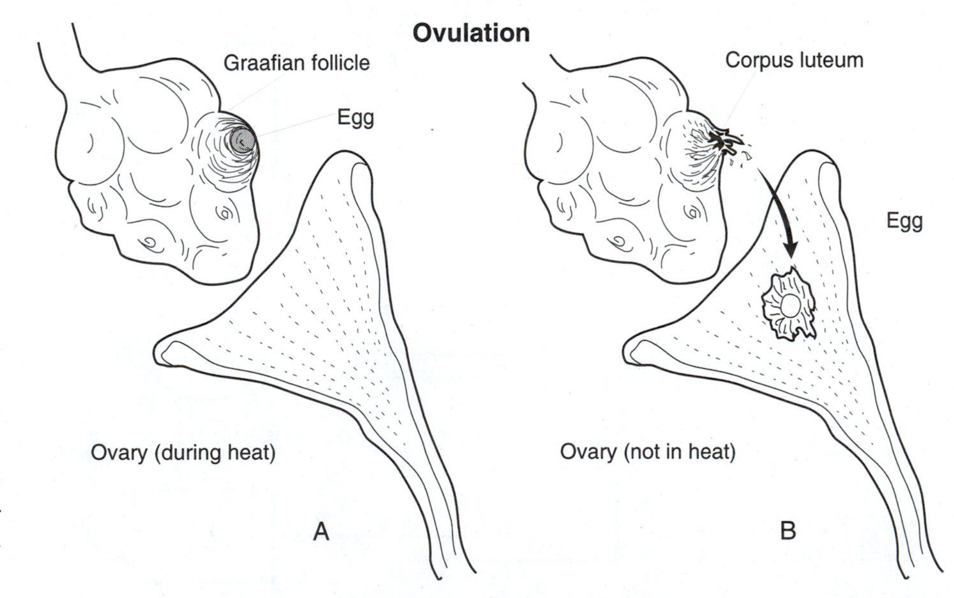

Figure 15-8. The process of ovulation. *Vocational Materials Service, Texas A & M University.*

secretes a hormone called **progesterone**. This hormone causes the walls of the uterus to thicken in preparation for receiving the fertilized egg. After **conception** (the uniting of the sperm and the egg) occurs, the corpus luteum continues to produce progesterone and the female remains pregnant. If conception does not occur, the corpus luteum recedes, the ovary returns to normal, and the cycle is begun again.

When the sperm from the male is deposited into the female's vaginal tract, the sperm cells make their way up through the cervix, through the uterus, and into the fallopian tubes. If a mature ovum (egg) is present, then conception may occur.

FERTILIZATION

Fertilization is the process by which the sperm is joined with the egg. Following this process, the embryo begins to develop. Since the sperm may only live for 20 to 30 hours, mating must take place at a time when the egg has matured and is released from the ovary. The entire process of gamete production is controlled by hormones. For example, the hormone estrogen, secreted by the follicle that develops on the ovary, causes the animal to enter a phase known as estrus or heat. During this time, which may last from a few hours to a few days depending on the species of animal, the female allows the male to approach her and to mate. Estrus is timed to occur as the follicle releases the egg. If fertilization does not occur and the female does not become pregnant, the whole process of egg production is started again. This cycle is referred to as the **estrus cycle**. It normally occurs every 21 days in hogs, cattle, and horses, and every 17 days for sheep, Figure 15–9.

During mating, a combination of sperm and fluid (semen) is deposited into the vagina of the female. The process is called **ejaculation** and the semen is sometimes referred to as ejaculate. As stated earlier, the fluid in the semen serves two purposes: to provide nourishment for the sperm and to provide a means for the sperm to move. In each ejaculation, millions of sperm are deposited. Each sperm is shaped like a tadpole and has a tail that causes the sperm

THE REPRODUCTION CYCLE IN FARM ANIMALS

Species	Length of Estrus Cycle (days)		Length of Estrus		Usual Time of Ovulation	Length of Gestation (days)		Age at Puberty (Months)
	average	range	average	range		average	range	
Mare	21	10–37	5-6 days	1-14 days	24–48 hours before end of estrus	336	310–350	10–12
Cow	19–21	16–24	16–20 hrs.	8–30 hrs.	10–14 hours after end of estrus	281	274–291	8–12
Ewe	16	14–20	30 hrs.	20–42 hrs.	1 hour before end of estrus	150	140–160	4–8
Sow	21	18–24	1–2 days	1–2 days	18–60 hours after estrus begins	112	111–115	5–7

Figure 15–9. The reproductive cycle in agricultural animals varies in length. *Vocational Materials Service, Texas A & M University.*

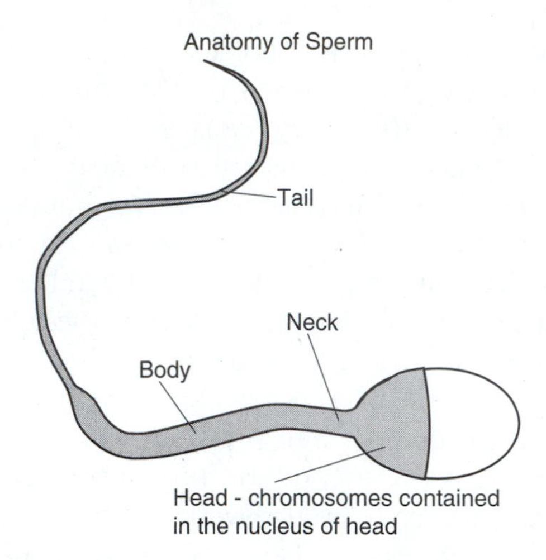

Figure 15-10. The anatomy of sperm. *Vocational Materials Service, Texas A & M University.*

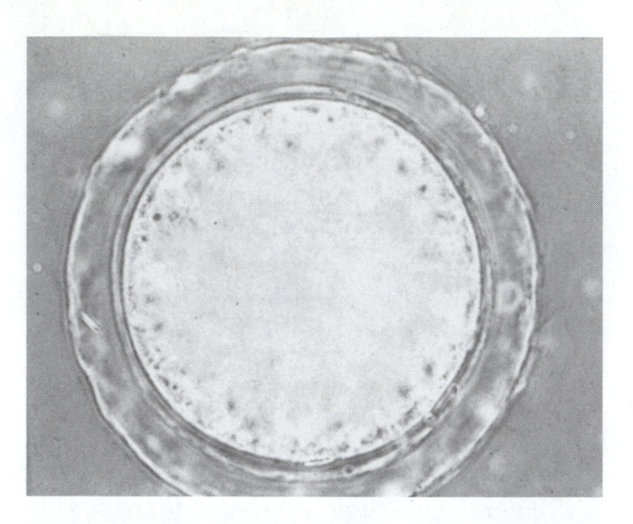

Figure 15-11. The outer membrane of the egg must be dissolved before the sperm can enter the egg. *Rick Jones, Cooperative Extension Service, The University of Georgia.*

to move in a whiplike action, Figure 15–10. Sperm that are able to move about freely are said to be **motile**. Through this means, the sperm begin a journey through the cervix, into the uterus, and into the fallopian tubes (oviduct) where fertilization occurs. The sperm are attracted to the egg by a chemical that is secreted by the egg. An obvious question is: If only one sperm is needed to fertilize the egg, why are millions of sperm deposited? This large number of sperm is needed for two reasons. First, not all sperm are hardy enough to make the long trip to the oviduct where the egg is. Therefore, a large number of sperm are necessary to ensure that some sperm reach the right destination. Second, even though only one sperm penetrates and fertilizes the egg, many sperm are needed for the process. The sperm swarm around the egg and secrete an enzyme that loosens the cells surrounding the egg, Figure 15–11.

The outer membrane of the egg is covered with a protective layer of a jellylike substance that must be dissolved before the sperm can enter the egg. The sperm release a chemical that works to dissolve the coating. One of the sperm forms a tubelike connection with the membrane of the egg. The nuclear material of the sperm then enters the egg and fertilization occurs. Only the nuclear material actually enters the egg, the tail of the sperm is left outside of the egg. When the nucleus of the sperm enters the egg, the egg releases carbohydrates and protein to form a layer around the egg in order to prevent any more sperm from entering. This new layer is called the **fertilization membrane**. Fertilization is completed when the nucleus of the sperm and the nucleus of the egg fuse together and the correct number of chromosomes for that particular species is restored. Remember that during meiosis only half of the number of chromosomes from the original cell are transmitted in the egg and half are transmitted in the sperm. At the completion of fertilization, the original number is reestablished. The resulting fertilized cell is referred to as a zygote, Figure 15–12.

Very soon after fertilization occurs, the zygote begins to divide. This type of cell division is called *mitosis* as compared to meiosis,

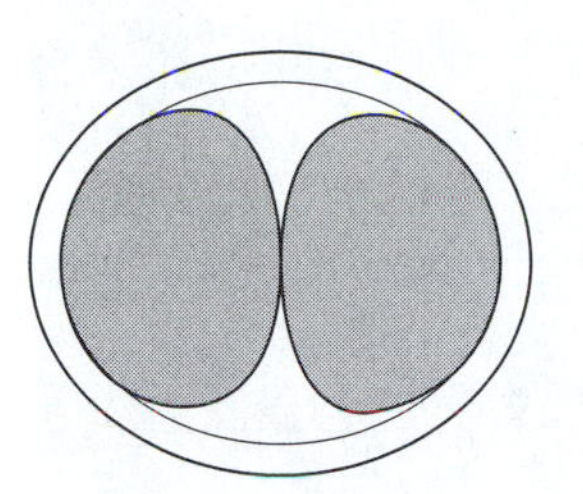
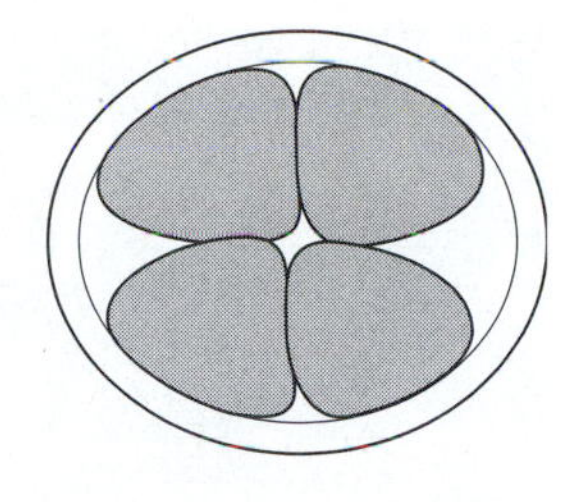

Figure 15-12. Fertilization occurs when the nucleus of the sperm and the nucleus of the egg fuse together. The cell multiplies and begins to differentiate. A zygote is formed.

which is the process by which the gametes are formed. Mitosis will be dealt with in Chapter 16. The division of the fertilized egg by mitosis is called **cleavage**. The cells continue to divide rapidly, and as they do, they make their way out of the oviduct into the uterus. As the cells enlarge and begin to differentiate, the mass of cells is called an *embryo*. Differentiate means that the cells begin to specialize; some begin to form cells that will become skin, some bones, some internal organs, etc. No one completely understands what causes differentiation, but we do know all parts of the animal's body come from the original fertilized egg. Once the embryo reaches the uterus, it attaches to the wall of the uterus and begins to develop. In the embryo development stage, the corpus luteum maintains a level of the hormone progesterone that causes the uterus to implant and nourish the embryo. As a result, the lining of the uterus remains intact and the embryo can continue to develop. In other words the female maintains pregnancy.

ARTIFICIAL INSEMINATION

As humans began to understand how reproduction works in animals, they began to use techniques that aided in natural reproduction. Healthier, faster-growing, more efficient animals could be produced with help from the humans who cared for the animals. One of the production practices that has been a real asset to animal producers is that of artificial insemination. Artificial insemination is not a new technology. Some say that the process goes all the way back to the Middle Ages when Arabs used the method to collect semen from stallions that belonged to their enemies and breed their own mares in order to produce superior foals. The first recorded use of artificial insemination was in 1780. Lazarro Spallanzani, an Italian scientist, was successful in artificially inseminating dogs. Perhaps the first large-scale use of artificial insemination was by the Russians shortly after the turn of the century. A Russian physiologist named Ivanoff used the process to help replenish the horse population in his country following World War I. Later, the technology was used for cattle and sheep on a large scale. Artificial insemination first began to be used in the United States in the 1930s. But as in the other countries, it had not reached its full potential because fresh semen had to be used. The life span of sperm is only about two to three days, so there were problems in obtaining semen when it was needed. In the 1950s, the technique of freezing semen was perfected. A **protectant** such as glycerine is added, and the semen is frozen at a specific steady rate until the temperature reaches -320°F, Figure 15–13. If the semen is kept at this temperature, it can remain viable for years. In fact, bull semen has been successfully stored for as long as 30 years. Semen from bulls, stallions, and rams can be frozen, stored, and thawed successfully. Semen from boars, however, is usually shipped immediately and used fresh because of problems with sperm viability when it is frozen.

Artificial insemination is used most widely in the dairy industry, Figure 15–14, but it is also used extensively for beef and to a lesser degree with horses, sheep, and swine.

Figure 15-13. A protectant such as glycerine is added to semen. The semen is frozen at a specific steady rate until the temperature reaches -320°F. *James Strawser, Cooperative Extension Service, The University of Georgia.*

Figure 15-14. Artificial insemination is widely used in the dairy industry. *James Strawser, Cooperative Extension Service, The University of Georgia.*

The advantages of using artificial insemination are numerous:

1. The producer may use a sire of higher quality than he/she could otherwise afford. A high-quality sire from any species is expensive; the cost of semen from artificial insemination is much lower.
2. Data from the progeny of sires used in artificial insemination are available to aid the producer in determining the quality of the sire. One bull may produce many thousands of offspring. The American Breeders Service reports that one of their superior bulls lived for 12 years and produced over 462,000 ampules of semen. Obviously, if production data for thousands of offspring can be compiled, the producer can get a clear idea of the type of animals to expect from the sire.
3. Artificial insemination allows the producer to select the type of sire needed for a particular group of females. For instance, a hog producer may need to increase the size of bone in the herd, or a beef producer needs a bull that will sire smaller calves at birth for calving ease. Through the use of sire data, the producer can select sires that are known for these characteristics.
4. Producers do not have to keep male animals. This can be an advantage not only because of expense but also from a safety standpoint. Mature male animals are by their nature aggressive and often dangerous. A large boar or bull can kill or seriously injure those who care for them.
5. The likelihood of disease is lessened. Many diseases are transmitted through direct contact with other animals. By using artificial insemination, contact between animals is avoided.
6. Sires from all over the world can be used. One of the largest problems associated with importing animals from other countries is strict **quarantine** laws that require the animals to be kept in isolation for a period of

SEMEN VOLUME & NUMBERS FOR FARM ANIMAL SPECIES				
Animal	Volume per Ejaculate (milliliter)	Sperm per milliliter (one thousand's)	Total Sperm per Ejaculate (billions)	Number of Females per Ejaculate
boar	150–250	100	15–25	10–12
bull	5–15	1,000	3–5	100–600
rooster	0.6–0.8	3,000	1.8–2.4	
ram	0.8–1.0	1,000	0.8–1.0	40–100
stallion	70–100	100	7–10	8–12

Figure 15-15. The volume of semen ejaculate varies with different agricultural animals. *Vocational Materials Service, Texas A & M University.*

time to make sure that they do not bring disease into the country. By using frozen semen, new genes can be brought into the United States with less risk of importing disease, and they can be brought in at much less cost.

7. Sires can be easily replaced. If a producer owns his/her own sires, the expense of changing sires is substantial. If the producer is not pleased with the offspring of the sire, then the old sire has to be sold and a new one bought. By using artificial insemination, all the producer has to do to change sires is to order semen from a different sire.

Semen Collection and Processing

Semen is collected through the use of an **artificial vagina**. The artificial vagina consists of a rigid tube that is lined with a smooth surface water jacket that is filled with warm water. At the end is attached a receptacle for collecting the semen. As the male approaches and mounts the dummy or live animal, the penis is guided into the artificial vagina where ejaculation occurs. The amount of the ejaculate or semen varies with the different species, Figure 15–15.

Once the semen is collected, it is examined in the laboratory under a microscope, Figure 15–16. It is checked for foreign material

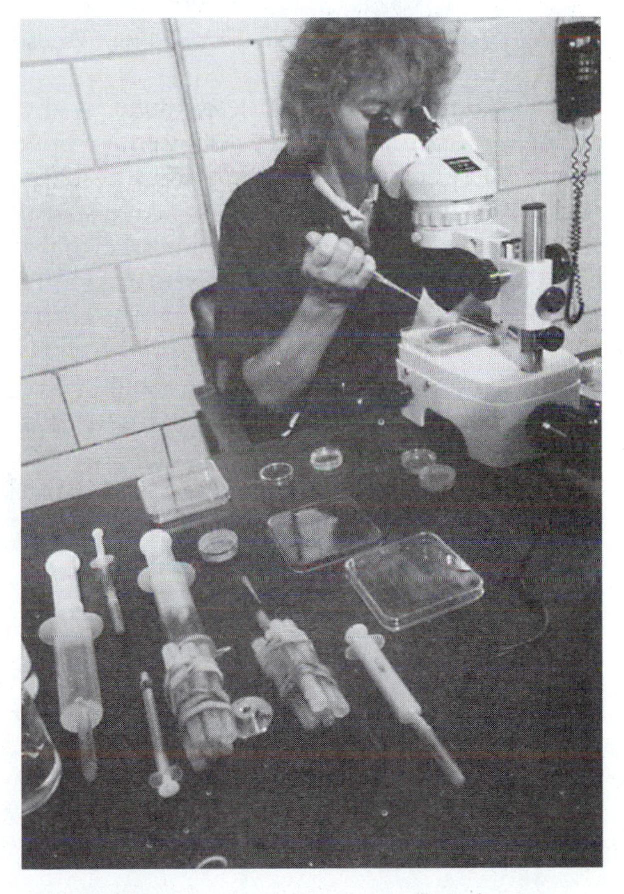

Figure 15-16. Once the semen is collected, it is examined in the laboratory under a microscope. *James Strawser, Cooperative Extension Service, The University of Georgia.*

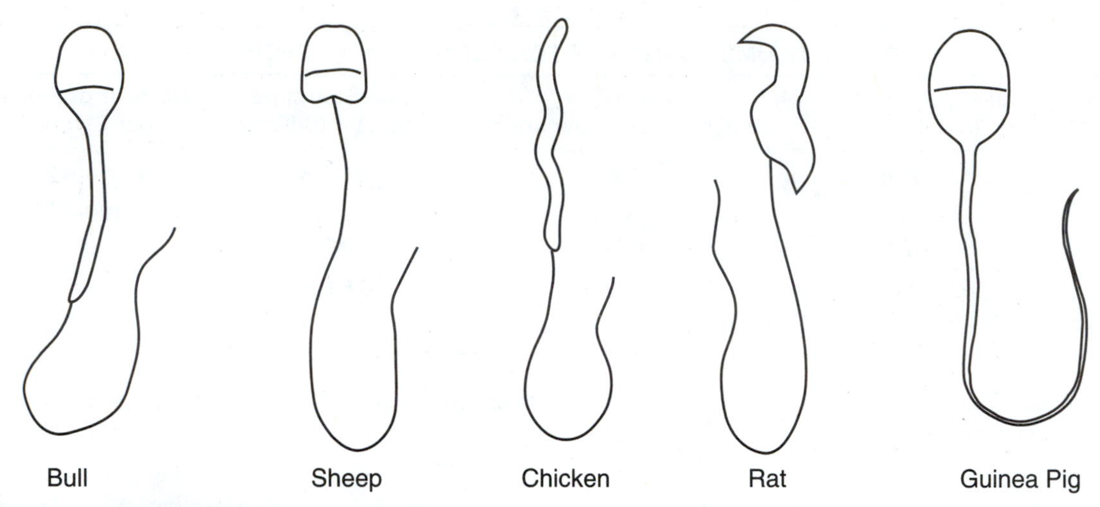

Figure 15-17. Species differ in sperm anatomy. *Vocational Materials Service, Texas A & M University.*

and for quality. Quality is determined by the number of sperm in a milliliter of semen, how active the sperm are (motility), and the shape of the sperm (morphology). Very active sperm are desirable because of the distance they must travel to reach the oviduct of the female. Sperm of different species are shaped differently, Figure 15–17. The sperm is checked for the normal shape of the sperm; a large number of sperm with an unusual shape is not desirable, Figure 15–18.

Once an ejaculate has been checked and determined to be of an acceptable quality, it is processed. Processing involves adding **extenders** such as milk, egg yolk, glycerine, and/or antibiotics. One purpose of the extenders is to provide a means of diluting the semen. The semen from one bull ejaculation may be divided into several units, depending on the number of sperm in the ejaculate. Another purpose served by the addition of extenders is to provide protection to the sperm during the freezing procedure. Extenders also provide nourishment for the sperm.

After the addition of the extenders, the semen is checked again to make sure that the

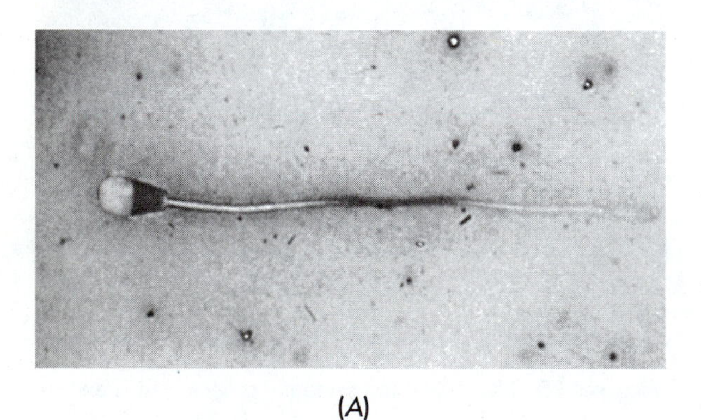

(A)

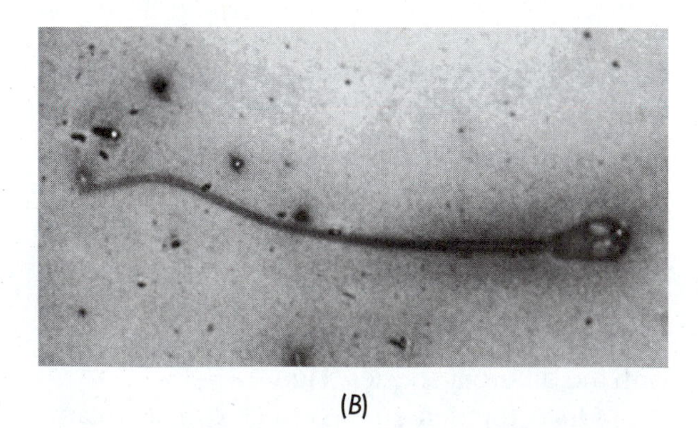

(B)

Figure 15-18. Sperm with abnormal shapes are not desirable. A: normal. B: defective. *Dr. Ben Bracket, Department of Physiology, School of Veterinary Sciences, The University of Georgia.*

sperm are still motile. The semen is then packaged in small hollow tubes called **straws**, sealed, and labeled with the name of the company, the date, and the name of the sire. The straws containing the semen are frozen at a specific rate to 320°F and are stored and transported in liquid nitrogen tanks. Boar semen does not freeze as well as bull semen. Although frozen boar semen is used, fresh semen is preferred because its use results in greater conception rates.

When the technician is ready to artificially inseminate an animal, the straws are carefully removed from the liquid nitrogen tank. Precautions have to be taken because liquid nitrogen can cause a frostbite-like injury if it contacts the skin. The straws of semen have to be thawed at the proper temperature and speed. Thawing may be accomplished through the use of a special apparatus that heats water to a certain temperature or through the use of a water-filled thermos bottle. The straw is placed into the water and left for not less than 30 seconds and not more than 15 minutes. Proper thawing assures that the thawed sperm will be healthy and motile.

Once the semen is properly thawed, the straw containing the semen is placed in a tube-like instrument that will be used to place the semen in the tract of the female, Figure 15–19. After the semen is placed in the female tract, the process of fertilization takes place just as in natural mating. The person doing the actual insemination must undergo special training before he/she can develop the skills necessary to properly thaw the semen and place it correctly.

Control of the Estrus Cycle

Another scientific advance that has greatly aided in improving the reproductive efficiency of agricultural animals is the use of **estrus** synchronization. Recall that estrus (the time the female allows breeding) is controlled by the production and secretion of hormones at the proper time. The estrus cycle is a chain of events that occur as certain hormones are released. If artificial **hormones** (from an outside source) are introduced into the female, the hormones will cause the same reaction as a naturally produced hormone. For example, remember that a hormone causes the follicle to develop on the ovary, and the egg is developed from the follicle. By injecting females with a hormone that stimulates the follicle, the female will begin the cycle at that point

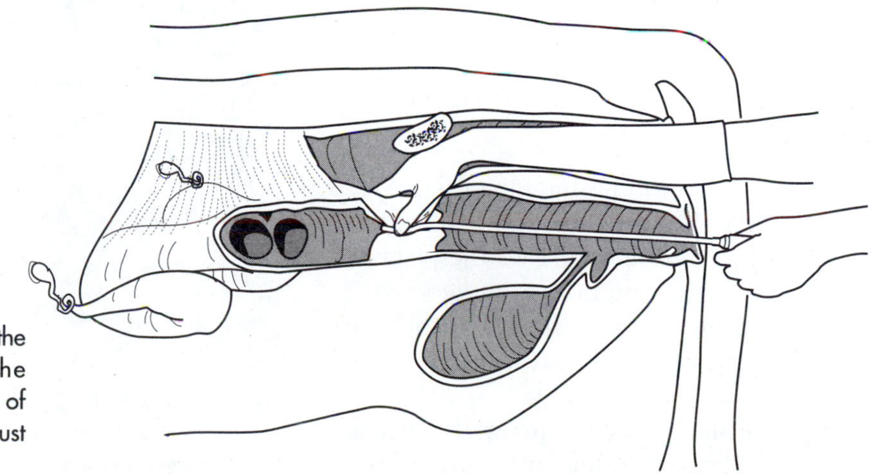

Figure 15-19. Once the semen is placed in the female tract, the process of fertilization takes place just as in natural mating.

and come into estrus. The advantage to the producer in inducing estrus is that by injecting all of the females in the herd at the same time, the cycle can be synchronized so that they all come into heat at about the same time. The obvious advantage is that a producer can have all of the animals artificially inseminated at the same time. Not only does this save time and resources at breeding time but it will also save because the females will all calve or farrow at about the same time. The crop of young animals will be of about the same age so they can be managed alike as they are grown out.

EMBRYO TRANSFER

Among the newest of the reproductive technologies is that of transferring embryos from one female to another. Just as artificial insemination has allowed genetic improvement from a single sire to be greatly increased, the transferring of embryos has increased the reproductive capacity of superior females. If a producer breeds a very valuable superior female, the animal will produce one offspring per year. Although a female is capable of producing many thousands of eggs during her lifetime, only a relative few develop into offspring. If the eggs are collected from a superior female and implanted in an inferior animal, the superior female has the capacity to produce many offspring in a year.

The transferring of embryos has many benefits.

1. The use of embryo transfer allows the rapid advancement of genetics from the dam. Just as artificial insemination allows the production of many offspring from a superior male, embryo transfer allows the production of many offspring from a superior female, Figure 15–20.
2. Embryo transfer allows the **progeny testing** of females. Progeny testing involves the gathering of data from the offspring of a particular animal. The data are analyzed to determine how valuable the animal is as a parent. Through the use of artificial insemination, a male can be progeny tested in a short time because of the tremendous number of offspring that can be born and raised at the same time. The problem with progeny testing with females is that her offspring are very limited in number and this does not allow sufficient numbers of offspring from which data can be collected. Through the use of embryo transfer, one female can produce many offspring in a short period of time, allowing for the testing of her progeny.
3. As in artificial insemination, embryo transfer permits the import and export of quality animals without the quarantine measures required of animals that are already born.
4. Embryo transfer allows the use of a dual production system. For example, by using embryo transfer, dairy cattle can produce calves that are pure beef animals. The reason dairy cows are bred is so they will continue to produce milk. But if they are bred naturally or by using artificial insemination, the calves will still be half dairy animals.

Figure 15-20. Embryo transfer allows the production of many offspring from a superior female. *Robert Newcomb, Cooperative Extension Service, The University of Georgia.*

When beef calves are preferred, embryo transfer is the method used.

5. By implanting two embryos into a recipient female, twin offspring can be produced.
6. A producer can rapidly convert his/her herd from grade animals to a purebred herd. By implanting a female of mixed breeding with purebred embryos, a producer can raise replacement animals that are both purebred and high quality.

Some argue that the use of embryo transplant has a big disadvantage. They say that if producers use embryo transfer and artificial insemination over a period of many years, the **genetic base** of the various breeds of animals will narrow. This means that in time there will be only a relative few animals that will eventually provide genetic material (egg and sperm) for the perpetuation of the breed. The fear is that producers, by demanding only embryos from the best animals, will eventually cause the loss of animals the producers feel are inferior. If the only animals of a breed that exist are related, there is reason to believe that this will lead to weakening rather than strengthening the breed. Others contend that through the use of embryo transfer, the importation of genetic strains from all over the world will prevent this problem from occurring. They contend that there are enough different strains in the world to make a narrowing of the genetic base highly improbable. Who is right? Only time will tell.

The Process of Embryo Transfer

The process of embryo transfer begins with the selection of **donor cows** and **recipient cows**. Cows selected as donors are usually animals that are of unusual value as breeding animals. They possess characteristics that are highly desirable to pass on to offspring, Figure 15–21. These characteristics might include high milking ability, growthability, or reproductive capacity. Or, the characteristics might be the type in demand for the show ring. In any case, the donor animals are too valuable to produce an offspring only one time a year.

Producers may purchase frozen embryos from one of many companies that specialize in the sale of genetically superior embryos. The producer selects the embryos he/she wishes to order by analyzing data that have been compiled about the donor and the sire. These data usually consist of production data about the animal's ancestors and their progeny. In this way, the producer can select for those traits that will be of most use in the producer's herd. On the other hand, recipient cows (those into which the embryo will be transferred) are usually cows of ordinary value. However, these animals are also carefully selected. They must be healthy animals that are able to reproduce efficiently. They must have the ability to maintain pregnancy and to deliver a healthy, growing calf at the end of the gestation period. Some producers like to use recipient cows that have at least some dairy breeding so they will produce adequate milk for the calf.

After the donor and recipient animals are selected, both groups must be synchronized so that they are at the same phase of their estrus cycle. This allows for the proper transfer of the embryo from one reproductive system to another. This synchronization is accomplished using the procedures discussed earlier. The only difference is that the donor animals undergo a process known as **superovulation** that causes them to release several eggs instead of just one. In this way as many as 12 to 15 eggs can be collected from one ovulation. Superovulation is accomplished by injecting the donor with a follicle-stimulating hormone (FSH). This hormone causes the ovaries to produce several follicles instead of one (the follicles provide a place for

Figure 15-21. Only very high-quality females such as this one are used as donor cows. *American International Charolais Association.*

the growing and maturing of the egg). During the process, the female is injected with **prostaglandin** to cause her to come into estrus. About 48 hours later, the female should be in estrus or heat. When this occurs, the cows are artificially inseminated or are bred naturally. Since there are multiple eggs to fertilize, more semen must be used than would be used in a regular artificial insemination.

Once fertilization occurs, the fertilized eggs (embryos) are allowed to grow for about a week before they are collected. In the earlier days of embryo transfer, collection was accomplished by removing the embryos surgically. This caused problems because of the scarring of the tissue in the reproductive tract of the donor female. Today, the embryos are removed by a process called flushing, Figure 15–22. In this procedure, a long, thin rubber tube called a **catheter**, Figure 15–23, is passed through the cervix and into the uterine horn. The catheter has an inflatable bulb about two inches from the end that fills like a balloon and seals the entrance to the uterus. A solution is then injected through the catheter into the uterus. When the fallopian tubes and the uterus are filled with solution, the flow of the solution is stopped and the solution is drained off into a collection cylinder. The fertilized eggs (embryos) are carried out of the uterus with the solution. An average of about six embryos are collected with each flush. After the embryos are flushed out, the uterus is flooded with another type solution that kills any embryos that were missed. The solution also helps to prevent infection.

Figure 15-22. Embryos are collected from the donor female by a process called flushing. *James Strawser, Cooperative Extension Service, The University of Georgia.*

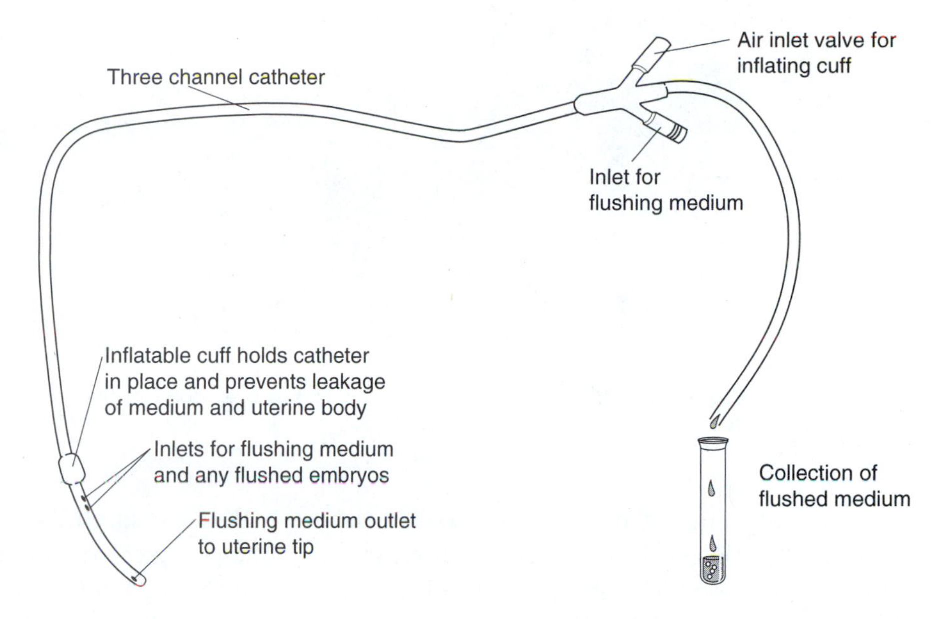

Figure 15-23. A Foley three-way catheter.

Once the embryos are collected in the solution, they are strained from the solution and are examined under a microscope to determine their quality. Only embryos that are in the proper stage of maturity and appear normal and undamaged are used for transferring, Figure 15–24. The embryos may be transferred directly to a recipient female or may be frozen and stored for implantation at a later date.

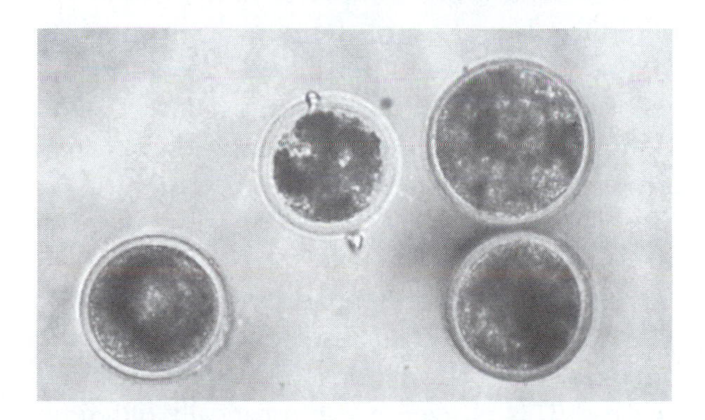

Figure 15-24. The embryo on the left is normal. The others are abnormal and not suitable for implant. *Dr. Ben Bracket, Department of Physiology, School of Veterinary Sciences, The University of Georgia.*

The recipient cow is made to come into estrus using injections of prostaglandins. When the corpus luteum reaches the proper stage, the embryo is placed in the uterus of the recipient cow. The pregnancy is allowed to progress as it would in a normal conception. Research has shown that pregnancies from embryo transfer are as likely to go full term and deliver a normal calf as are pregnancies from natural conception.

New Technology in Embryo Transfer

In the not-too-distant future, exciting possibilities for the use of embryo transfer exist. In fact, much of the required knowledge to make these practices a reality is already known, if not feasible economically. For example, one of the techniques that producers would like to use is that of being able to determine the sex of an embryo. Being able to transplant an embryo that they would be sure would produce a ram, a heifer, a boar, or a filly would be a tremendous benefit to livestock producers. For example, dairy producers might like for a majority of the calf crop to be female to be used as

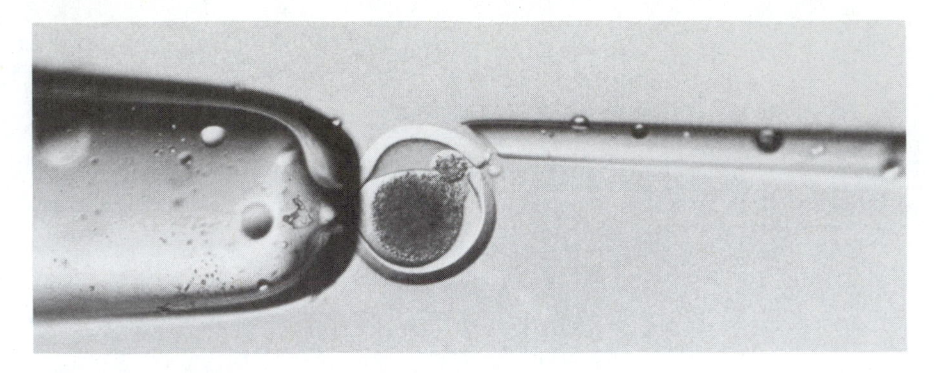

Figure 15-25. In the cloning process, microscopic tools are used to divide embryos. *Dr. Carol L. Keefer, ABS Specialty Genetics.*

replacements in the herd. If the embryos could be separated before implantation, all female embryos could be used. If a purebred breeder needs to produce mostly males to sell as herd sires, the process could be beneficial in his/her operation. Of course, scientists are already able to determine the sex of an embryo before it is implanted in the recipient female. The goal is to make the process inexpensive enough to make sex determination profitable.

Another technology that holds promise is that of cloning. A **clone** is an animal that is genetically identical to another animal. Cloning involves the splitting of embryos into two or more parts that will produce genetically identical offspring. The ability to clone not only exists but is rapidly becoming economically feasible. In fact, cloning is now done routinely on a commercial scale by several embryo transfer companies.

The process of splitting embryos for cloning involves the use of a device called a **micromanipulator**, which allows the control and maneuvering of tiny instruments that are small enough to slice an embryo, Figure 15–25.

The embryo is placed in a small dish and is covered with **paraffin**. By looking through a microscope set at a magnification of 100, the technician carefully divides the embryo using a microsurgical blade attached to the micromanipulator. Each half of the split embryo is removed from the dish and may be implanted or frozen. One advantage of splitting embryos is that one half may be implanted and the other half frozen. If the implanted half results in a superior animal, then the other half is transplanted. If the first implanted half does not result in a superior animal, then the other half may be discarded without the expense and time involved in producing a calf.

Another means of cloning is to place DNA material from a superior embryo into an unfertilized egg. All of the genetic material is removed from the unfertilized egg, and one cell is selected from the embryo, Figure 15–26. The embryo cell is fused to the egg and creates a new embryo. This process is repeated for all the cells in the embryo, creating many new embryos from the original.

A new type of cloning was accomplished in 1997 when Scottish research scientists produced a sheep born through a new cloning process. This process was dramatically different because instead of splitting embryos, the scientists used a single cell from an adult sheep to create an entirely new sheep. Remember from

NUCLEAR TRANSFER PROCEDURE

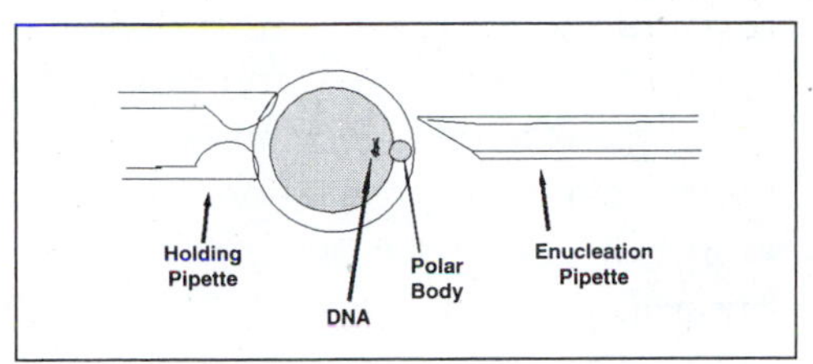

STEP 1: Remove DNA from unfertilized egg (enucleation).

Why: We do not want the genetic information from the egg (unknown genetics).

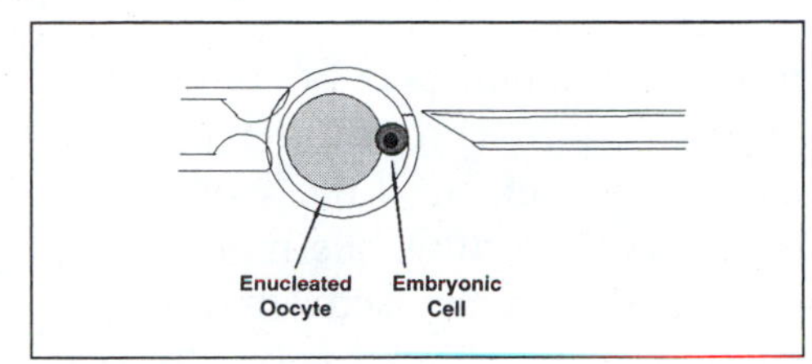

STEP 3: Place embryonic cell next to egg from Step 1 (enucleated egg).

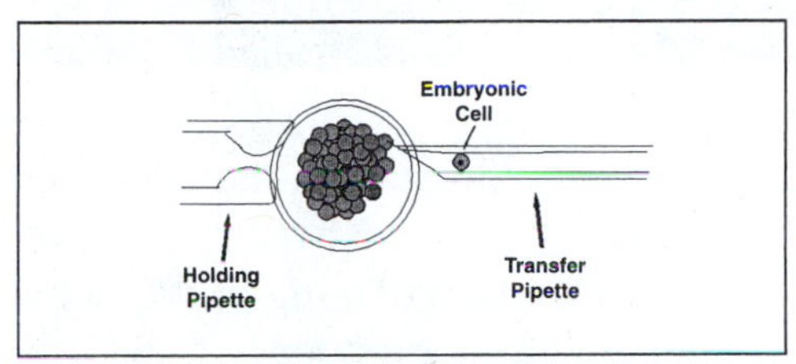

STEP 2: Pick up one cell from a good embryo.

Why: We want to use the good genetics of the embryo.

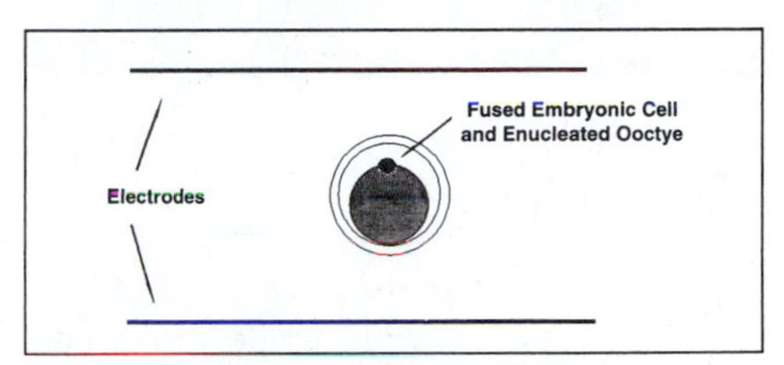

STEP 4: Fuse egg and cell by brief electrical pulse

WHY: Embryonic cell will supply genetic information and egg will supply cytoplasm to form new embryo (copy).

STEP 5: Repeat Steps 1-4 until each cell of the embryo has been used.

Why: Each cell has potential to form a copy of the original embryo. A 32 cell embryo could produce up to 32 copy embryos (clones) during one cycle of the cloning process.

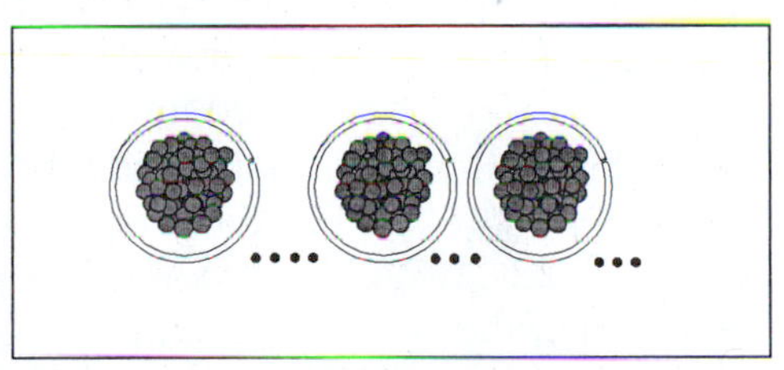

STEP 6: Culture clones for 5 to 6 days until they reach the morula or blastocyst stages. Transfer or freeze clones.

Why: The best stages of embryo development for embryo transfer or freezing are that of morula and blastocyst.

Figure 15-26. Nuclear transfer procedure. *Dr. Carol L. Keefer, ABS Specialty Genetics.*

the chapter on genetics that each cell in the body contains the complete genetic code for the animal. By unlocking the mechanism that controls cell differentiation, the scientists were able to use a single cell to create a new sheep with the identical genetic make-up as the cell. This was a tremendous breakthrough that has implications for all aspects of biology and agriculture. The innovation was not only that a new genetically identical sheep had been produced but also that the process for cell differentiation and the trigger for genetic coding was unlocked for the first time. As the process becomes better understood, the possibilities are almost limitless.

SUMMARY

The most important aspect of animal agriculture is reproduction. Through this process, new animals are brought into the world. The natural process of breeding has many years been controlled by humans with increasingly more dramatic results. The use of artificial insemination and embryo transplantation have revolutionized the entire industry of animal agriculture. Selective breeding has brought about gains in efficiency and the type of animal better suited for production. The exciting new world of gene manipulation and cloning will bring about changes we can't yet comprehend.

Review Exercises

TRUE OR FALSE

1. An organism that reproduces asexually originates from one parent, while one that reproduces sexually arises from two parents. ______.
2. Zygotes divide by a process called meiosis. ______.
3. The material that gametes have that determines all of the characteristics of the new animal is called deoxyribonucleic acid or DNA. ______.
4. Each chromosome from the male parent is matched with a chromosome from the female parent. ______.
5. The egg of the female and the sperm of the male each have half the chromosomes that normally occur in other cells of the body. ______.
6. Conception takes place in mammals in the uterus. ______.
7. Testicles are suspended away from the body to increase the temperature, which is necessary for sperm production. ______.
8. The pituitary gland produces a hormone that stimulates the ovary to produce a follicle. ______.
9. The male organ that deposits sperm in the female tract and serves as the means of expelling urine from the body is called the penis. ______.
10. The external covering of the penis is called the sheath or prepuce. ______.
11. The life span of sperm is only two to three hours. ______.
12. All species of mammals have the same amount of ejaculate. ______.
13. The female reproductive system is much more simple than that of the male. ______.
14. When the gamete is matured, the follicle becomes soft and expels the egg into the fallopian tube. ______.
15. After conception occurs, the corpus luteum discontinues producing progesterone and the female aborts the zygote. ______.
16. Embryo transfer allows for the production of many offspring from a superior male. ______.
17. Embryo transfer also allows dairy cows to produce calves that are pure beef animals. ______.
18. Collecting embryos by removing them surgically often causes problems due to the scarring of the tissue in the reproduction tract of the donor female. ______.
19. Only embryos in the proper stage of maturity that appear normal and undamaged are used for transferring. ______.

20. Scientists cannot determine the sex of an embryo before it is implanted in the recipient female. ______.

FILL IN THE BLANKS

1. The production of the male gamete or ______________ is called ______________, while female gamete production is referred to as ______________.
2. The testicles produce the hormone ______________, which controls the animal's ______________ and stimulates the development of ______________ characteristics.
3. The three accessory glands in the male reproductive tract are the ______________ ______________, the ______________ gland, and the ______________ gland.
4. Replicated chromosomes are called ______________. When they come together and are matched up in pairs, it is called ______________.
5. In sperm production ______________ new sperm cells were produced from the original cell, but in egg production only ______________ egg cell is produced.
6. A long tube called the vas deferens leads from the ______________ to the ______________ vesicle located at the upper end of the ______________.
7. The ovaries have several functions: they produce the ______________ or ______________, they provide a place for ______________ to take place, and they produce the hormones ______________ and ______________.
8. The fluid in the semen serves two purposes: to provide ______________ for the ______________ and to provide a means for the sperm to be ______________.
9. Large numbers of sperm are needed because not all sperm are able to make the ______________ ______________ to the ______________ where the ______________ is, and many are needed to secrete an ______________ that loosens the ______________ surrounding the ______________ and they also release a ______________ that works to dissolve the ______________ around the egg.
10. Leading from the ovaries are two tubes known as the fallopian tubes, which serve to transport the ______________ from the ______________ to the ______________.
11. When the nucleus of the single sperm enters the egg or ______________, the ______________ then releases ______________ and ______________ that form a layer around the ______________ to ______________ any more ______________ from entering.
12. Once the embryo reaches the ______________, it ______________ to the wall of the ______________ and begins to ______________.
13. When a mass of cells begins to differentiate, the cells begin to specialize to form ______________, ______________, ______________ ______________, etc..
14. An artificial vagina consists of a ______________ tube that is lined with a ______________ surface ______________ ______________, that is filled with ______________ ______________.
15. Some producers say that embryo ______________ and ______________ insemination over a period of years will narrow the ______________ base of the various ______________ of animals.

16. Quality of semen is based on the __________ of ________ in a __________ of semen, how ______________ the sperm are, and the __________ of the sperm.
17. Some producers feel that embryo transfer will allow the ________________________ of _______________________ strains from all over the world.
18. In embryo transfer, flushing involves a long, thin rubber ________________, called a ______________________, which is passed through the ________________ ________________________ and into the ________________ ________________.
19. Cloning involves the ________________ of embryos into two or more __________ that will produce ________________ identical ________________.
20. A micromanipulator allows the ______________ and ______________ of tiny ______________ that are small enough to slice an ______________________.

DISCUSSION QUESTIONS

1. What are the two different methods by which organisms reproduce?
2. What is a gamete?
3. What function does a gene serve?
4. Why do chromosomes always come in pairs (in higher ordered animals)?
5. Explain why mules cannot reproduce.
6. What is the difference between meiosis and mitosis?
7. List and explain the steps in the production of spermatozoa.
8. Name and explain the purpose of the hormone produced by the male's testicles.
9. What purposes do the accessory glands of the male reproductive tract serve?
10. Name and give the functions of three hormones that assist in the control of the female reproductive cycle.
11. What function does the cervix serve?
12. Explain how the female gamete (the egg) is produced in the ovary.
13. Where does fertilization take place?
14. Why is a large number of sperm needed to ensure that fertilization takes place?
15. What is the difference between a zygote and an embryo?
16. What are the advantages of using artificial insemination?
17. How is semen stored in order to last a long period of time?
18. What is estrus synchronization?
19. Why would a producer want to synchronize the estrus cycles of his/her herd?
20. What are the benefits of embryo transfer?
21. How are embryos collected from the female?
22. What potential problem do some people see in the continued use of embryo transfer?
23. What is cloning and how is this process achieved?

STUDENT LEARNING ACTIVITIES

1. From a local slaughterhouse obtain reproductive tracts of beef and pork animals. In the laboratory, dissect and examine the tracts. In order to protect yourself from any disease organisms that might be present in the tracts, be sure to wear rubber gloves at all times.

2. Obtain samples from a local veterinarian of frozen semen and embryos. Following the veterinarian's instructions, carefully thaw the semen and embryos and examine them under the microscope. Look for sperm or embryos that may be damaged or low quality.
3. Using the catalogs of breeder companies, compare the costs of semen and embryos of different animals. List some reasons why there would be a price difference. Explain why a producer might want to buy the cheaper or more expensive semen or embryos.
4. Conduct a survey of livestock producers in the area. Determine how many of the producers use artificial insemination and/or embryo transfer. Determine their reasons for or against using the techniques.

Chapter 16

Animal Growth and Development

Student Objectives in Basic Science

As a result of studying this chapter, you should be able to

1. Describe how an animal grows.
2. Distinguish between prenatal and postnatal growth.
3. Define mitosis.
4. Explain the three phases of prenatal growth.
5. Define cleavage.
6. List the layers of the blastula and the organs that are derived from each layer.
7. Describe the function of the placenta.
8. Explain how muscle cells are different from most body cells.
9. Discuss the sequence in which fat tissue is deposited in an animal's body.
10. Define the role of fat cells.
11. Explain the effects of hormones in the growth process.
12. Describe the aging process in animals.
13. Distinguish between chronological and physiological age.

Student Objectives in Agricultural Science

As a result of studying this chapter, you should be able to

1. Explain why animal growth is so important to producers of agricultural animals.
2. Explain why selection for muscling in breeding cattle is important.
3. Explain the phases of an animal's life in which the most rapid growth occurs.
4. Discuss the effect castration has on the growth of animals.
5. Define the lean-to-fat ratio. Explain the effects aging has on the productivity of animals.

Key Terms

prenatal
postnatal
hyperplasia
hypertrophy
morula
blastula
placenta
amniotic fluid
ectoderm
mesoderm
endoderm
morphogenesis
cartilage
ossification
organic
energy
lean-to-fat ratio
barrows
aging
physiological age
vertebrae
collagen

All animals must go through the process of growth in order to develop from a fertilized egg to a mature adult. During their lives, animals go through several stages of development in which their growth takes place in slightly different ways. A vast amount of both basic and applied research has been completed to better understand how growth occurs and how to make the growth process more efficient. Animal producers earn their living based on the amount of growth that occurs in animals. This is true whether the producers raise cattle for beef, sheep for wool, or horses for pleasure riding. Without growth, animal products would be nonexistent, Figure 16–1.

Growth is generally defined as an increase in the size or volume of living matter. An animal begins as a microscopic speck and grows within its mother until a certain mass of body weight and degree of maturity are achieved. A young newborn calf may weigh 85 pounds at birth, but by the time the animal is mature, it may weigh over a ton, Figure 16–2. Growth occurs in two major phases of the animal's life: **prenatal** and **postnatal**. Prenatal refers to occurrences that take place before the animal is born; postnatal refers to occurrences after the animal is born. During prenatal growth, all of the organs of the animal's body will be formed. When the animal is born, the organs begin to function in order for the animal to live and grow outside the mother's womb.

During postnatal growth, the animal increases in size and the systems of the body mature and develop. During both pre- and postnatal growth, the increase in size is a result of the cells increasing in size or in number. The increase in the number of cells is referred to as **hyperplasia**, and the increase in the size of the cells is referred to as **hypertrophy**. Growth in the size of cells is usually due to the accumulation of materials such as protein or calcium in the cytoplasm of a cell. Cytoplasm is

Figure 16-1. Producers depend on the growth of animals in order to make a living. *North American Limousin Association.*

Figure 16-2. Calves begin small but may grow to weigh over a ton. *Cooperative Extension Service, The University of Georgia.*

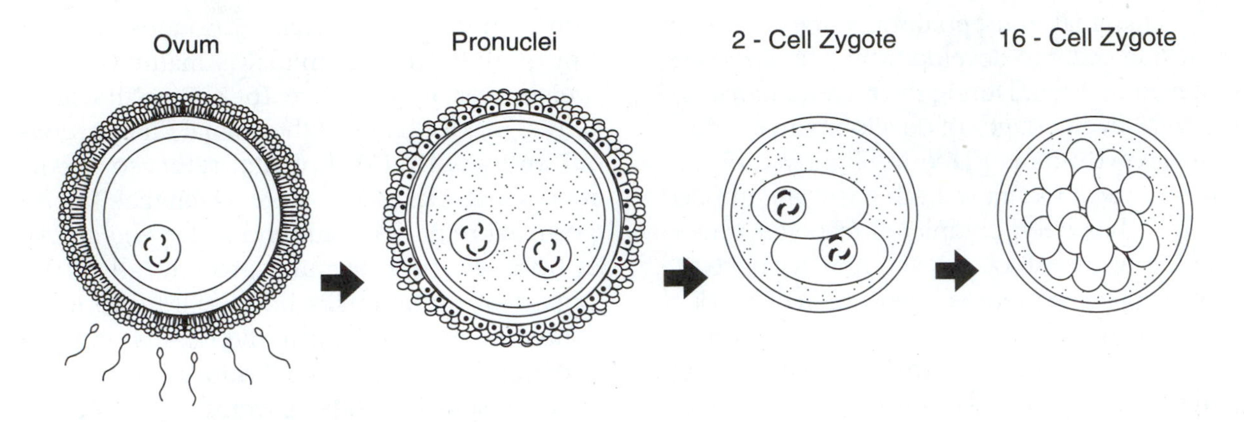

Figure 16-3. The ovum stage is the first phase of life. The single cell is the beginning of animal life.

the material within the cell wall that does not include the nucleus.

PRENATAL GROWTH

Prenatal growth may be divided into three phases: ovum, embryonic, and fetal. The ovum phase lasts from fertilization of the ovum by the sperm until the mass of cells attaches to the wall of the uterus. The ovum phase lasts for about ten days in sheep and eleven days in cattle. During this period little change takes place, and the shape of the cell mass remains approximately spherical.

The single cell (fertilized egg) is the beginning of animal life, Figure 16–3. Remember from Chapter 15 that the sperm cell and the egg cell are developed through a process called *meiosis*. Each of these cells contains half the number of chromosomes needed to create a new animal. When the two gametes (the egg and the sperm) unite at fertilization, they combine to supply the correct number of chromosomes for the new animal.

After fertilization, the fertilized cell must reproduce itself in order to begin to grow; this process of cell division is called *mitosis*. Through mitosis the nucleus of a cell divides and the DNA is replicated so that each cell has the exact same genetic information. As the number of these cells increases, the fertilized egg develops into a tight ball containing a large number of cells. This process is called *cleavage*.

As the cells multiply and divide, they form into a spherical mass called the **morula**. The cells of the morula are arranged to form an outer layer and a central core, Figure 16–4. As

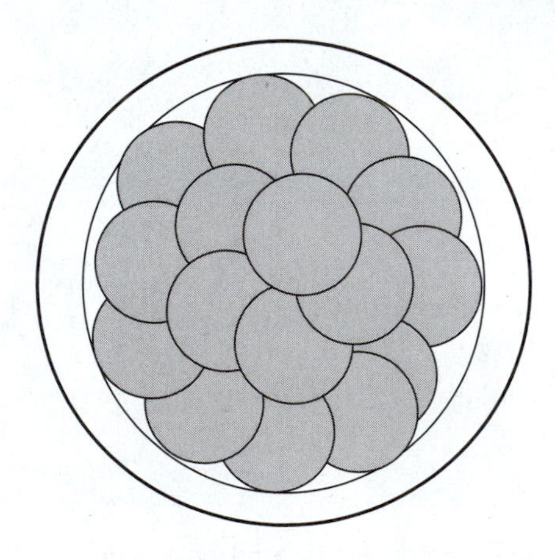

Figure 16-4. As the cells multiply, they form into a spherical mass called the *morula*.

the mass of cells continues to grow, a fluid-filled cavity develops in the center of the group of cells. This mass of cells with a cavity in the center is called a **blastula**. From the blastula, the cells begin to differentiate.

The Embryonic Phase

During the embryonic phase, the major tissues, organs, and their major systems are differentiated. The embryo is attached to the uterus, and a saclike pouch develops around the embryo. This enclosure is called the **placenta** and it contains fluid known as the **amniotic fluid**. The purpose of the placenta is to give the embryo (fetus) nutrition and oxygen from the mother. Also, the waste materials are absorbed by the placenta and disposed of through the mother's lungs and kidneys. The amniotic fluid in the placenta serves as a mechanism to absorb shock and to protect the fetus. It also provides a lubricant when the fetus is moving through the birth canal.

The embryonic phase lasts from the tenth day to the thirty-fourth day in sheep and from the eleventh day to the forty-fifth day in cattle. During the embryonic phase, the body undergoes a series of successive changes without much weight gain.

The blastula goes through a process that begins the development of all the organs and tissues in the animal's body. This process begins with the cells dividing into three layers, Figure 16–5. The outer layer is called the **ectoderm**, the middle layer is called the **mesoderm**, and the inner layer is called the **endoderm**. All parts of the animal's body are developed from these layers. This process of cell development into different tissues and organs is known as **morphogenesis**.

After the layers are formed, the outer layer (the ectoderm) develops into the tissues that are on the outside or near the surface of the animal's body. These include the skin, hair, hooves, and certain endocrine glands (ductless glands that secrete hormones into the blood-

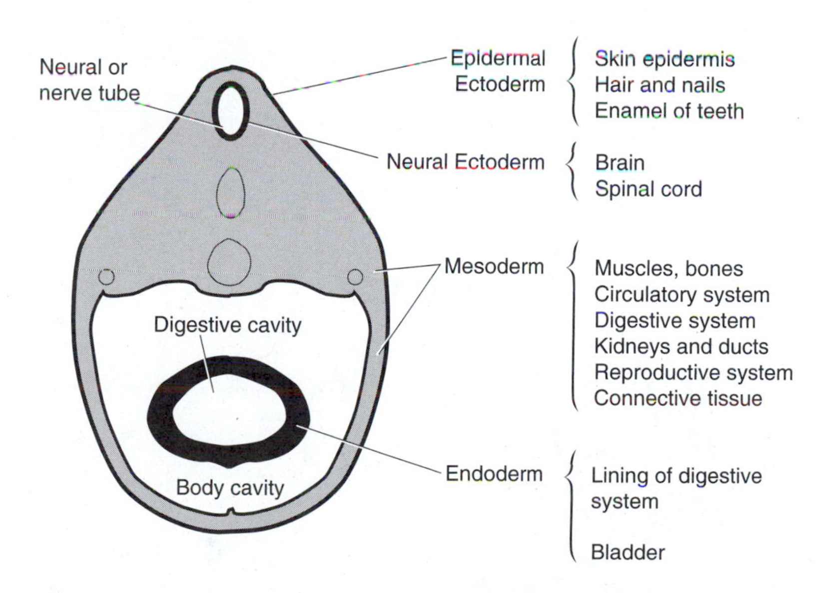

Figure 16-5. From the blastula, the cells begin to differentiate into distinct types of tissue.

stream). In addition, this layer forms the central nervous system—the brain, the spinal cord, and all of the nerve branches. The development of the brain and spinal cord begins by the formation of the neural tube. The neural tube is formed by a thickened strip of ectodermal cells that fold together to form a tube. Another group of endodermal cells form a rodlike structure (notochord) that gives rise to the vertebral column. From this beginning, the brain and nervous system of the animal are formed.

The mesoderm layer develops into the skeletal and muscular system of the animal, including the bones, muscles, cartilage, tendons, and ligaments. The heart, veins, arteries, and other parts of the circulatory system are also developed from this layer. In addition, the reproductive system is formed from the mesoderm layer, Figure 16–6. The inner organs, such as the liver and digestive system, are developed from the endoderm layer. The endoderm forms the digestive system, lungs, the liver, and other endocrine glands.

The animal begins to take on a recognizable form in the late embryonic stage, Figure 16–6. Scientists do not fully understand what triggers cells into taking on different forms in the differentiation process. They believe the answer is in the genetic code of the DNA. As more and more research is completed about genetics, scientists are realizing that this code is tremendously more complicated than they once thought.

The Fetal Stage

Animals of different species grow at different rates during the fetal stage. During this stage, the new animal develops all of the organs of the body, Figure 16–7. Organs, such as the liver, heart, and kidneys, have functional importance during fetal growth. These organs undergo a greater proportion of their growth in the early stages, while the digestive tract undergoes a far greater proportion of its growth in the later stages. The reason for this is that the animal must have fully functioning organs, such as the

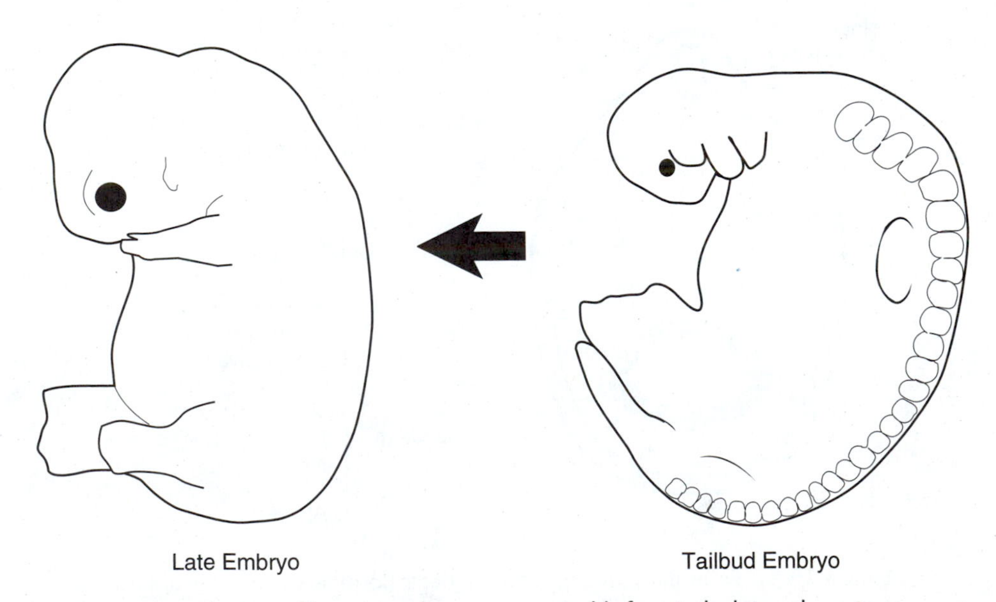

Figure 16-6. The animal begins to take on a recognizable form in the late embryonic stage.

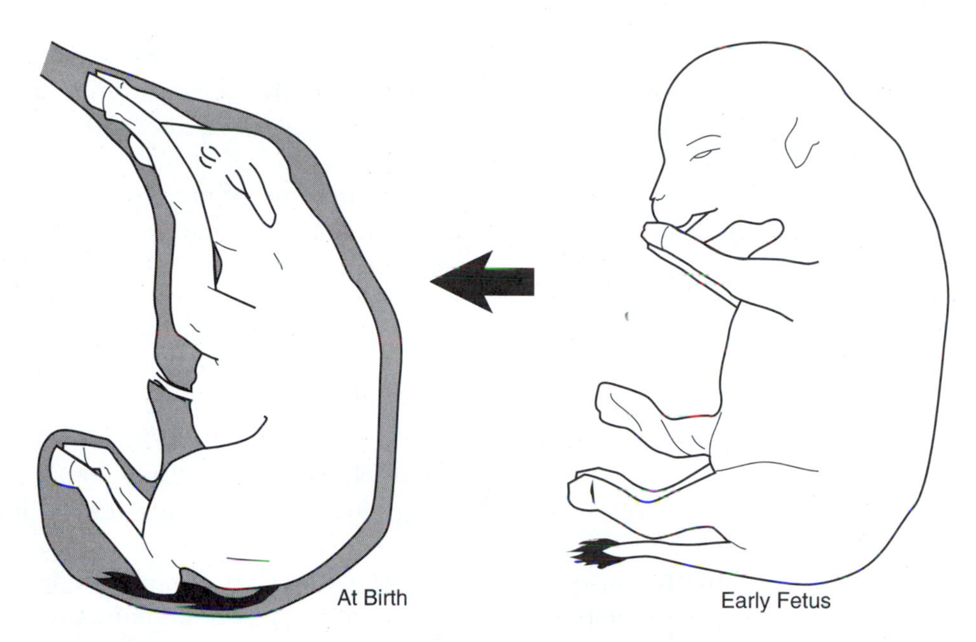

Figure 16-7. The animal's vital organs are developed in the fetal stage.

heart and lungs, before it is born. Nutrients are passed from the mother's bloodstream, and so the digestive system of the fetus is not as vitally important.

Growth of the various parts of the fetus and the organs is characterized by changing rates of growth. This means that various organs grow faster at times than other organs. Continuous changes in the fetus occur during the fetal stage. As the fetus matures, it is less and less dependent on the mother. When the fetus is able to live on its own, a hormone called *oxytocin* stimulates the muscles of the uterus into contracting. The birth canal is relaxed and the new animal is expelled from the uterus. As the animal hits the ground, the lungs are stimulated into functioning and the animal lives on its own, Figure 16–8.

Postnatal Growth and Development

After an animal is born, the parts of its body do not grow and develop at the same rate, nor is the development rate of different species the same. However, the order in which parts and systems of the animal develop is much the same in all species. Tissues are generally developed in the order of the importance to the survival of the animal.

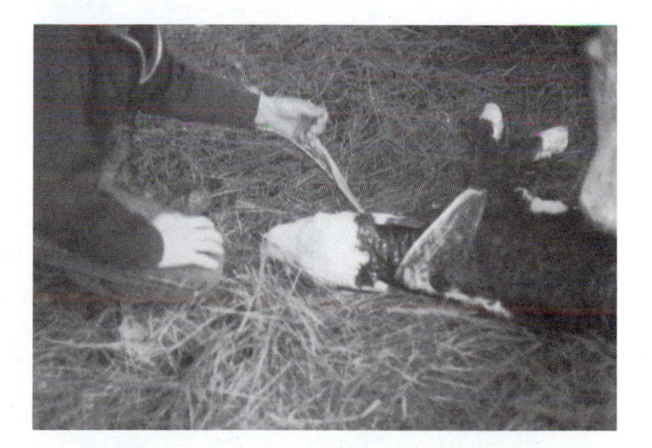

Figure 16-8. At birth, the animal is expelled from the uterus. When the animal hits the ground, its lungs are stimulated into operating. *Cooperative Extension Service, The University of Georgia.*

The heads of most species of animals comprise a larger portion of the body at birth than at any time during the animal's postnatal development. The head contains the brain, and this organ directs not only the growth but also the functions of all the other systems and organs.

The animal's legs tend to be a larger portion of the body at birth than at later stages. Well-developed legs are necessary at an early stage because young animals must be able to stand and nurse or be able to get away from predators, Figure 16–9.

The brain, central nervous system, heart, and circulatory system are all well developed at birth. These organs are essential because the animal is expelled from its mother's womb and must survive on its own.

The respiratory system and the digestive system of animals tend to develop early after birth. The first milk from the mammary glands of the mother is called *colostrum.* This milk is rich in nutrients and it passes immunity from the mother to the newborn. Also, the colostrum cleanses the digestive tract and stimulates it into functioning.

Figure 16-9. Strong legs are necessary at an early stage so animals can stand and nurse.

Animals grow rapidly from the time they are born until they reach sexual maturity. In the brief period just after birth, the animal grows relatively slowly as the organs adjust to functioning outside the placenta of the mother. After that period, the animal begins a stage of rapid growth when bone and muscle tissue grow steadily. This period continues until the animal has reached sexual maturity. During this phase, the animal achieves the greatest growth rate and the greatest feed efficiency.

The size of an animal is dependent for the most part on the size and amount of the bones and muscles in the animal's body. All of the animal's muscle cells are in place by birth. Muscle cells are different from most other cells in that they are long and relatively thin and contain many nuclei, Figure 16–10. After the animal is born, the growth in the muscle system of an animal is due to increases in the size of but not in the number of cells. This is why it is so important to select parent animals that will pass the genetic ability for developing muscle cells of an adequate quantity.

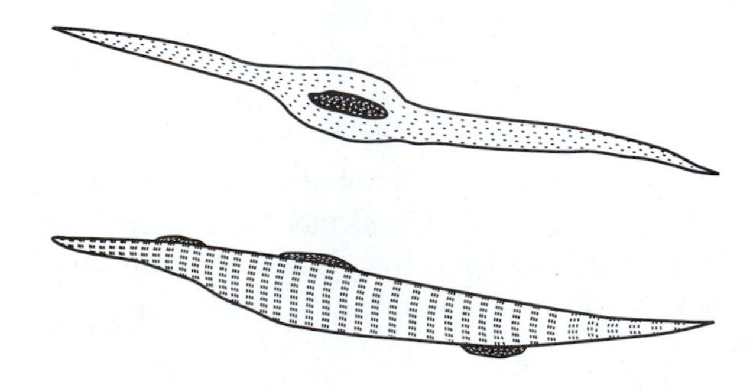

Figure 16-10. Muscle cells are different in that they are elongated and contain many nuclei.

Bone tissue cells multiply both before and after birth. Bones grow longer by the hardening of **cartilage** tissue at the end of the bones. Cartilage is tissue that is softer than bone and solidifies as the bone matures. Once the cartilage solidifies all the way to the end of the bone, the growth of the bone ceases. This process is called **ossification**. When bones mature, they are made up of about half minerals; the two main minerals are calcium and phosphorus. The other half is made up of **organic** matter such as protein.

After the animal reaches sexual maturity, it continues to grow, although it grows at a slower pace until the growth of muscle and bone stops. The animal then begins to lay down layers of fat. Fat is present in all phases of the animal's growth if there is adequate nutrition in the diet to allow for the storage of fat. Fat is nature's way of storing **energy** for the animal to use when it is not getting as much feed as it normally does.

An animal's body contains two types of fat: white fat stores energy; brown fat functions to maintain the animal's body heat. The animal first deposits fat in the abdominal cavity, where the fatty tissue serves to cushion the internal organs. After a sufficient amount is deposited in the abdominal cavity, the fat is then placed between the muscles; this tends to give the animals a sleek, smooth appearance. The last place for storage of fat is inside the muscles. In retail cuts of meat, these deposits of fat are called marbling.

THE EFFECTS OF HORMONES ON GROWTH

Hormones are chemical substances formed by the endocrine glands and secreted into the bloodstream. They are carried to other parts of the body where they stimulate organs into action or control bodily processes. The growth of an animal is regulated by hormones. The more important of the glands that secret hormones affecting the growth of animals are the pituitary, thyroid, testicles, ovaries, and adrenal glands. Figure 16–11 shows the hormones that are responsible for the stimulation and regulation of animal growth. These hormones regulate the ultimate size of the animal. If there is an excessive amount, the animal will be abnormally large or malformed. An absence of the hormones can cause the animal to be a dwarf.

Producers make use of artificial hormones to cause animals to grow more rapidly or more efficiently. These hormones or hormonelike substances are implanted in the animal (usually in the ear). The implants slowly release a very small amount of the growth stimulant over a period of time. The Food and Drug Administration (FDA) closely regulates the amount of these substances that are allowed to be present in the meat from implanted animals (see Chapter 22).

Castration of males slows growth and increases fat deposition in cattle, sheep, and swine. As shown in Figure 16–11, the testicles produce testosterone that stimulates growth to a certain degree. When the testicles are removed, the production of this hormone stops and growth is somewhat affected. The degree of this effect varies within the species. It also affects the amount of lean and fat in the animal's body. This is referred to as the **lean-to-fat ratio**. In cattle and sheep, the female has the lowest lean-to-fat ratio, Figure 16–12. A larger portion of their bodies contain fat than contain lean. The uncastrated males have the highest lean-to-fat ratio, and the castrated males fall somewhere in between. However, **barrows** (castrated male pigs) have lower lean-to-fat ratios than gilts because they reach maturity earlier and

GLAND	HORMONE	EFFECTS ON SKELETON	EFFECTS ON PROTEIN METABOLISM
Pituitary	Somatotropin (STH)	Stimulates growth and closure of long bones	Increases nitrogen retention and protein synthesis
Thyroid	Thyroxin	Stimulates growth of long bones; essential for STH effect	Controls body metabolism by increasing energy production and oxygen consumption by tissues
Testicles	Testosterone	High dose is weak stimulator of epiphyseal closure and inhibits STH; low dose increases the width of epiphysis and helps effect of STH	Stimulates nitrogen retention; promotes muscle growth and development of sex characteristics
Ovaries	Estrogens	Inhibits skeletal growth; promotes epiphyseal closure	Increases nitrogen retention and protein synthesis in ruminants
Adrenal	Glucocorticoids	Decreases growth of epiphysis; decreases stimulation of epiphysis by STH	Increases protein and amino acid degradation; inhibits protein synthesis and increases fat deposition

Figure 16-11. Hormones have quite an affect on the way animals grow.

have more fat deposition than gilts. Animals with the highest lean-to-fat ratio (intact males) usually have the least marbling. This is especially true for bulls.

The Aging Process in Animals

Aging, from the time of birth to the time of death, involves a series of changes that lead to the deterioration of the animal and eventually to death. There are two different terms for aging in animals. Chronological age refers to the actual age of the animal. **Physiological age** refers to the stage of maturity the animal has reached.

In determining the physiological age of an animal after it has been slaughtered, the most common method is that of examining the bones. As mentioned earlier, as bones mature, they harden. Soft cartilage tissue, such as that on the rib cage and the **vertebrae**, continues to solidify throughout the animal's life. Physiological age can be determined by the degree to which these cartilage tissues have solidified.

LEAN-TO-FAT RATIO	CATTLE	SHEEP	SWINE
Highest	Bull	Ram	Boar
Intermediate	Steer	Wether	Gilt
Lowest	Heifer	Ewe	Barrow

Figure 16-12. The gender of an animal affects the lean-to-fat ratio of its body.

It is said that as an animal is born, it begins to die. This is true from a physiological sense because after the formation of the embryo, many tissue cells stop dividing and only the cells essential to life (skin, blood, etc.) continue to divide. The animal's life span is directly related to the rate of growth or the length of time to reach maturity, Figure 16–13. Sheep mature in about a year and have a life expectancy of about ten years. Cattle require two to three years to reach maturity and have a life expectancy of 20–25 years.

Physiological functions of animals decrease with age. Muscular strength and speed decline, reproductive organs secrete lower levels of hormones, and the time of recovery from substance imbalances becomes longer with age. There is a gradual breakdown of separate nerves, the nervous system, and glandular control involved in aging. Wrinkles form as **collagen** and protein become less elastic.

Production by agricultural animals is often limited by age. The reason for this is because animals tend to lose their teeth as they age and become inefficient. Mothers produce less milk as they age; their young do not grow as rapidly. Some animals, such as the sow, reach excessive size, causing management problems. Therefore, cows are usually culled by age 10–11 years, ewes are culled at about age seven to eight years, and sows are culled at two to three years of age. The factors affecting how long an animal may live or be productive may be genetic or environmental. How the animals are cared for determines to a large extent how well they do at an advanced age.

SUMMARY

Animals begin to grow at birth and in some ways continue this process until they die. All of the body's systems must develop, grow to the proper size, function, and replace cells as these are lost or wear out. Growth and development is what the animal industry is all about. Ensuring that animals grow properly is the job of all connected with the industry from conception to the slaughter process.

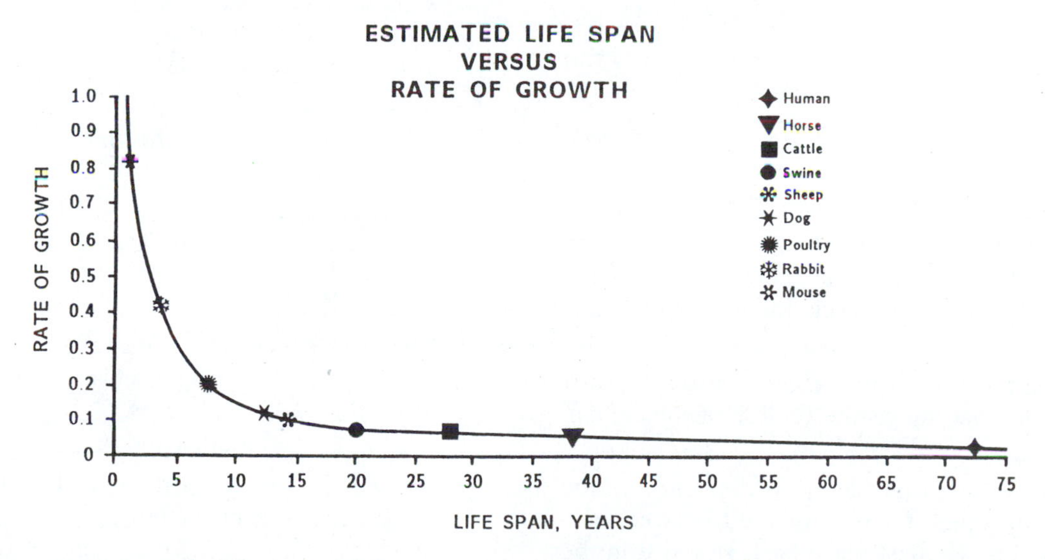

Figure 16-13. Animals differ in the length of their lives. *Vocational Materials Service, Texas A & M University.*

Review Exercises

TRUE OR FALSE

1. Prenatal refers to occurrences that take place before the animal is born; postnatal refers to occurrences after the animal is born. ______.
2. Prenatal growth may be divided into three phases: ovum, embryonic, and trimester. ______.
3. Differentiate means that the cells take on different forms in order to produce different tissues. ______.
4. The placenta serves as a mechanism to absorb shock and to protect the fetus. ______.
5. During the embryonic phase, the body undergoes a series of weight gains without much new development. ______.
6. The inner organs, such as the liver and digestive system, are developed from the endoderm layer. ______.
7. The differentiation process has been studied and researched so extensively that there is little left to find out. ______.
8. All species of mammals grow at nearly the same rate. ______.
9. Although the order in which parts and systems of the animal develop is different for different species, the rate is the same. ______.
10. After the organs have adjusted to functioning outside the mother, the animal begins a stage when bone and muscle tissue grow rapidly. ______.
11. At sexual maturity, another growth spurt occurs that affects the fertility of the animal. ______.
12. After enough fat has been deposited inside the muscles, fat is deposited between the muscles. The last place fat is stored is in the abdominal cavity. ______.
13. Hormones stimulate organs into action or control bodily processes. ______.
14. A low lean-to-fat ratio means there is more fat than muscle in the animal's body. ______.
15. Production by agricultural animals is never limited by age. ______.

FILL IN THE BLANKS

1. Growth is generally defined as an ______________ in the ______________ or ______________ of ______________ matter.
2. The increase in the number of cells is referred to as ______________, and the increase in the size of the cells is referred to as ______________.
3. The blastula is a mass of ______________ with a ______________ -filled cavity in the ______________.
4. The differentiation process begins with the ______________ forming three basic layers: an outer layer called the ______________, a middle layer called the ______________, and an inner layer termed the ______________.
5. The development of the brain and spinal cord begins by the formation of the ______________ tube, which is formed by a ______________ strip of ______________ cells which fold together to form a ______________.
6. The mesoderm layer develops into the ______________ and muscular system of the animal, such as bones, ______________, ______________, tendons and ______________.
7. As soon as they are born, animals must have certain organs such as the ______________ and ______________ that begin to function immediately.

8. When the fetus is able to live on its own, a hormone called ________________ stimulates the muscles of the ________________ into contracting, and the birth canal is ________________.
9. The brain, ____________ ____________ system, heart, and ______________ system are all ____________ developed at birth.
10. Colostrum is the __________ milk from the mammary glands of the mother and passes ________________ from the mother to the ____________________.
11. Muscle cells are different from most other cells in that they are ___________ and relatively ________________ and contain many ______________.
12. Fat is ________________ way of storing ____________ for the animal to use when it is not getting as much ____________ as it normally does.
13. Hormones are ____________ substances formed by the ____________ glands and ___________ into the ____________.
14. Two terms for aging in animals are chronological age, which refers to the __________ _________ of the animal, and physiological age, which refers to the __________ of ___________ the animal has __________.
15. With age, muscular strength and _________ decline, ______________ organs secrete lower levels of ___________, and the time of __________ from substance ___________ becomes longer.

DISCUSSION QUESTIONS

1. Why is animal growth essential to producers?
2. By what two methods does growth occur?
3. What is the difference between the morula and the blastula?
4. What is the purpose of the placenta?
5. Name three layers of the blastula.
6. What organs arise from the ectoderm?
7. What is meant by differentiation?
8. What is oxytocin?
9. Why is it essential that newborn animals have well-developed legs?
10. At what stage in an animal's life does the most rapid growth occur?
11. What is meant by ossification?
12. List the sequence in which an animal's body deposits fat.
13. List five glands that secrete hormones that affect growth.
14. What is meant by the lean-to-fat ratio?
15. Tell the difference between chronological and physiological age.

STUDENT LEARNING ACTIVITIES

1. Obtain from a slaughterhouse fetuses from animals that have been slaughtered. Try to determine the stage of maturity for each. Be sure to wear latex gloves when handling the material.
2. Visit with a producer and determine how he/she determines when animals have reached the level of maturity desired for market. Try your hand at making the determination of animals in the producer's herd.

Chapter 17

Animal Nutrition

Student Objectives in Basic Science

As a result of studying this chapter, you should be able to

1. Explain why animals must have nutrients.
2. List the six nutrients essential to life.
3. Discuss the role water supplies in supporting life.
4. Discuss the relationship between proteins and amino acids.
5. Distinguish between a carnivore, omnivore, and a herbivore.
6. Discuss the importance of protein in the diet of animals.
7. Discuss the importance of carbohydrates.
8. List the types of common sugars.
9. Distinguish between a starch and a sugar.
10. Discuss the importance of fats in the diet of animals.
11. Discuss the role minerals play in sustaining life.
12. List the vitamins that are important in the diet of an animal.
13. Discuss the function of vitamins.
14. Distinguish between a monogastric and a ruminant digestive system.
15. List and define the function of the organs of the monogastric digestive system.
16. List and define the function of the organs of the ruminant digestive system.

Student Objectives in Agricultural Science

As a result of studying this chapter, you should be able to

1. Explain the sources from which protein is obtained for feeds.
2. List the common grains that are used as a source of carbohydrates.
3. Distinguish between a concentrate and roughage.
4. List the sources of fats in animal rations.
5. List the sources of minerals in animal feeds.
6. List sources for the various vitamins that are of use to animals.
7. Explain the differences in the feed used by monogastrics and feed used by ruminants.

Key Terms

maintenance ration
feedstuff
anabolism
catabolism
amino acids
essential amino acids
nonessential amino acids
crude protein content
carnivores
herbivores
tankage
cellulose

Key Terms *(continued)*

monosaccharides
disaccharides
glucose
fructose
galactose
sucrose
lactose
lipids
inorganic
macrominerals
microminerals
trace minerals
free choice
carotene
toxic
rumen
monogastric
gastrointestinal tract
alimentary canal
absorption
cecum
esophagus
pepsin
duodenum
jejunum
ileum
semipermeable membrane
diffusion
boluses
reticulum
omasum
abomasum
mucous membranes
bloat

In order for an animal to go on living, growing, reproducing, and performing all of the bodily functions, it must have nourishment. All movement and body processes of the animal require the use of energy. There are only two places where an animal can obtain energy: one is from the food it ingests; the other is from the energy stored by its body in fat cells. Obviously even for the animal to store energy in the fat cells, it must have an intake of food.

In the wild, animals must spend most of their time in search of food to sustain themselves. Most agricultural animals are given their food every day. The nutrients obtained from their feed can go into growing and producing the products desired by the producers. Therefore, producers carefully balance the diets to fit the needs of the animals.

A certain level of nutritional needs, known as the **maintenance ration**, must be met first. This is the level of nourishment needed by the animal to maintain its body weight and not lose or gain weight. Nourishment over that amount can be used for growing, gestating, and producing milk or other products, Figure 17–1.

In the wild, most animals eat a variety of foods. This variety gives the animals the nutrients they need to support their bodily functions. In

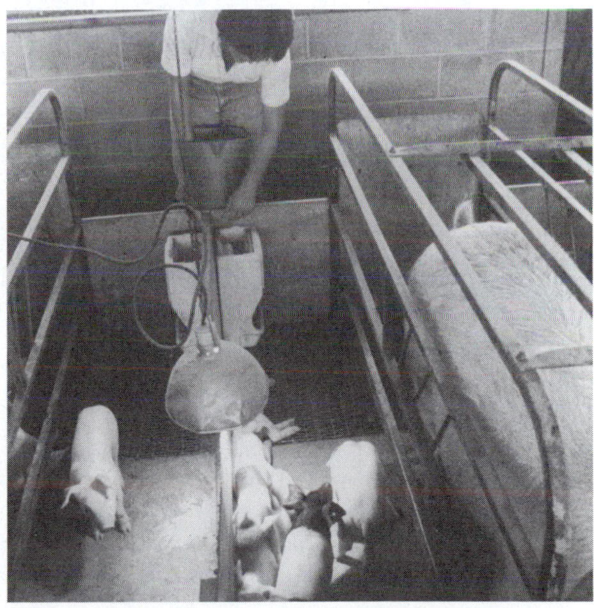

Figure 17-1. Animals must have nourishment above the maintenance level in order to grow, gestate, and/or produce milk. *Rick Jones, Cooperative Extension Service, The University of Georgia.*

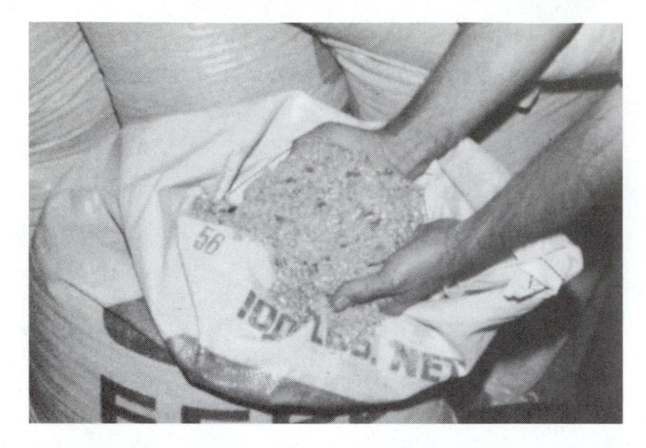

Figure 17-2. A lot of research has gone into developing feeds that provide animals with the proper nutrients. *Dr. Frank Flanders, Agricultural Education, The University of Georgia.*

agricultural operations, producers balance the feeds of their animals to ensure that the proper nutrients are consumed. In confinement, the animals have to eat what the producer gives them. A lot of research has gone into the development of feeds that give animals exactly what they need to remain healthy and to perform at their peak, Figure 17–2. One type of feed may supply several of the needed nutrients, but usually a certain **feedstuff** contains a certain concentration of a particular nutrient. A feedstuff is generally a feed component that producers would not normally give by itself, but combined with other types of feedstuffs, it helps comprise the animal's feed.

Animals must have nutrients in each of six major classes: water, protein, carbohydrates, fats, vitamins, and minerals. Each of these classes of nutrients serves a specific function in the metabolism of the animal. Metabolism refers to all of the chemical and physical processes that take place in the animal's body. These processes provide energy for all of the functions and activities of the animal's body. Metabolism that builds up tissues is called **anabolism**. Examples of anabolism are the maintenance of the body, growth, and tissue repair. Metabolism that breaks down materials is called **catabolism**. An example is the breakdown of food within the digestive system.

WATER

Water is the most abundant compound in the world. Over two-thirds of the world's surface is covered with water. Since this nutrient is essential for sustaining life, animals must have frequent intakes of water to remain alive, Figure 17–3. Even animals such as llamas and camels,

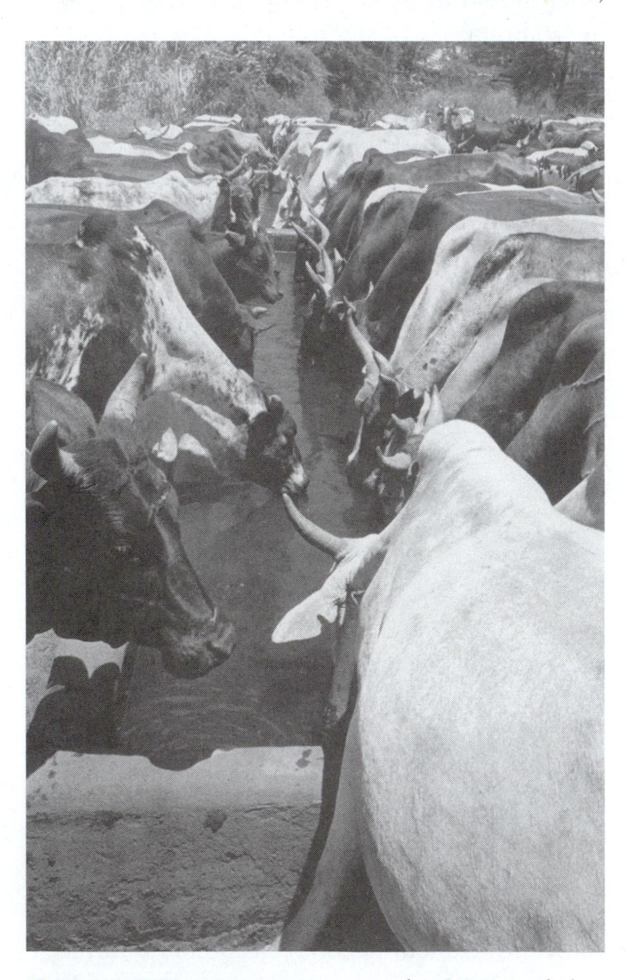

Figure 17-3. Water is an essential nutrient needed to provide all the bodily fluids. *United States Agency for International Development.*

which can go for long periods of time without drinking, have to have water. This nutrient provides the basis for all of the fluid of the animal's body. The bloodstream must be a liquid in order for circulation to occur. Digestion requires moisture for the breakdown of the nutrients and the movement of the feed through the digestive tract. Water is needed to produce milk. It is needed to provide fluid for the manufacture of all the bodily fluids. It provides the cells with pressure that allows them to maintain their shape. It helps the body maintain a constant temperature. Another vital function of water is that of flushing the animal's body of wastes and toxic materials. This nutrient is so vital that over half of the animal's body is composed of water. A loss of 20 percent of this water will result in the death of the animal.

Using ballpark figures, animals generally need about three pounds of water for every pound of solid feed they consume. Some of this water comes in the feed itself. For instance, animals that graze obtain water from the succulent green forages they eat. Some water can be obtained in feeds such as silage that have a relatively high water content. However, most of the water an animal needs comes from the water it drinks. Since water is so essential, producers make sure that animals are given a constant supply of clean water.

Animals may require more water at some periods than at others. A horse that is working hard in hot weather will sweat profusely and will need more water intake to replenish the fluid lost from its body. Likewise, a sow that is nursing a litter of 12 piglets requires a lot of water to produce milk for the young, Figure 17–4.

Figure 17-4. A sow nursing a litter of piglets needs a lot of water.

PROTEIN

Protein can make up to around 15–16 percent of an animal's ration and may be the most costly part of the ration. Proteins are composed of compounds known as **amino acids**. Amino acids are often said to be the building blocks of life because they go into the formation of tissues that provide growth for the animal. Muscle production in particular is dependent on the amino acids found in proteins. All of the enzymes and many of the hormones in the bodies of animals are composed of protein. To a certain degree, protein is also used to provide energy.

Like water, some animals need larger amounts of protein than others. Young, rapidly growing animals need more protein than mature animals. This is because the amino acids in the protein are needed to build muscles, skin, hair, bones, and all of the other cells that go into the growth process, Figure 17–5. A cow that is giving large amounts of milk needs more protein than an animal that is not lactating.

In all, there are over 20 different types of amino acids that an animal's body uses. Of these, there are ten essential amino acids that the animal must obtain from its feed. The other amino acids can be synthesized by the animal's

Figure 17-5. Animals use protein to grow muscle, hair, and other tissues. *USDA photo.*

digestive tract. This means that the 13 nonessential amino acids can be made from the ten that the animal consumes. In this sense, the ten are essential in that they cannot be manufactured by the animal and must be consumed. The following table lists the **essential amino acids** and **nonessential amino acids**.

Essential	*Nonessential*
Arginine	Alanine
Histidine	Aspartic acid
Isoleucine	Citrulline
Leucine	Cystine
Lysine	Glutamic acid
Methionine	Glycine
Phenylalanine	Hydroxyproline
Threonine	Proline
Tryptophan	Serine
Valine	Tyrosine

Animals may not be able to digest all of the protein in a particular feed. The total amount of protein in a feed is called the **crude protein content**. The amount of crude protein in a feed is calculated by analyzing the nitrogen content and multiplying that percentage by 6.25. Digestible protein is the protein in a feed that can be digested and used by the animal. The digestible protein is usually about 50–80 percent of the crude protein.

Of all the ingredients in an animal's feed, protein is usually the most costly. Although protein can be found in most of the feedstuffs, some have a much lower content than others. For example, yellow corn has a protein content of around eight percent. A growing pig may need a ration that consists of 16 percent protein. Being fed corn alone will not give the pig an adequate amount of protein to provide for the building of the body cells to sustain growth. This means that a feedstuff that is higher in protein content will have to be added.

Protein can come from basically two sources: animal and plant. **Carnivores** (animals that eat other animals), such as dogs, cats, and foxes, get almost all of their protein from meat. After all, the muscles in an animal's body are primarily composed of protein and can serve as food for another animal. Omnivores (an animal that eats both plants and animals), such as humans and pigs, can get protein from both plants or animals. Animals that eat only plants are called **herbivores**, and they must get protein exclusively from plants.

Most feedstuffs that are rich in protein come from plant sources. Pigs were once fed slaughterhouse by-products such as **tankage** and blood meal. Recent research has shown that these protein sources are inferior to plant sources in terms of protein that is usable to the animal. Some dried fish meal is fed to hogs as a supplement. Much of plant protein that goes into the feed of animals comes from the vegetable oil industry. Cooking oil is usually pressed from cottonseed, soybeans, peanuts, or corn, Figure 17–6. These seeds are run through huge presses where the oil is squeezed out. The material that is left is in the form of a cake composed of the seeds minus the oil. It is dried and ground into a meal for feed. This material is usually 40–45 percent crude protein and can greatly increase the percentage of protein in a feed. This feedstuff is then mixed with the other feedstuffs in the proper ratio to give the desired protein content for the feed.

Figure 17-6. Soybeans provide protein for animal rations. *John Deere and Company.*

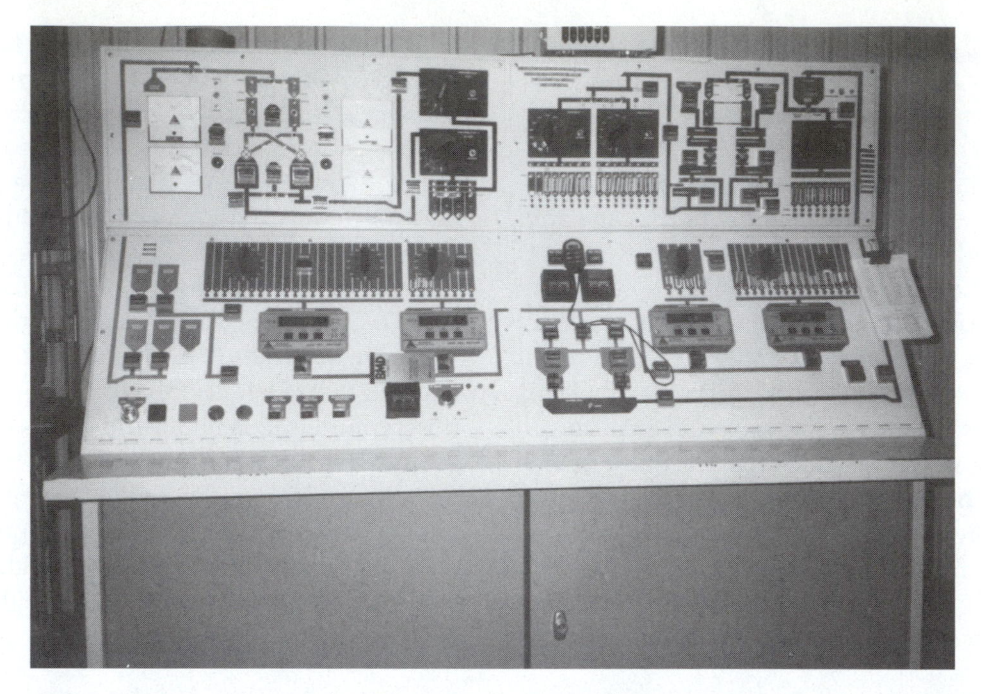

Figure 17-7. Feeds are balanced using computers. This is a computer control panel of a large, modern feed mill. *Dan Rollins, ConAgra Feed Mill, Falkville, Alabama.*

The protein source used is often determined by the animal that is being fed. For instance, pigs are not fed cottonseed meal because this feedstuff contains a substance known as gossypol that is toxic to them. Other needs of the animal also determine the protein source that is used. Modern livestock operations no longer just balance a feed ration based on the percentage of protein. Now the feed formulas are based on the types and amounts of amino acids that are needed by a particular group of animals. The process of balancing feed rations based on amino acid content is so complicated that it is done by computers. Large modern feed mills have computers that control the formulating and blending of the different types of feed, Figure 17–7. Two feedstuffs may have the same percentage of protein and have different percentages of the essential amino acids. Different amino acids are needed for different body functions. For example, a different amino acid is needed for growth than is needed for milk production. Growing animals need a different type of protein supplement than a lactating dam needs. Feeds balanced on the type of amino acids needed by the animal are more cost-efficient, Figure 17–8.

CARBOHYDRATES

The main source of energy from animals comes from carbohydrates. Carbohydrates are compounds made up of carbon, hydrogen, and oxygen. They include sugars, starches, and **cellulose** and are the major organic compounds in plants. Almost all carbohydrates come from plants and are developed by photosynthesis. By weight, plants are composed of about 75 per-

Figure 17-8. Feed samples are constantly monitored to ensure the proper balance and quality of the feed. *Dan Rollins, ConAgra Feed Mill, Falkville, Alabama.*

cent carbohydrates. As will be discussed later, some animals are more efficient than others at making use of these carbohydrates.

Starch is generally found in grain. It is used by the plant as energy storage for the seed. Grains such as wheat and corn contain a lot of starch and therefore a lot of energy for the animal to use. Starches are composed of sugars, and as digestion occurs, the starch is broken down into the component sugars.

There are several different types of sugars. Two broad groups are **monosaccharides** (the simple sugars) and **disaccharides** (the more complex sugars). Simple or complex refers to the chemical composition of the sugar and the different ways the molecules are formed.

There are several common simple sugars (monosacchrides); among these are **glucose**, **fructose**, and **galactose**. Glucose is the simplest of all the sugars and is found in a low concentration in plant materials. It is also the major energy source found in an animal's blood. The animal's body breaks down some of the other sugars into glucose.

Fructose is found in fruits and honey and is the sweetest of all the sugars. Common table sugar (**sucrose**) is a disaccharide composed of fructose and glucose. Galactose is obtained from the breakdown of the disaccharide **lactose** (milk sugar).

Cellulose is the portion of cell walls that gives the plant its rigid structure. The enzymes in an animal's digestive system cannot break down cellulose. However, some animals have microorganisms in their digestive system that break down the cellulose fiber so the enzymes can digest the material. (This will be discussed later in this chapter.)

The most important source of carbohydrates for agricultural animals is grains. Most of the millions of tons of corn grown in this country each year go into the production of livestock feed, Figure 17–9. Other grains such as wheat, oats, and barley are also used. Feeds that are high in grain content are known as concentrates because of the high concentration of carbohydrates.

For horses and ruminant animals, forages grown for grazing and for hay are valuable sources of feed. These food sources are referred to as roughages because of the amount of fiber in the diet. Sometimes a combination of grain and forage is used in the form of silage or similar types of feed, Figure 17–10.

FATS

Fats are part of a group of organic compounds known as **lipids**. These compounds will not dissolve in water but will dissolve in certain organic solvents. Besides fats and oils, lipids also include cholesterol. Fats are found in

Figure 17-9. Most of the millions of tons of corn grown each year go into the production of animal feeds. *John Deere and Company.*

Figure 17-10. Corn is chopped to create silage, which is a combination of roughages and concentrates. *John Deere and Company.*

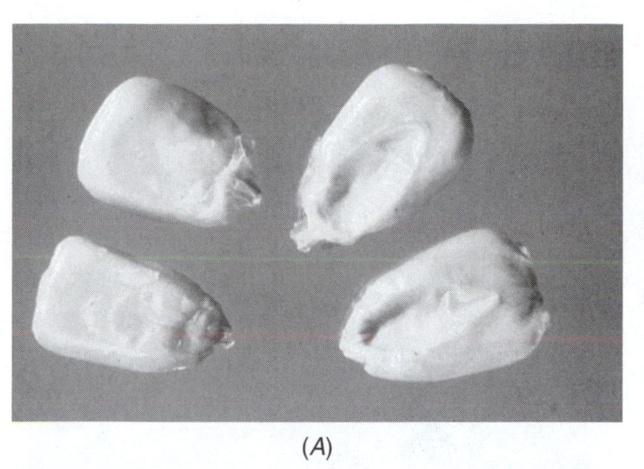

(*A*)

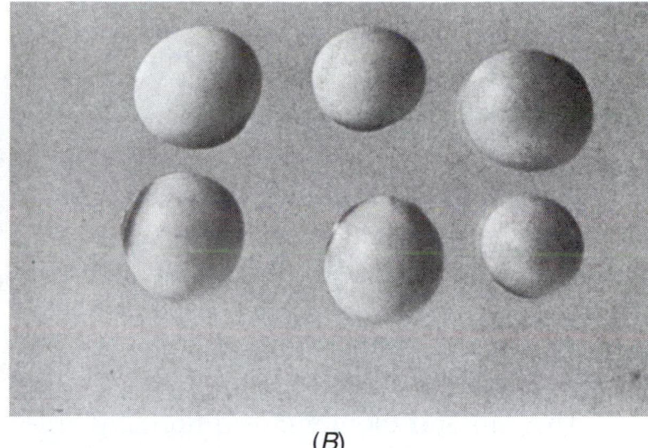

(*B*)

Figure 17-11. Oil within seeds such as corn (*A*) and soybeans (*B*) is an example of plant fats. *American Association of Feed Microscopists.*

both plants and animals. They serve as concentrated storage places for energy. Oil within seeds such as peanuts and soybeans is an example of plant fats, Figure 17–11.

Fats serve the purposes of providing energy for the animal and of storing excess energy. When an animal consumes more energy (especially in the form of fats) than it needs to provide for all the needed bodily functions, the excess is stored in the form of fat. When the body does not take in enough energy to perform the normal bodily functions, these reserves of fat are used.

Certain acids, referred to as the essential fatty acids, are also derived from fats. These acids are necessary in some animals for the production of some hormones and hormone-like substances.

The most important sources of fats in feeds for agricultural animals are the grains that contain oil. Corn and most of the other feed grains contain oil that is used as a fat source by the animals. Some types of animals, pigs for example, may have problems if fed too much oil. Hogs fattened on oily feeds such as whole peanuts may produce soft, oily pork that is not acceptable to the consumers.

MINERALS

Minerals are the only group of nutrients besides water that are **inorganic**. Although they provide only a small portion of the total feed intake, they are vitally important. This group serves the important role of providing structural support for the animal. Bones are formed by a combination of calcium and phosphorus. Another example is eggshells, which are mainly composed of calcium, Figure 17–12.

Figure 17-12. The shells of eggs are mostly composed of the mineral calcium. *Dr. Nicholas Dale, Cooperative Extension Service, The University of Georgia.*

Eggshells are ground up and added to chicken feed as a source of calcium. Animals must have a sufficient intake of these inorganic materials to provide the building materials for their body structure.

In addition to building bones, minerals provide other essential needs. They aid in the construction of muscles, blood cells, internal organs, and enzymes. Animals with a deficiency in minerals never develop properly and are more susceptible to disease.

The mineral elements required by animals include seven **macrominerals** (required in relatively large amounts in the diet) and nine **microminerals** or **trace minerals** (required in very small amounts in the diet). The macrominerals are calcium, chlorine, magnesium, phosphorus, potassium, sodium, and sulfur. The microminerals are cobalt, copper, fluorine, iron, iodine, manganese, molybdenum, selenium, and zinc. These inorganic, crystalline, solid elements make up 3–5 percent of the body on a dry-weight basis, with calcium (approximately one-half the body mineral) and phosphorus (approximately one-fourth the body mineral) accounting for the largest portion of the total mineral content.

Minerals are usually added to the feed of animals in their chemical form. Calcium is sometimes added from other animal sources. For example, ground up oyster shells and eggshells are fed to laying hens to provide materials for their bodies to create strong eggshells.

Minerals are often fed **free choice**. This means that the animals are given free access to the minerals and are allowed to eat all they wish. For cattle, this is done by a mineral box or trough, or by the use of a salt block, Figure 17–13. Essential minerals are in the block, and the animals get them as they lick the block for salt.

Figure 17-13. Cattle are fed minerals "free choice." This mineral box turns with the wind to help keep out the rain.

Vitamins

Vitamins are considered to be micronutrients. This means that the body needs them in very small amounts. Even though only small amounts are required, vitamins are essential for life. They are essential for the development of normal body processes of growth, production, and reproduction. They are also vitally important in providing the animal with the ability to fight stress, disease, and to maintain good health.

Some animals are able to synthesize certain vitamins in their body tissues. Other vitamins cannot be created by the animal from other nutrients and must be obtained from the diet or by microbial synthesis in the digestive system.

There are 16 known vitamins. The B vitamins and vitamin C are water soluble. Fat-soluble vitamins are A, D, E, and K.

Vitamin A is not found in feeds, but it is converted by the animal's body from the provitamin **carotene**, which is found in green, leafy forages from pastures, hay, silage, and dehydrated legumes (alfalfa). Other sources include yellow corn, fish liver oils, and whole milk. Vitamin A can be stored in fats and the liver for several months, to be used when forage quality is low or stress conditions increase the body's demand for vitamin A. Supplementation is usual for ruminants and swine.

Vitamin D is sometimes referred to as the "sunshine vitamin" since both animal and plant sources depend on ultraviolet light to make a form of vitamin D. This form of the vitamin is converted by the liver and kidneys to forms that are usable. Animals make their own vitamin D, and diets of sun-cured forages, yeast, and certain fish oils provide the basis for synthesis, Figure 17–14. Commercial vitamin D is available and generally is made from irradiated yeast. Excessive amounts of vitamin D can reduce an animal's efficiency and is **toxic** in some incidences. Animals in total confinement often receive supplements of vitamin D.

The cereal grains, germ oils, and green forage or hay supply vitamin E. Vitamin E is found in several forms of a complex organic compound called *tocopherol.* Commercially

Figure 17-14. Sun-dried forages can provide the basis for the synthesis of vitamin D. *John Deere and Company.*

produced vitamin E is available for supplemental feeding. There are no known toxic effects from excessive levels in the diet.

Vitamin K is utilized to form the enzyme prothrombin, which in turn helps to form blood clots. Deficiencies rarely occur because vitamin K is synthesized in the **rumen** and in the **monogastric** intestinal tract. Green forages, good-quality hays, fish meal, and synthetic forms of vitamin K can be used to increase the level in the diet.

The B Vitamins

Thiamine is essential as a coenzyme in energy metabolism. Dietary sources of thiamine include green forage or well-cured hays, cereal grains (especially seed coat or bran), and brewer's yeast. Heat processing of grains reduces the amount of available thiamine. Thiamine is usually synthesized in the rumen. The diet of monogastric animals usually provides enough thiamine. However, thiamine is commercially available in vitamin premixes.

Riboflavin is important as a part of two coenzymes that function in energy and protein metabolism. Sources include green forages, leafy hays, or silage; milk and milk products; meat; fish meal; and distiller's or brewer's by-products, Figure 17–15. Commercially available, riboflavin is generally added to swine rations and may be needed in ruminant rations.

Pantothenic acid is a component of coenzyme A, which is important in fatty acid and carbohydrate metabolism. Sources include brewer's yeast, liver meal, dehydrated alfalfa meal, molasses, fish solubles, and most feedstuffs. Commercial sources should be included in the diets of confined animals.

Niacin is part of an enzyme system essential in the metabolism of fat, carbohydrates, and proteins. Niacin is found in animal by-products, brewer's yeast, and green alfalfa. There is some present in most feeds, but the niacin in grains is largely unavailable to nonruminants and supplementation is often needed.

Figure 17-15. Green forages are a good source of vitamins such as vitamin K, thiamine, and riboflavin. *Kentucky Horse Park.*

Pyridoxine is important as a coenzyme component needed for fatty acid and amino acid metabolism. Most feedstuffs are fair-to-good sources of pyridoxine, including cereal grains and their by-products, rice and rice bran, green forages and alfalfa hay, and yeast. Supplementation in animal diets is usually not needed.

Biotin is widely distributed. It is found in large quantities in egg yolk, liver, kidney, milk, and yeast. Biotin is a part of an enzyme involved in the synthesis of fatty acids. It can be readily synthesized by animals and is not deficient in normal farm animals.

Folic acid is needed in body cell metabolism. Folic acid is found in green forages, such as alfalfa meal, and in some animal proteins.

The animal body synthesizes some folic acid, and although it is available in synthetic forms, supplements are not greatly needed in farm animals.

Choline is found as a component of fats and nerve tissues and is needed in greater levels than other vitamins. Most commonly used feeds are good sources of choline. Choline is synthesized in the animal body when other vitamins such as B_{12} are abundant.

B_{12} functions as a coenzyme in several metabolic reactions and is an essential part of red blood cell maturation. Synthesis of vitamin B_{12} requires cobalt. Sources of B_{12} include protein feeds of animal origin and fermentation products. Most swine rations are supplemented with vitamin B_{12}.

Inositol is found in all feeds and synthesized in the intestinal tract, so is not generally needed as a supplement.

Although the function is not well known, paraaminobenzoic is synthesized in the intestine and is usually not deficient in livestock rations.

Vitamin C is essential in the formation of the protein collagen. Vitamin C is found in citrus fruits; green, leafy forages; and well-cured hays. Animals normally can synthesize sufficient quantities of vitamin C to meet their needs.

The Digestion Process

Animals use feed nutrients on a cellular basis—all of the different nutrients that an animal takes in must be converted to a form that the cells in the body can use. Once this conversion (digestion) is completed, the nutrients must be transported to the cells where they are needed. The system that performs this task is referred to as the digestive system. The organs that make up this system are known as the **gastrointestinal tract** (GI tract).

The gastrointestinal tract is also referred to as the **alimentary canal**. This is the tract reaching from the mouth to the anus, through which feed passes following consumption and where it is exposed to the various digestive processes. The digestive system includes the various structures, organs, and glands involved with the procuring, chewing, swallowing, digestion, **absorption**, and excretion of feedstuffs. Although there are similarities in the digestive systems, farm animals are often classified according to the nature of their digestive system. There are basically two types of digestive systems. One is known as monogastric (single-compartment stomach), Figure 17–16; and the other is known as ruminant (multicompartment stomach). Following are descriptions of

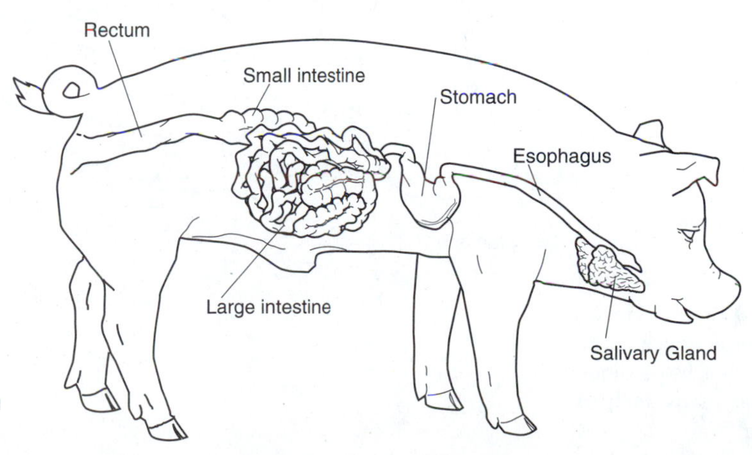

Figure 17-16. The pig has a monogastric digestive system.

processes in the two digestive systems in the order of occurrence.

Monogastric Digestive Systems

Monogastric systems are those that have only one-compartment stomachs. These include the pig, horse, dog, cat, and birds. The horse has an enlargement, known as a **cecum**, that enables it to utilize high-fiber feeds by means of microbial fermentation, much as do ruminants, Figure 17–17. This means that the horse is not typical of monogastric animals. Simple-stomach (monogastric) animals are not capable of digesting large amounts of fiber and are usually fed concentrate feeds.

The digestive process begins in the mouth, which is the first organ of the digestive tract. Within the mouth, the tongue is used for grasping the food, mixing, and swallowing. The teeth are used for tearing and chewing the feed. This is the first step in the process of breaking down the feed into fine particles. The mouth also contains salivary glands, which consist of three pairs of glands that excrete saliva. Saliva contains several substances: water to moisten, mucin to lubricate, bicarbonates to buffer acids in the feeds, and the enzyme amylase to initiate carbohydrate breakdown.

The **esophagus** is a hollow, muscular tube that moves food from the mouth to the stomach. This is accomplished by muscular contractions that push the food along.

The stomach is a hollow muscle that causes further breakdown of foods by physical muscular movement. The food is pressed together and massaged by the movement of muscles in this area. In the stomach, food is also broken down by chemical action. The walls of the stomach secrete hydrochloric acid that begins to dissolve the food. Another secretion, **pepsin**, begins to break down proteins into the amino acids. The secretion rennin acts to curdle the casein in milk. Gastric lipase causes the breakdown of fats to fatty acids and glycerol.

The small intestine is a long, hollow tube that leads from the stomach to the large intestine. This organ is made up of several parts: the **duodenum**, the **jejunum**, and the **ileum**.

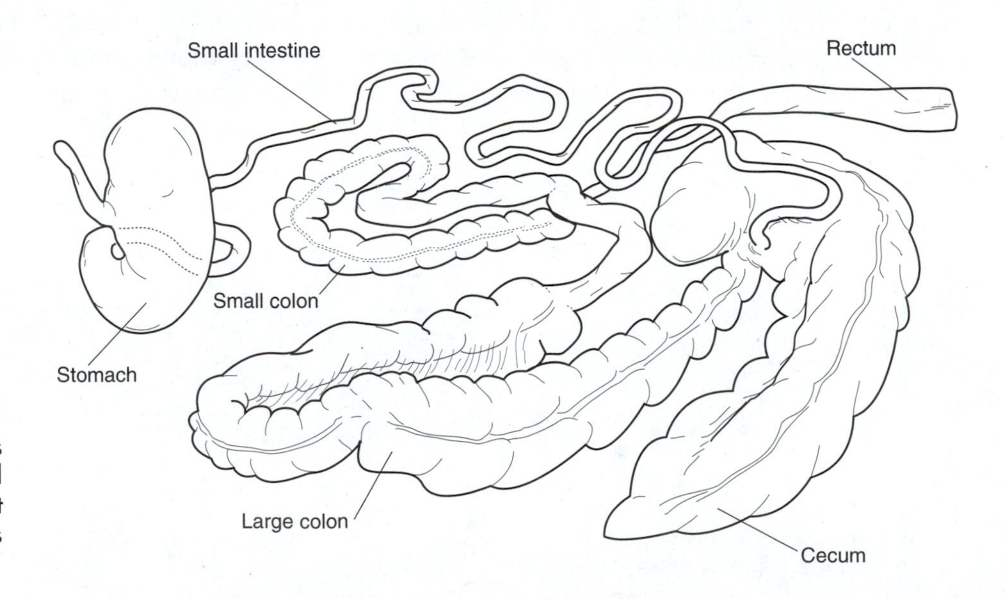

Figure 17-17. Horses have a large pouch called a *cecum* that helps digest high-fiber feeds such as hay.

The entrance to the small intestine is controlled by a sphincter muscle that helps move food into and through the tract.

The first segment of the small intestine is the duodenum. The duodenum receives secretions from the pancreas, which act to break down proteins, starches, and fats. Here the intestinal walls secrete intestinal juices that contain enzymes that further the process of breaking down the food.

The next segments of the small intestine are the jejunum and the ileum. These are the areas of nutrient absorption. Absorption is the process by which the nutrients are passed into the bloodstream. The villi (small fingerlike projections) in these areas facilitate absorption into the bloodstream and/or lymph system through membranes that surround the villi, Figure 17–18. This type of membrane is called a **semipermeable membrane**. This means that the membrane will allow particles to pass through in a process called **diffusion**.

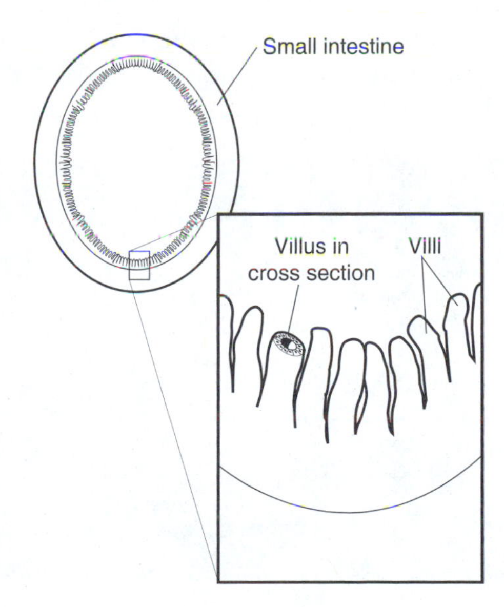

Figure 17-18. The small intestine is lined with fingerlike projections, called *villi*, that absorb nutrients into the bloodstream.

The last organ of the digestive tract is the large intestine. This organ is divided into three sections. The first section is the cecum, which is a blind pouch. A cecum is of little function in most monogastric animals. However, in some animals, such as the horse, this area is where fibrous food such as hay and grass is broken down into usable nutrients.

The second segment of the large intestine is the colon, which is the largest part of the organ. Its function is to provide a storage space for wastes from the digestive process. Here water is removed from the wastes and some microbial action begins on fibrous materials.

The rectum is the final segment of the large intestine and the final part of the digestive system. It serves to pass waste material through to the anus where it is eliminated.

Ruminant Digestive System

Animals such as cows and sheep have multicompartment stomachs that allow them to use high-fiber feeds such as grasses and hays, Figure 17–19. These animals are often called "cud chewers" because they regurgitate **boluses** of feed, consumed earlier, and remasticate (chew) and reswallow. The digestive systems of ruminants differ from monogastric systems in several ways.

In the mouth of ruminants there are no upper front teeth. Instead there is a dental pad that works with the lower incisors for tearing off forages and other feedstuffs. The upper and lower jaw teeth (molars) enable the animal to chew on one side of the mouth at a time. Large quantities of saliva are produced. This saliva is highly buffered and provides phosphorus and sodium for the rumen microorganisms. Unlike most monogastric animals, there are no enzymes in the saliva, but there is some urea released that provides nitrogen for the bacteria in the rumen.

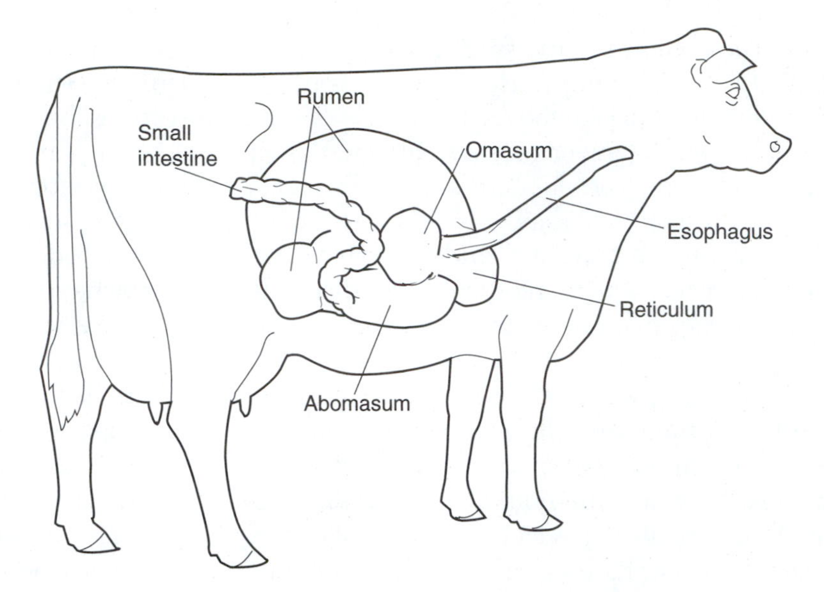

Figure 17-19. Cattle have four compartments to their stomachs. This allows them to use large amounts of roughages.

The stomach has four compartments. It consists of the **rumen** (paunch), **reticulum** (honeycomb), **omasum** (many piles), and the **abomasum** (true or glandular stomach). In the young ruminant animal, an esophageal groove or heavy muscular fold allows milk from the suckling animal to bypass the rumen and reticulum to the omasum.

Compartments of the Ruminant Stomach

The esophagus leads to both the reticulum and the rumen. As ruminants graze, they tend to pick up hard, indigestible objects such as small stones, nails, and bits of wire. These heavy materials fall into the reticulum. The walls of the reticulum are made up of **mucous membranes** that form subcompartments with the appearance of a honeycomb. These small compartments trap and provide a storage place for "hardware" that does not float, Figure 17–20. This prevents dangerous objects from proceeding through the rest of the digestive tract.

The reticulum also functions to store, sort, and move feed back into the esophagus for regurgitation or into the rumen for further digestion. The process of breaking down roughages begins with a contraction of the reticulum and muscles in the esophagus to move roughage and fluid to the mouth. Excess fluid is squeezed out and the material is reswallowed.

After the material is reswallowed, it moves to the rumen. The rumen functions as a storage vat where food is soaked, mixed, and ferment-

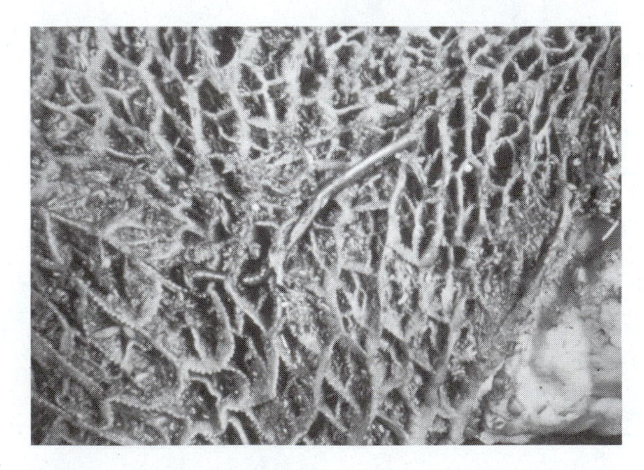

Figure 17-20. The reticulum collects hard indigestible objects. Note the nail. *North Dakota Department of Education.*

ed by the action of bacteria. The hollow, muscular paunch fills the left side of the abdominal cavity and contains two sacks, each lined with papillae (nipplelike projections) that aid in the absorption of nutrients. Bacteria thrive in the rumen environment and function to break down fibrous feeds. Carbohydrates are broken down into starches and sugars. Volatile fatty acids are released as the carbohydrates are broken down, and these fatty acids are absorbed through the rumen wall to provide body energy.

Bacteria also use nitrogen to form amino acids and eventually proteins. The bacteria can also synthesize water-soluble vitamins and vitamin K. By-products of the microbial activity include methane and carbon dioxide. A small portion of these gases is absorbed by the blood, but much of the gas is eliminated by belching. Belching occurs when the upper sacs of the rumen force gases forward and down so the esophagus can dilate and allow gases to pass. If gases are not eliminated due to froth or foam blocking the esophagus, a condition called **bloat** (an inflation of the rumen) will sometimes occur.

After leaving the rumen, the food material passes through to the omasum. The omasum is a round organ on the right side of the animal and to the right of the rumen and reticulum. The omasum grinds roughage using blunt muscular papillae that extend from many folds of the omasum walls.

The last compartment of the ruminant's stomach is the abomasum. The abomasum is the only glandular (true stomach) stomach of the ruminant. The abomasum is located below the omasum and extends to the rear and to the right of the rumen. This compartment functions similarly to the stomach of monogastric animals. By the time food materials reach the abomasum, the fiber of the roughages have been broken down to the extent that they can be handled by the abomasum. The small and large intestines of the ruminant animal function much the same way as they do in the monogastric animal.

SUMMARY

In order for animals to grow and thrive, they must have the proper nutrition. An understanding of how the digestive process works in different types of animals is essential in order to formulize the proper diet for animals. Factors such as species, age, sex, and environmental conditions all must be taken into account in formulating rations. Today's feeds are scientifically balanced to give animals exactly what they need in each particular circumstance. Research scientists are continuing to find new and better ways of providing nutrients to animals.

◆ Review Exercises

TRUE OR FALSE

1. A feedstuff is generally a feed component that producers would not normally give to an animal exclusive of other feedstuff. ______.
2. Water, although an essential nutrient, is needed only in small amounts and makes up only a tiny fraction of the animal's body. ______.
3. Protein, composed of amino acids, may be the most costly part of the animal's ration. ______.
4. Digestible protein, that protein in a feed that can be digested and used by the animal, makes up about 20–25 percent of the crude protein. ______.

5. Carnivores get almost all their protein from meat, omnivores can get protein from meat and plants, and herbivores get protein exclusively from plants. ______.
6. The main source of tissue-building proteins is carbohydrates. ______.
7. Photosynthesis accounts for almost all carbohydrate development. ______.
8. Of all the millions of tons of corn grown in this country each year, only a small amount is used for feed grain. ______.
9. Fats can only be obtained from animals. ______.
10. Animals with a deficiency in minerals never develop properly and are more susceptible to disease. ______.
11. Both animal and plant sources depend on ultraviolet light to make a form of vitamin D. ______.
12. It is possible for animals to ingest too many vitamins, and harm can result. ______.
13. The terms gastrointestinal tract and alimentary tract refer to the same system. ______.
14. The small intestine consists of three parts called the duodenum (which breaks down the food), the jejunum, and the ileum (both of which are areas of absorption). ______.
15. Ruminant animals do not have small and large intestines. ______.

FILL IN THE BLANKS

1. The maintenance ration is the level of ____________ needed by the __________ to maintain ____________ weight and not ____________ or gain weight.
2. Metabolism refers to all of the ____________ and ____________ processes that take place in the ____________ body.
3. Metabolism that builds up tissue (such as growth and __________ repair) is called ____________; metabolism that breaks down material (such as the breakdown of ____________ within the ____________ system) is called ________________.
4. Young, rapidly growing animals need more protein than ____________ animals because the __________ acids are needed to build muscles, __________, ___________, bones and all of the other cells that go into the __________ process.
5. Feed formulas are now based on the __________ and __________ of __________ acids that are needed by a particular group of ____________.
6. Carbohydrates are compounds made up of ____________, hydrogen, and __________ and include sugars, ______________, and ____________.
7. Simple or complex sugars refer to the ____________ composition of the sugar and the different __________ the ____________ are formed.
8. When an animal consumes more energy (especially in the form of ___________) than it needs to ______________ for all the necessary bodily ______________, the excess is stored in the form of ______________.
9. Although minerals provide only a small portion of the total ___________ intake, they are vitally important in providing ____________ support for the ________________.
10. Although vitamins are ________________—needed only in small amounts—they are essential in the development of normal body ____________ of health, ____________, production, and ______________.
11. Good sources for pyridoxine include ___________ grains and their by-products, __________ and rice bran, __________ forages and _________ hay, and ___________.
12. Choline is found as a component of _______ and ____________ tissues and is needed in _________ levels than other ____________.

13. Simple-stomach or ______________ animals are not capable of ______________ large amounts of ___________ and are usually fed _______________ feeds.
14. Ruminants such as the cow and ___________ have ________________ stomachs that allow them to use high- _____________ feeds such as grasses and ______________.
15. The esophagus in a ruminant animal leads both to the reticulum (which serves to store _________________ and functions to store, sort, and ____________ feed back into the esophagus for ________________ or into the rumen for further ______________) and the rumen (which functions as a storage __________ where food is soaked, _________, and ____________).

DISCUSSION QUESTIONS

1. What is the difference between a feed and a feedstuff?
2. Define metabolism.
3. List at least four functions water plays in sustaining life.
4. Explain why a young animal needs more protein than a mature animal does.
5. List the sources of protein for livestock feeds.
6. From what source do all carbohydrates come?
7. What purpose do fats serve?
8. What essential needs do minerals provide?
9. List the vitamins that are needed by animals.
10. List the parts of the monogastric digestive system and briefly describe the function of each.
11. List the compartments of the ruminant stomach and describe the function of each.

STUDENT LEARNING ACTIVITIES

1. Locate and report on an article telling about research that has been completed on livestock feeds. Explain what practical differences you feel this research will make.
2. Obtain the tag from a bag of feed. Make a list of the ingredients and tell which nutrients are derived from each component.
3. From a slaughterhouse, obtain the digestive tract of a monogastric animal (pig) and a ruminant (sheep or cow). Dissect the tract and identify all of the parts. Be sure to wear latex gloves when the organs are handled.

Chapter 18

Meat Science

Student Objectives in Basic Science

As a result of studying this chapter, you should be able to

1. Describe physiological processes that take place in an animal's body at death.
2. Describe the process of ossification.
3. List the different types of tissue that make up muscles.
4. Explain the factors that affect the sensation of taste.
5. Explain why meat is so highly perishable.
6. Discuss the types of microbes that cause spoilage.
7. List the factors that favor the growth of microbes.
8. Discuss the scientific principles behind the preservation of meats.

Student Objectives in Agricultural Science

As a result of studying this chapter, you should be able to

1. Explain the steps in the slaughter of meat animals.
2. Distinguish between quality grading and yield grading of carcasses.
3. List the wholesale cuts of beef, pork, and lamb.
4. Discuss the factors that affect the palatability of meats.
5. Discuss the various methods used to preserve meats.

Key Terms

wholesale cuts
immobilization
kosher
exsanguination
chine
shroud
rigor mortis
aging
primal cuts
adipose
mastication
elastin
oxidation
rancid
microbes
psychrophiles
thermophiles
mesophiles
aerobic organisms
anaerobic organisms
facultative
dry curing
injection curing
combination curing
blast freezing
radiation
irradiation
E. Coli.

THE MEAT INDUSTRY

Americans are a nation of meat eaters. Each year the average person in this country consumes 48 pounds of beef and veal, 29 pounds of pork, and 24 pounds of poultry. Very few nations in the world even come close to us in the per capita consumption of meat.

In addition, when compared to the rest of the world, Americans spend a small percentage of their annual income for food. This means that they can afford to buy the type of food they prefer and, obviously, they prefer meat.

Lean meat is very dense in nutrients and a pound of meat may equal or surpass the nutritive content of the feed that it took to produce the meat. Meat is among the most nutritionally complete foods that humans consume. Foods from animals supply about 88 percent of the vitamin B_{12} in our diets because this nutrient is very difficult to obtain from plant sources. In addition, meats and animal products provide 67 percent of the riboflavin, 65 percent of the protein and phosphorus, 57 percent of the vitamin B_6, 48 percent of the fat, 43 percent of the niacin, 42 percent of the vitamin A, 37 percent of the iron, 36 percent of the thiamin, and 35 percent of the magnesium in our diets.

In recent years, the trend in meat consumption has changed rapidly. Modern consumers want meat products that are already cooked or are ready to place in the oven or microwave. Meat products such as chicken wings that were once considered to be almost by-products are now packaged fully cooked, frozen, and ready to place in the microwave oven. Individually wrapped chicken parts, chops, and steaks that are ready to go on the grill are all gaining in popularity with consumers. The meat industry is changing rapidly to accommodate the wishes of the consumer.

Figure 18-1. Animals are processed into meat in large, automated plants. *Dr. Estes Reynolds, Cooperative Extension Service, The University of Georgia.*

The vast majority of the agricultural animals raised in this country are produced for meat. Meat is defined as the edible flesh of animals and may include the muscles, certain internal organs, such as the liver, and other edible parts. A huge industry has developed around processing animals into edible products, Figure 18–1.

In this country there are over five thousand plants that slaughter animals. The trend in recent years is for the slaughter plants to become larger. Eighty to ninety percent of the animals slaughtered under federal inspection are processed in less than 200 plants. These plants may slaughter over 100 thousand animals in a single year.

In the larger plants, the slaughter process is highly organized and automated. Like an assembly line, workers perform only a few tasks with the animal or carcass, and the product goes on down the line to the next person.

When the animals are brought in, they must be inspected and slaughtered. The head, entrails, hair or hide, and feet must be removed.

Figure 18-2. The carcass is the part of the animal left after the head, entrails, hide (or hair), and feet are removed. *Cooperative Extension Service, The University of Georgia.*

What is left is referred to as the carcass, Figure 18–2. The carcass is then cut into **wholesale cuts** and sold to grocery stores and other outlets where meat cutters divide the wholesale cuts into retail cuts. The consumer buys the retail cuts that are ready for cooking.

The Slaughter Process

After the live animals are inspected (see Chapter 22), they are brought down a chute onto the slaughter floor.

The first of several steps in the slaughter process is **immobilization**. The purpose of immobilization is to render the animal unconscious so that it feels no pain. This must be done in such a way as to allow the heart to continue to pump in order to drain the animal's body of blood. Regulations for the immobilization process are set by the Federal Humane Slaughter Act. Immobilization may be accomplished in several ways. The animal may be placed in a chamber of carbon dioxide until it goes to sleep from lack of oxygen. Some slaughter plants use electric shock to render the animal unconscious. Others use a cartridge or a mechanical bolt that is driven by compressed air to stun the animal.

Livestock slaughtered for **kosher** markets are exempt from the requirement of stunning under the Humane Slaughter Act. Kosher means that the animal is slaughtered under the regulations of Jewish religious laws. In any case, humane slaughter must be performed quickly and with as little stress as possible.

After immobilization, the animal must be bled quickly to keep it from regaining consciousness and to prevent hemorrhaging due to a rise in blood pressure. This process, known as **exsanguination**, is usually done by severing the jugular vein with a sharp knife. Unless bleeding is accomplished within a few seconds of immobilization, the blood pressure may cause hemorrhaging (escape of blood from a ruptured blood vessel), resulting in blood spots in the meat. As soon as the blood pressure begins to drop, the heart speeds up to maintain pressure. This action fills many of the organs with blood and thus blood loss at exsanguination is only about 50 percent of the total volume of blood.

The animal is hoisted on a rail and the hide and entrails are removed, Figure 18–3. Pigs that have been slaughtered may be dipped into scalding water and placed on a machine that scrapes the hair from the hide; other slaughter plants may skin the pigs. When the internal organs are removed, care is taken to preserve those that are to be used for food.

The liver is the most commonly eaten of the internal organs, Figure 18–4. The brains, the pancreas (sweetbread), intestines (chitterlings or tripe), and the heart are all used for human food. In some parts of the world the kidneys are also used.

During the slaughter process, federal inspectors are on hand to inspect the internal

Figure 18-3. The animals are hoisted onto a rail and the entrails are removed. *Dr. Estes Reynolds, Cooperative Extension Service, The University of Georgia.*

organs and the carcass to detect any health concern over the meat from the animal, Figure 18–5. If a problem is found, the carcass may be condemned as unfit for human consumption. Each carcass that is to be sold for human consumption must be inspected.

Beef carcasses are generally sent into the cooler in halves called *sides of beef.* After slaughter, the carcass is sawed down the backbone (called the **chine**) and the carcass is divided in two pieces. In very large carcasses, each side may be divided into two quarters. To do this the side is divided between the twelfth and thirteenth rib into the fore and hind quar-

Figure 18-4. Livers are an example of internal organs that are used for food. *Dr. Estes Reynolds, Cooperative Extension Service, The University of Georgia.*

Figure 18-5. Federal inspectors examine the internal organs of each carcass. *Dr. Estes Reynolds, Cooperative Extension Service, The University of Georgia.*

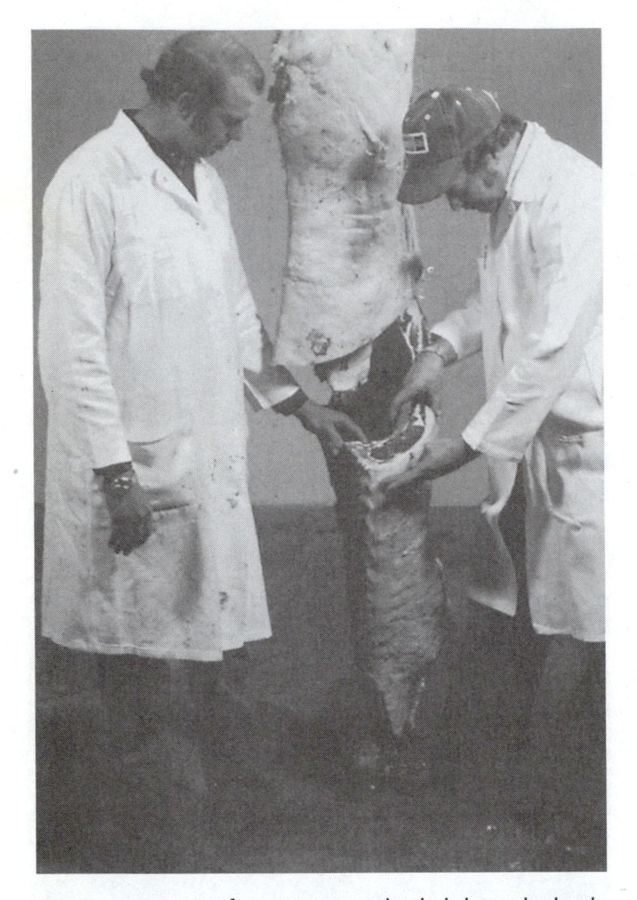

Figure 18-6. Beef carcasses are divided down the backbone and between the 12th and 13th ribs. *James Strawser, Cooperative Extension Service, The University of Georgia.*

ter, Figure 18–6. Lamb carcasses are generally sent to the cooler whole. Hog carcasses are usually divided in half down the backbone. The carcasses of beef and pigs that have been skinned are covered in a heavy cloth soaked in salt water. The purpose of this covering (called a **shroud**) is to prevent the carcass from drying out.

After slaughter, the carcasses need to be cooled down rapidly. Here the carcasses undergo a process known as **rigor mortis** in which the muscles lock into place and the carcass becomes stiff. The physiology of rigor mortis is similar to muscle contractions in a live animal, except in the carcass the muscles do not relax. The onset of rigor mortis usually takes from six to twelve hours for beef and lamb; thirty minutes to three hours for pork. As enzymes and microorganisms begin to break down the muscle tissue, rigor mortis is partially relaxed.

It is usually desirable to reduce muscle temperature as quickly as possible after death to minimize protein degradation and inhibit growth of microorganisms. Pork and lamb carcasses are usually cooled for 18 to 24 hours before they are cut into wholesale cuts. Larger beef carcasses may require 30 or more hours of cooling before they are ready to be cut, Figure 18–7.

Figure 18-7. After slaughter, the carcasses are chilled down by hanging in the cooler. *Dr. Estes Reynolds, Cooperative Extension Service, The University of Georgia.*

Higher quality beef carcasses may be aged in the cooler for as much as a week. The carcasses undergo a period of **aging** to allow enzymes and microorganisms to begin the process of breaking down the tissue. This improves tenderness and flavor, but adds to the expense of processing the meat.

An alternative to aging is the use of electrical stimulation of the muscles. A current of 600 volts is sent through the carcass immediately after slaughter and before the hide is removed. The stimulation speeds up natural processes that occur after death, like the depletion of energy stores from the body. Although this process affects the tenderness of the meat only slightly, there are other benefits such as improved color, texture, and firmness. This process also makes the hide easier to remove.

GRADING

After the carcasses are cooled, they are graded according to USDA standards. Federal Meat Grading was officially established in 1925 and is administered by the Agricultural Marketing Service (AMS) of the United States Department of Agriculture (USDA). The meat grade certifies the class, quality, and condition of the agricultural product examined to conform with uniform standards. Quality grades are a prediction of the eating quality (palatability) of properly prepared meats. Yield grades indicate expected yield of edible meat from a carcass and the subsequent wholesale cuts from that carcass. Grading is voluntary and is paid for by the meat packer.

Quality grades of beef are as follows: prime, choice, select, standard, commercial, utility, cutter, and canner. Grades are determined by the age of the animal at slaughter and the amount of fat intermingled with the muscle fibers. The age is determined by the maturity of the cartilage and bones in the carcass. Remember that as an animal ages, the cartilage hardens and turns to bone. Graders inspect the rib cage and vertebrae of the carcass for the degree of bone and cartilage hardening (called *ossification*), Figure 18–8.

As the animal ages, vertebrae in the lower end of the backbone tend to fuse or grow together. By determining the degree of ossification, graders are able to classify the animal according to its maturity. Animals that appear to be older than about 42 months in age cannot receive the highest two grades (prime and choice). Younger animals are usually more tender than older animals.

The amount of fat amongst the muscle fibers is determined by examining the longissmus dorsi muscle that runs the length of the vertebrae on each side. The carcass is separated between the twelfth and thirteenth ribs to expose this cross section. The fat, known as

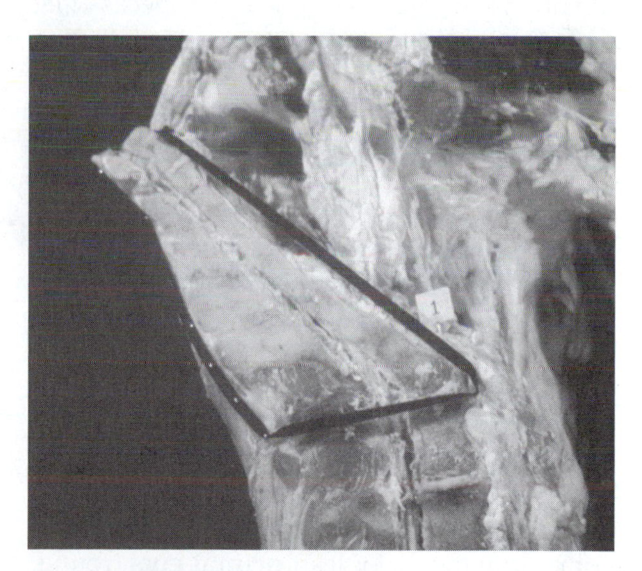

Figure 18-8. Graders determine the age of an animal by examining the vertebrae. Note how the bones of the vertebrae are fused together. This indicates the carcass of an older animal. *Dr. Estes Reynolds, Cooperative Extension Service, The University of Georgia.*

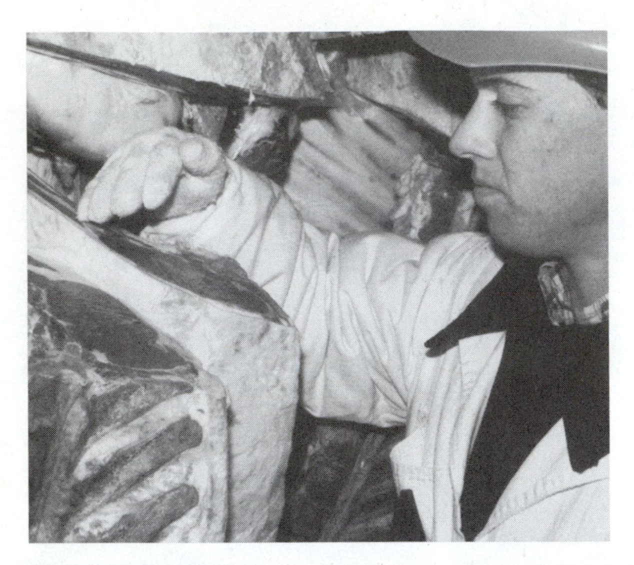

Figure 18-9. A USDA grader inspects the amount of marbling in the rib eye of a beef carcass. *North American Limousin Association.*

Figure 18-10. A USDA grader uses a rib eye grid to measure the area of the rib eye. *North American Limousin Association.*

marbling, shows up as specks of white across the rib eye, Figure 18–9. The more specks of fat that are visible, the higher the grade (assuming maturity is the same).

Beef that grades prime has the highest degree of fat in the muscle. Fat is what gives meat its flavor and juiciness. Fat is expensive to put on animals, so the leaner grades are usually less expensive. Most feedlot owners want their animals to grade a low choice at slaughter. Those that feed animals to grade prime usually cater to the restaurant trade. Most beef bought in the grocery store is choice grade, although a few market chains are selling the leaner select grade as a low-fat meat.

Yield grade is an estimate of the percentage of boneless, closely trimmed retail cuts that come from the major lean **primal cuts** (round, loin, rib, and chuck). USDA yield grades for beef are as follows:

1. over 52.3 percent lean primal cuts
2. 50.0–52.3 percent lean primal cuts
3. 47.7–50.0 percent lean primal cuts
4. 45.4–47.7 percent lean primal cuts
5. less than 45.4 percent lean primal cuts

The yield grade is determined by a formula used by the grader. Factors used in the formula are the chilled carcass weight, the amount of internal (kidney, pelvic, and heart) fat, the size of the rib eye area, and the amount of backfat on the carcass, Figure 18–10 and Figure 18–11.

The Wholesale Cuts

After the carcasses have been chilled for the proper time, they are taken out of the cooler and cut into pieces that are sold to retail outlets. These cuts, known as *primal cuts,* are packaged in vacuum packs, placed in boxes, and shipped to the retailers. One advantage of this is that around 30 percent of the carcass weight is trimmed away with the excess fat and bone. This saves in the shipping costs. Boxed primal cuts are also easier to handle than are

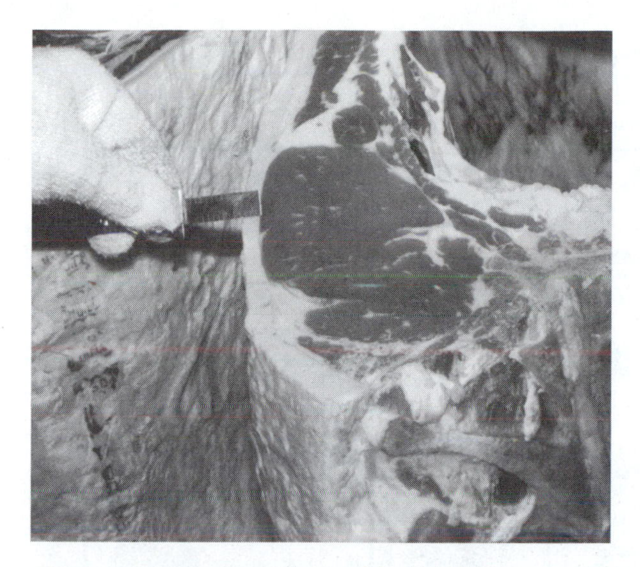

Figure 18-11. The thickness of the backfat is measured at the rib eye. *North American Limousin Association.*

sides of beef or pork. Retailers also see this as an advantage because they can order the cuts of meat they need without getting parts they have no market for, Figure 18–12.

Wholesale or primal cuts of beef usually consist of the chuck, loin, rib, and round. Pork cuts include shoulder, loin, sides, and ham. Lamb cuts include shoulder, rib, loin, and leg. Primal cuts are sometimes divided into smaller units called *subprimal cuts.*

The wholesale cuts are divided into the retail cuts. These are the cuts of meat that the consumer buys at the grocery store. They are sized into portions that can be easily cooked and eaten without further cutting or trimming, Figure 18–13a-c. The most expensive retail cuts usually come from the loin area, Figure 18–14. This muscle group is usually the most tender of the muscle groups. This is the area from which chops and steaks such as T-bones come.

As the retail cuts are made, there is always muscle that is trimmed or portions that do not make good retail cuts of meat. These portions or trimmings are made into sausage or ground meat, Figure 18–15. Often the sausage is spiced and preserved by drying or smoking. Ground beef becomes hamburger meat.

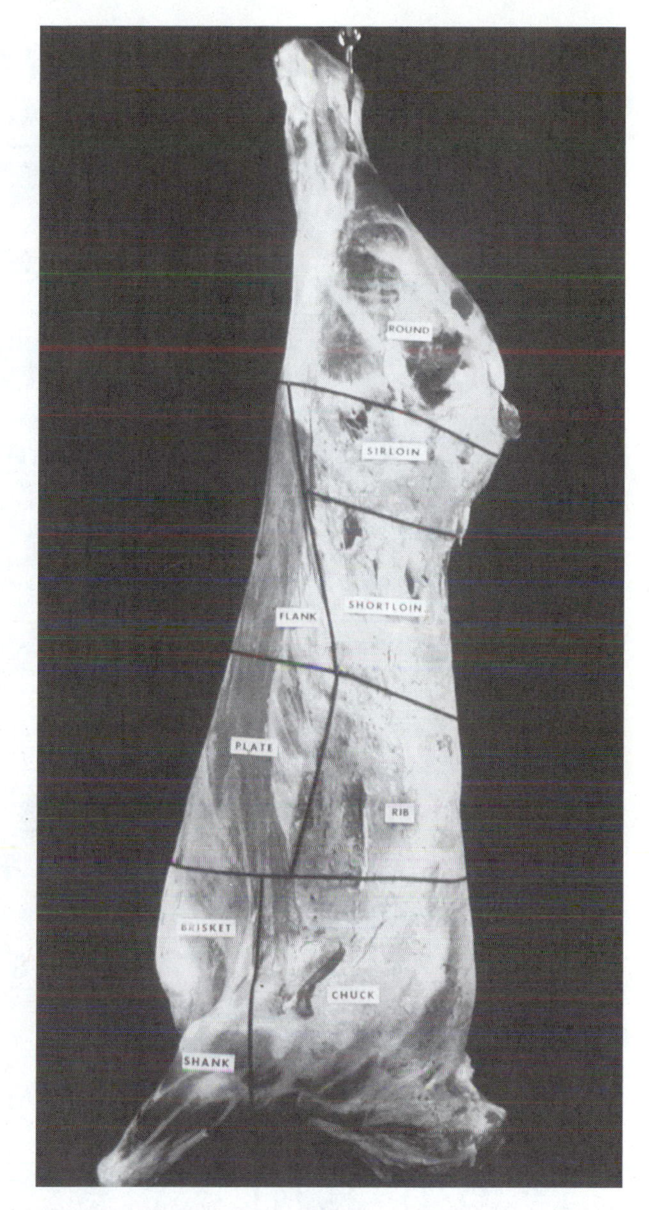

Figure 18-12. Carcasses are divided into wholesale cuts. *Dr. T. F. Price, Michigan State University.*

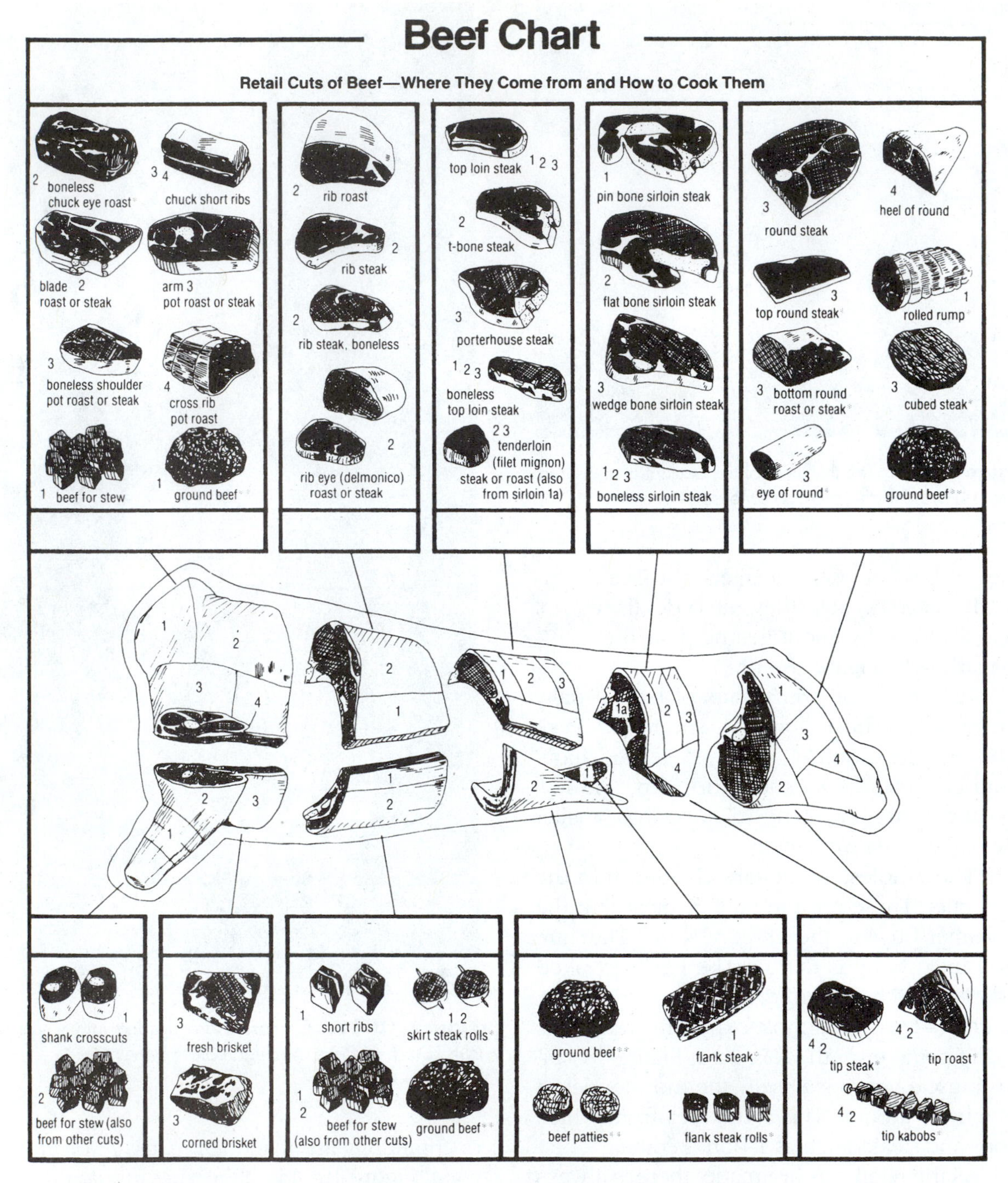

Figure 18-13A. Retail cuts of beef—where they come from and how to cook them.

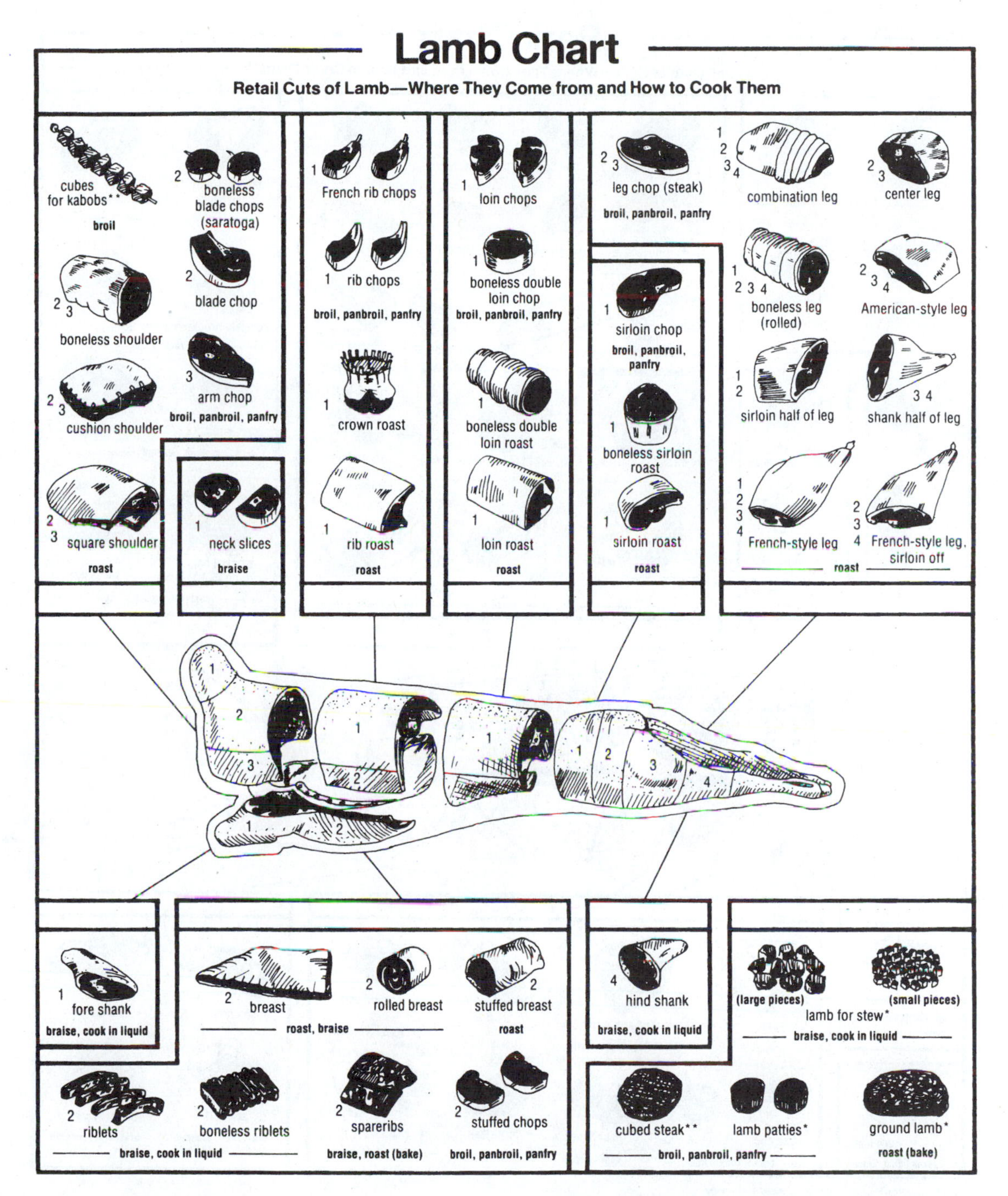

Figure 18-13B. Retail cuts of lamb—where they come from and how to cook them.

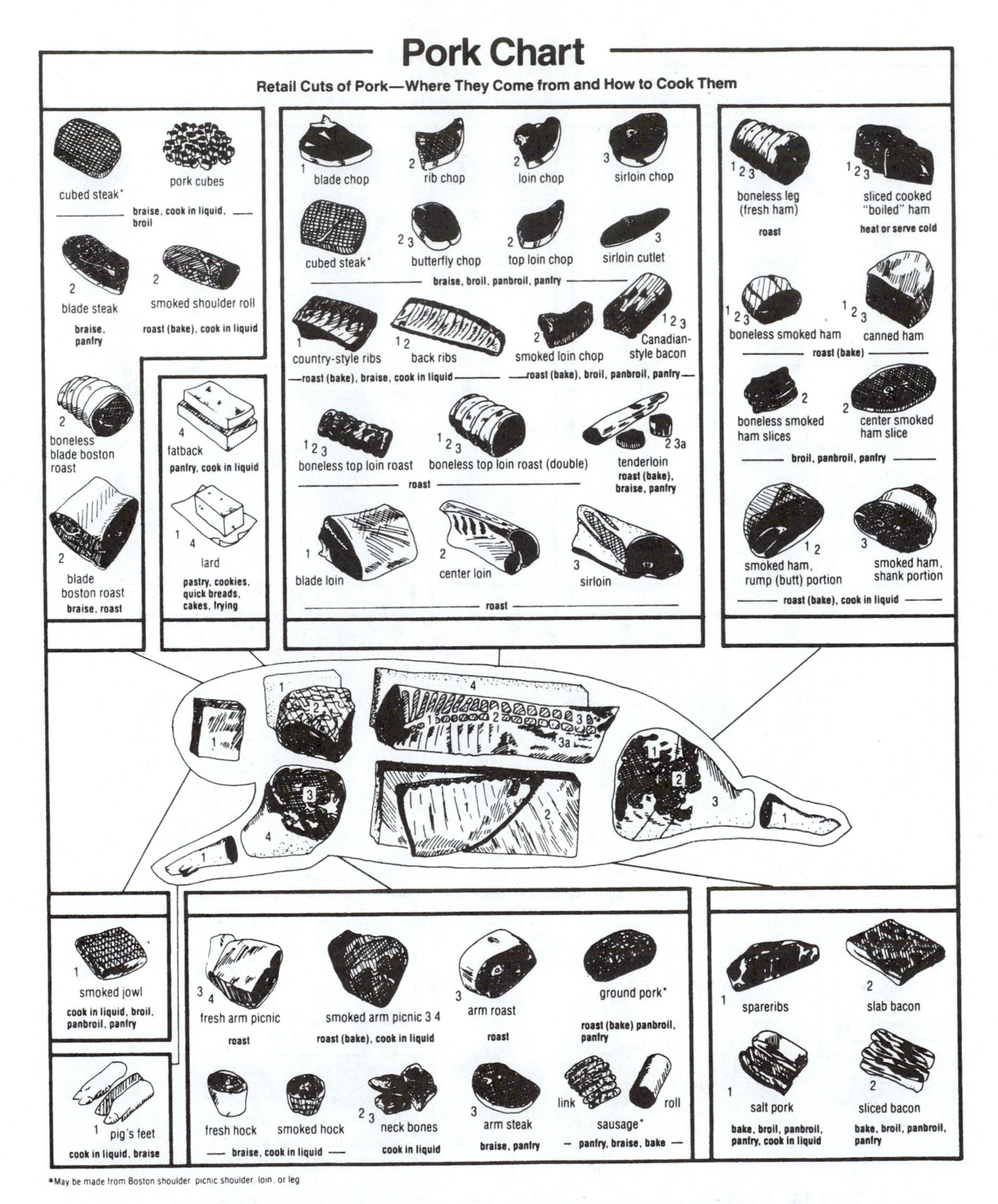

Figure 18-13C. Retail cuts of pork—where they come from and how to cook them.

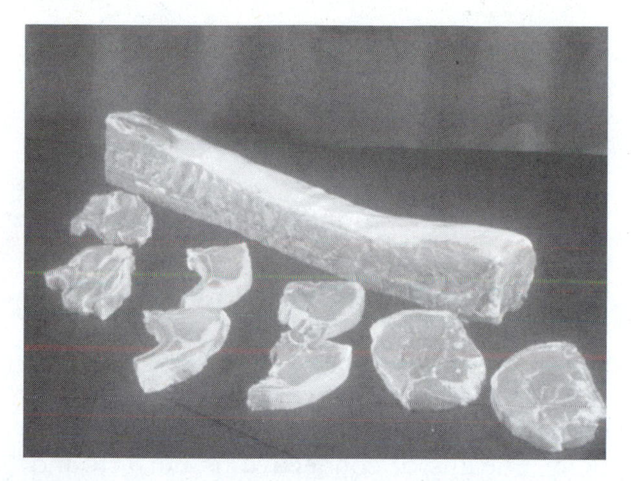

Figure 18-14. This is a wholesale cut (the loin) and the retail cuts that come from it. *Dr. Estes Reynolds, Cooperative Extension Service, The University of Georgia.*

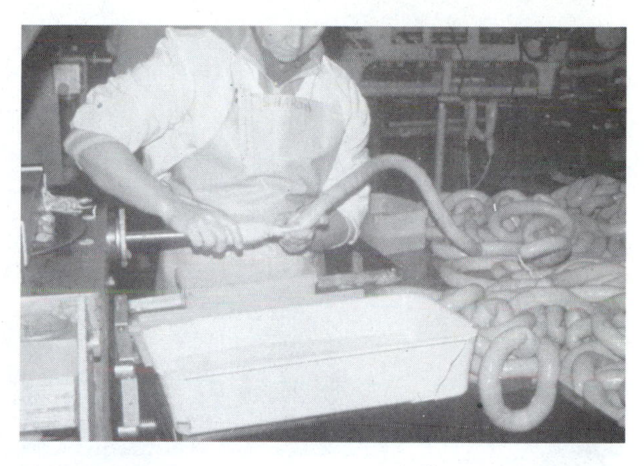

Figure 18-15. Sausages are made from ground meat. Here tubes made from intestines are being filled with sausage meat. *Dr. Estes Reynolds, Cooperative Extension Service, The University of Georgia.*

FACTORS AFFECTING PALATABILITY

Palatability refers to how a food appeals to the palate (a portion of the roof of the mouth). Meat palatability depends on such qualities as appearance, aroma, flavor, tenderness, and juiciness. How palatable a piece of meat is depends upon a combination of the qualities listed above and the way in which it was cooked. However, consumers buy most retail meat in the uncooked form.

Processed meats are an exception. These meats are composed of those parts of the carcass that do not make good retail cuts (much the same as sausages and ground meat). These meats undergo a processing that may include pressing, forming, and slicing. These products are usually fully cooked and are used as cold cuts and sandwich material. Bologna, hot dogs, processed ham, and salami all fit this category, Figure 18–16.

Appearance is the factor that first influences the expectations of quality. Beef, pork, and lamb all vary in the shades of red color. Darker meat tends to be associated with either a lack of freshness or meat from older animals. Bright red meat gives the appearance of being fresh and wholesome.

Fat that is yellow instead of a creamy white is less appealing to consumers. Yellow fat is generally associated with certain breeds that are unable to convert carotene (yellow to red pigment found in vegetables, body fat, and egg yolks) to vitamin A. Cattle that have been grass

Figure 18-16. There are many types of processed meat that are made from various parts of the carcass. *National Livestock and Meat Board.*

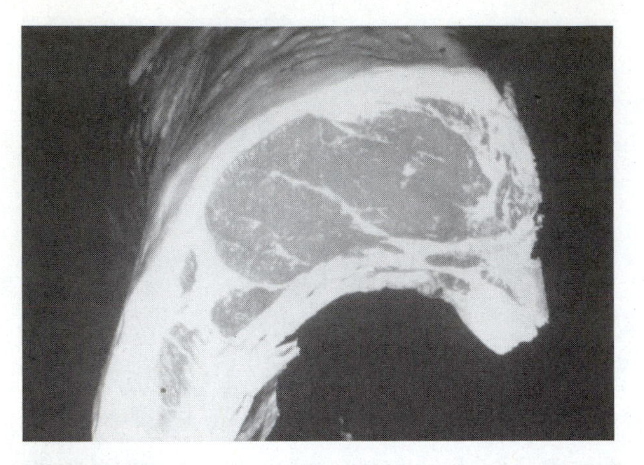

Figure 18-17. The amount of fat in proportion to the muscle affects consumer approval. This steak has too much fat. *Cooperative Extension Service, The University of Georgia.*

fed and have consumed an excess of carotene may have yellow fat. Grain-fed beef is generally considered to taste better than grass-fed beef.

The amount of fat and bone in proportion to muscle also affects the appearance and consumer appeal of meat, Figure 18–17. Consumers realize that fatter meat with bones has a smaller portion of edible meat or more plate waste (meat scraps, bone, and fat that is not consumed).

Tenderness is a sensation that has several components and has been the object of considerable study. Tenderness is difficult to measure. Terms that have been ascribed to tenderness during chewing include: resistance to tooth pressure, softness to tongue and cheek, ease of fragmentation, mealiness, adhesion or stickiness, and residue after chewing. Not only is it difficult to describe eating tender versus tough meat, but also it has been difficult to use mechanical devices that measure the many aspects of chewing and taste.

Components of muscle that contribute to tenderness are connective tissue, state of the muscle fibers (what degree of contraction), and the amounts of **adipose** (fat) tissue. However, the amount of adipose tissue or fat, particularly intramuscular (within the muscle) fat known as marbling, probably does not influence tenderness to any great degree. Research has not been able to relate marbling to tenderness except that marbling may tend to act as a lubricant during **mastication** (chewing) and swallowing, and that marbling is related to high-energy feed and production systems.

Connective tissue connects various parts of the body and is distributed throughout the body. Sheaths of connective tissue surround muscle bundles, nerve trunks, blood vessels, tendons, and fat cells (adipose tissue). Connective tissue consists of a structureless mass (called *ground substance*), embedded cells, and extracellular fibers (fibers outside the cell). Extracellular fibers include collagen and **elastin**. Collagen is the most abundant protein in the animal and is found in all tissues and organs. The presence of collagen in skeletal muscle is generally proportionate to the physical activity of the muscle—the more activity, the more collagen. Muscles of the limbs contain more collagen than the muscle along the spinal column. As the animal grows older, the collagen becomes less soluble. An example of collagen solubility is the gelatin material in the pan after cooking a pot roast in moist heat. Therefore, the muscles of younger animals and the muscles responsible for less physical activity, such as those along the spinal column, are more tender.

Elastin is an elasticlike protein found throughout the ligaments, arterial walls, and organ structures. Elastin fibers are easily stretched and they return to their natural state when the tension is released. Cooking has no appreciable effect on elastin fibers. Cuts of meat with high elastin content tend to appear tougher.

When selecting fresh cuts of meat in the retail case, the consumer should avoid extremes in apparent juiciness. A dark, dry appearance usually indicates either age in the retail case or the age of the animal from which the cut was obtained. Extremely moist-appearing meat (moisture oozing out of the retail cut) indicates the pale, soft, and exudative (PSE) condition associated with some meats.

The juiciness of cooked meat is important in the perception of palatability to consumers. The juice is made up of water and melted intramuscular fats. As the meat is chewed, juices are released that stimulate the flow of saliva, thus further increasing apparent juiciness. The juices contain flavor components, and they assist in lubricating, softening, and fragmenting the meat during chewing.

Flavor changes often occur in meat after extended storage periods. Chemical breakdown of nucleotides (flavor compounds) give a desirable aged flavor, while **oxidation** of fatty acids (oxidative rancidity) results in a **rancid** flavor and sharp unpleasant aroma.

Aroma is detected from numerous gaseous aspects of meat that stimulate nerve endings in the linings of the nasal passages. The total sensation is a combination of taste (gustatory) and smell (olfactory). The meaty flavor and aroma stimulate the flow of gastric juices and saliva that aid in digestion and increase the apparent juiciness of the meat.

Preservation and Storage of Meat

Meat is a highly perishable product that can spoil in a very short time. It is preserved by creating conditions that are unfavorable for the growth of spoilage organisms and the development of off-flavors, which are usually due to chemical oxidation of fatty acids or proteins. Historically, methods of preservation have included drying, smoking, salting, refrigeration, freezing, canning, and freeze-drying. Degradation of the animal tissue begins upon slaughter due to chemical, biological, and physical reactions. Microorganisms are an integral part of this process as they thrive on the nutrients supplied by meat.

Meat provides an ideal medium for the growth of many **microbes** (microorganisms), Figure 18–18. Molds, yeast, and bacteria are all found on or in meat. Molds are multicellular, multicolored organisms that have a fuzzy or mildewlike appearance. Molds are spread by producing spores that float in the air or are transported by contact with objects. Yeast are large, unicellular bud and spore forms that are spread by contact or in air currents. Most yeast colonies are white to creamy in color and are usually moist or slimy in appearance or to the touch.

Among the factors affecting the growth of microbes are temperature, moisture, oxygen, pH, and the physical form of the meat. Temperature can influence the rate and kind of microbial growth. Some microbes grow well in

Figure 18-18. Meat provides an ideal medium for the growth of microbes. This package of processed meat has spoiled. Note the ballooning of the wrapper caused by gases released by the spoiling meat. *Dr. Estes Reynolds, Cooperative Extension Service, The University of Georgia.*

cooler temperatures of 0°–20°C (32°–68°F). These are known as **psychrophiles**. Microbes that grow at higher temperatures between 45°C and 65°C (113°–150°F) are called **thermophiles**. Microorganisms with a growth optimum between the psychrophiles and the thermophiles are called as **mesophiles**. Temperatures below 5°C (40°F) greatly retard the growth of spoilage microorganisms and prevent the growth of most pathogens.

Moisture and relative humidity greatly affect the growth of certain microorganisms. Most microbes must have moisture to reproduce and grow.

Oxygen availability determines the type of microorganism that grows. Microorganisms requiring free oxygen in order to grow are called **aerobic organisms**. Those organisms growing in the absence of oxygen are called **anaerobic organisms**. Microbes that can grow with or without free oxygen (reduced oxygen availability) are termed **facultative**. Molds, yeast, and many of the bacteria commonly associated with meat are aerobic. Many of the bacteria found in meats, such as lactobacillus, are anaerobic or facultative and will, over time, cause spoilage in vacuum-packaged meat products. In vacuum packaging, the meat is placed in a plastic bag and all the air is removed, Figure 18–19. This creates a package that does not let air or moisture pass through. The vacuum packaging of meat and meat products increases the storage time of these products by inhibiting the growth of aerobic organisms.

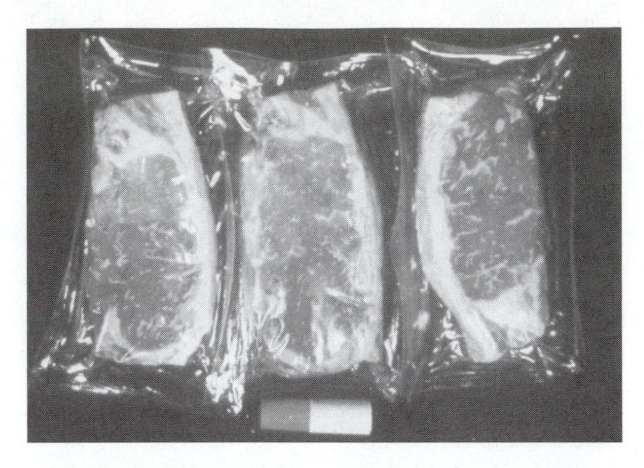

Figure 18-19. Vacuum packaging of meat improves the storage time. *Dr. Estes Reynolds, Cooperative Extension Service, The University of Georgia.*

Most microorganisms have an optimum pH near neutrality (pH 7.0). Molds survive in a wide range of conditions (pH 2.0 to 8.0). Yeasts favor a slightly acid condition of pH 4.0 to 4.5. Most bacteria favor a range of pH 5.2 to 7.0. Meat and meat products generally range from pH 4.8 to 6.8, with the norm being approximately pH 5.4 to 5.6. Therefore, meat conditions favor the growth of molds, yeast, and the acidophilic (acid-loving) bacteria.

The physical form of the meat affects the growth of microorganisms. As carcasses are separated in different primal, subprimal, and retail cuts, more surface area is exposed. This surface area provides nutrients, moisture, and oxygen for the microorganisms to grow. When meat is ground for such products as sausage and hamburger, the maximum surface area is exposed and the microorganisms are spread throughout the product. Therefore, sanitation and keeping the temperature as low as possible are the major considerations for maintaining acceptable numbers of microbes.

Meat that has been frozen and thawed is more susceptible to microbial growth due to ruptured cells and increased surface moisture. Refreezing meat that has been thawed does not cause serious deterioration or breakdown in itself, but refreezing will not reverse the deterioration done by microbes. Refreezing with less than optimum conditions, such as often exist in home refrigerator/freezer units, will cause decreases in juiciness and flavor due to the formation of large ice crystals that rupture the cells. This causes drying and the oxidation of fats.

Curing and Smoking

Meat processing developed soon after people became hunters in prehistoric times. The salting and smoking of meat has been documented as far back as 850 B.C. by Homer and as far back as the thirteenth century B.C. by the Chinese. In these early times, the smoking (drying) and salting (curing) of meats were the only methods of preservation known. Today, the curing and smoking of meat is a method of imparting a particular flavor to the meat, Figure 18–20. Very few people in this country still rely on curing and smoking as methods of preservation. There are just about as many cured products today as there are regions of the world, many of these being descendants of the ancient curing methods.

Salt and nitrite are the two main ingredients used to cure meats. Some of the most frequently used are sugar, ascorbate, erythorbate, phosphates, and delta gluconolactone. Today, salt is used at levels that generally impart a flavor to the product (one to three percent) instead of the amounts that were used to preserve the meat item being cured (9–11 percent).

Figure 18-20. Smoking imparts flavor to meat. These sausages are being smoked. *Dr. Estes Reynolds, Cooperative Extension Service, The University of Georgia.*

Nitrates (saltpeter) and nitrites are used to impart the "cured meat" color and flavor and to inhibit bacteria action. The use of nitrites or nitrates is not permitted to result in more than 120 parts per million (ppm) of nitrite in the finished product.

The oldest of these methods is known as **dry curing**. In this method, the cure ingredients are rubbed onto the surface of the product and allowed to move into the product by osmosis. This method takes the longest amount of time to infiltrate the meat. Over time, technology has improved on this simple procedure.

A more modern method of curing meat is **injection curing**, Figure 18–21. This method involves pumping the curing solution (brine) into the meat product. This shortens the curing process since the curing ingredients are

Figure 18-21. These hams have been injected with curing solution and are about to be placed in the smoker. *Dr. Estes Reynolds, Cooperative Extension Service, The University of Georgia.*

dispersed directly into the meat. There are three methods of placing the cure into the product. The first of these uses the artery system in the meat (primarily used in hams) to disperse the curing adjuncts. The second of these methods involves placing a hollow needle (stitch) into the major muscle masses of the product to inject the curing solution.

The third method of curing is known as **combination curing** and is simply a combination of dry curing and injection curing.

Refrigerator Storage of Meat

Carcass temperatures upon exiting the slaughter floor generally range between 30°C and 35°C (85°–95°F). Chill coolers generally operate at about -3°C to 1°C (27°–34°F), and carcasses need to be chilled to less than 5°C (40°F). The time required for the chilling process is affected by the carcass size, amount of fat on the carcass (fat reduces heat dissipation), and the initial heat of the carcass. Rapid air movement can reduce the chilling time up to 25 percent. After 12 to 24 hours in the chill cooler (often called a *hot box*), beef carcasses are moved to aging or holding coolers at 0° to 3°C (32°–37°F) until they are fabricated or shipped.

The storage life of carcasses or meat products in refrigeration depends on factors such as initial numbers of microbes on the meat, temperature and humidity conditions during storage, use of protective coverings or packaging, animal species, and the type of product being stored. As a rule of thumb, fresh meat under good home refrigeration conditions should be consumed within four days of purchase.

Freezer Storage of Meat

Freezing acts as a preservation method because microbial and enzymatic activity is stopped at about -10°C (14°F). However, some changes do still take place, such as the development of rancidity and surface discoloration due to dehydration. Factors that affect the quality of frozen meat include freezing rate, length in freezer storage, packaging materials used, and the variability of the freezer temperatures.

The most common method of commercial freezing is to utilize high-velocity air and temperatures of -10°C to -40°C (14°-40°F). This method is commonly termed **blast freezing**. Methods of freezing utilizing condensed gases in direct contact with the product are termed *cryogenic*.

Problems associated with relatively slow freezing include the formation of large ice crystals and loss of moisture during thawing. Temperature fluctuations during frozen storage may also cause these problems.

The length of time meat can be stored in a freezer varies with freezer temperature, temperature fluctuations, species, type of product, and the type of wrapping material. The product must be packaged using vapor proof materials to keep oxygen out and to keep moisture in the package. Oxygen causes oxidative reactions such as rancidity. Moisture loss causes dehydration and a condition known as freezer burn.

Storage time may be extended by lowering storage temperatures. Although not economically feasible, maintaining a temperature of -80°C (-112°F) stops most chemical changes. Most commercial and home freezers are maintained at approximately -18°C (0°F) or less. Even at this temperature, fluctuations in temperature may cause migration of water and increased moisture loss upon thawing.

Meats from different species differ in the time that freezer storage will maintain acceptable quality. The difference is primarily due to differences in fat composition. Softer fats, such as those found in pork, are more susceptible to oxidative changes and subsequent loss of flavor. Due to the differences in fat composition, recommendations for length of freezer storage at -18°C (0°F) or less are as follows: beef—six to twelve months; lamb—six to nine months; pork, four to six months; cured meats—one to two months. This can vary with packaging material, cuts of meat, and freezer temperature fluctuation.

Processing of the meat into sliced, ground, or cured products influences the acceptable freezer storage time. Exposure to oxygen or the addition of salts enhances the development of rancidity and reduces flavor acceptability.

Preservation of Meat by Drying

Since moisture is a critical element for microbial growth, the removal of moisture from meat is an effective means of preservation. Low-moisture foods are those containing less than 25 percent moisture. Beef jerky is an example of a low moisture food, Figure 18–22.

Intermediate-moisture foods have less than 50 percent moisture. Dry salami is an example of an intermediate dry meat product that is shelf stable (requires no refrigeration) but still subject to mold growth unless treated with a mold inhibitor. Most meat products that are dried also contain some salt, which assists in lowering the moisture of the product.

Figure 18-22. Beef jerky is an example of a low-moisture food. *Dr. Estes Reynolds, Cooperative Extension Service, The University of Georgia.*

Irradiated Meat

In 1997, the Food and Drug Administration (FDA) granted approval to use **radiation** to help preserve red meat. This technology has been around for several years and has been used on a wide variety of foods ranging from wheat to onions. In 1990, the technology was approved for use on the poultry to control salmonella and on pork to stop trichina but the approval for beef was delayed. In fact, **irradiation** is currently used in over 40 countries to preserve food, Figure 18–23.

The technology makes use of low levels of radiation to kill pathogens in food products. Of course, additional preservation techniques such as refrigeration are required because the food is not permanently preserved However, the rate of spoilage is greatly reduced and dangerous bacteria such as salmonella and **E. coli** are killed. This makes meat much safer for the consumer.

The greatest hindrance to the use of irradiation on meats and other foods is consumer acceptance. There seems to be an aversion to eating anything that was submitted to radiation. However, the FDA has declared the process to be safe and effective. A symbol indicating that the product has been treated by irradiation is required on all packages. After treatment, the food is no more radioactive than your teeth are

Figure 18-23. Food treated with irradiation must be labeled with this identifying logo.

after a dental x-ray. Perhaps after the irradiated foods have been sold for a while, they will be more readily accepted by consumers.

SUMMARY

Americans eat a lot of meat and this has spawned a tremendous industry that is constantly changing in an attempt to meet the needs of the consumer. We want products that are safe, tasty, and affordable. The government, through the USDA and the FDA, regulate the growing, processing, grading, packaging, and sale of all meat products. These products require a wide variety of preservation methods to keep the meat from spoiling. Because of our high-tech agricultural food systems, we enjoy the best, cheapest, and safest food supply in the world.

◆ Review Exercises

TRUE OR FALSE

1. Regulations for the immobilization process in the slaughterhouse are set by the Cattlemen's Association. ______.
2. Exsanguination refers to the practice of rendering an animal unconscious so that it feels no pain during slaughter. ______.
3. The temperature must be raised immediately after death to kill microorganisms and minimize protein degradation. ______.
4. The meat grade, which certifies class, quality, and condition of the carcass, is mandatory for all meat packers and is paid for by the U.S. Department of Agriculture. ______.
5. The amount of fat in the muscle fibers is referred to as marbling. ______.
6. Tenderness is one of the easiest qualities of meat to test. ______.
7. The more collagen and elastin meat contains, the less tender it is. ______.
8. Aroma is a combination of taste (gustatory) and smell (olfactory). ______.
9. Preserving meat involves creating conditions that are unfavorable for the growth of spoilage organisms and the development of off-flavors. ______.
10. The best temperature to keep meat from spoilage is below 40°F. ______.
11. Vacuum packing meat prevents any bacteria from contaminating meat by keeping oxygen and moisture out. ______.
12. Meat that has been frozen and thawed is more susceptible to microbial growth due to the rupture of the cells and increased surface moisture. ______.
13. Meat processing is a recent development that became popular in the 1940s. ______.
14. Once meat is frozen, no more changes take place (such as microbial breakdown of the meat and the development of rancidity). ______.
15. Low-moisture foods, such as dry salami, contain less than 25 percent moisture; intermediate moisture food, such as beef jerky, contains less than 50 percent moisture. ______.

FILL IN THE BLANKS

1. Meat is defined as the ____________ flesh of animals and may include the ____________________, certain internal ___________, such as the _________ and other edible parts.

2. Federal inspectors are on hand during the slaughter process to inspect the ______________ ______________ and the ______________ to detect any ______________ concern over the meat from the ______________.
3. The use of electrical stimulation instead of aging involves a current of ______________ ________volts that is sent through the ______________ immediately after ______________ and before the ______________ is removed.
4. Factors affecting the yield grade are the ______________ carcass weight, the amount of ______________ fat, the size of the ______________ ______________ area, and the amount of ______________ on the carcass.
5. Primal cuts, which are sold to ______________ outlets, are packaged in ______________ packs, placed in ______________, and shipped to the ______________.
6. Meat palatability depends on such qualities as appearance, ______________, flavor, ______________, and ______________.
7. Connective tissue consists of a ______________ mass called ______________ substance, ______________ cells, and ______________ fibers.
8. The juices of cooked meat contain ______________ components, and they assist in ______________, softening, and ______________ the meat during ______________.
9. Meat provides an ideal ______________ for the growth of many ______________ such as ______________, yeast, and ______________.
10. As carcasses are separated in different primal, ______________, and retail cuts, more ______________ area is exposed, providing ______________, moisture, and ______________ for the ______________ to grow.
11. Refreezing meat with less-than-optimum conditions, such as often exist in ______________ refrigerator/freezer units, will cause decreases in ______________ and ______________ due to the formation of large ______________ ______________ that rupture the ______________.
12. There are just about as many cured ______________ today as there are regions of the ______________, many of these being descendants of the ______________ curing ______________.
13. In dry curing, the cure ingredients are ______________ onto the ______________ of the product and allowed to move into it by ______________.
14. The storage life of carcasses depends on such factors as the initial numbers of ______________ on the meat, temperature and ______________ conditions during ______________, use of protective ______________ or ______________, animal species, and the type of ______________ being stored.
15. Softer fats, such as those found in ______________, are ______________ susceptible to ______________ changes and subsequent loss of ______________.

DISCUSSION QUESTIONS

1. Define the term *meat*.
2. What are the methods used to render an animal immobile during the slaughter process?

3. What parts of the animal are edible besides the muscles?
4. What is rigor mortis?
5. What is the difference between quality grading and yield grading?
6. List, in order, the quality grades of beef.
7. List three factors that affect the palatability of meat.
8. List six methods of preserving meats.
9. Name three types of microbes that cause meat spoilage.
10. What problems are encountered when meat is frozen slowly?
11. Discuss the greatest hindrance to preserving meat with irradiation.

STUDENT LEARNING ACTIVITIES

1. Visit a local grocery store and make a list of all the cuts of meat in the meat counter. Develop a chart illustrating the wholesale cut from which each came.
2. Interview a meat buyer for the grocery store. Determine how and where the meat is bought, the quality and yield grade bought, and which wholesale cuts the buyer buys the most of.
3. Go to the grocery store and look for food products that have been treated with irradiation. Share your list with the class.

Chapter 19

Parasites of Agricultural Animals

Student Objectives in Basic Science

As a result of studying this chapter, you should be able to

1. Explain symbiotic relationships.
2. Distinguish between mutualism, commensalism, and parasitism.
3. Discuss how parasitism causes harm to host animals.
4. Explain the process of metamorphosis.
5. List the phases in the life cycle of an insect.
6. Distinguish between a roundworm and a segmented worm.
7. Explain how scientific research is used in the eradication of parasites.

Student Objectives in Agricultural Science

As a result of studying this chapter, you should be able to

1. List the types of parasites that infest agricultural animals.
2. Explain how production losses are incurred because of parasites.
3. List the conventional means of controlling parasites on agricultural animals.
4. Discuss how the life cycle of a parasite can be used to control the parasite.

Key Terms

symbiosis
mutualism
commensalism
parasitism
host
life cycle
anemia
internal parasites
external parasites
metamorphosis
stomach worms
strongyle
colic
ascarids
intermediate host
flukes
warm-blooded
nymph
systemic pesticides
bolus
biological control

Animals of different species that live in close association with each other are said to live in a symbiotic relationship. **Symbiosis** can take at least three different forms.

Mutualism is a relationship that is beneficial to both species of animals. For example, as explained in the chapter on nutrition, certain bacteria live in the rumen of cattle. The cattle provide the bacteria with food and a place to live. The bacteria help the cattle break down fibers into a form that can be digested. Another example is that of tick birds that light on the back of cattle and other animals. The birds obtain food from eating the ticks on the animals, and the animals benefit by having an annoying pest removed.

Commensalism is the relationship of animals when one benefits and the other is not harmed. An example of commensalism is the relationship between cattle and houseflies. The housefly must lay its eggs in the feces of animals. The fly benefits from the fecal material deposited by the cattle, but the cattle are not harmed by the housefly.

The third form of symbiosis is that of **parasitism**. Parasitism is a relationship that is beneficial to one animal and harmful to another. It accounts for the vast majority of the incidences of symbiosis. All agricultural animals are susceptible to parasites, and measures must be taken by producers to deal with parasitism. The animal that lives off the other animal is called a *parasite*; the animal that the parasite lives on or in is called the **host**.

According to the USDA, parasitism causes almost a billion dollars worth of damage to agricultural animals each year. Parasites that live in and on livestock are generally insects that live out one or more of the phases of their **life cycle** at the expense of the agricultural animal. The damage they cause comes about in several ways.

Most parasites live off the blood of the host animal. The continual loss of blood causes the animal to develop a condition known as **anemia**. One of the major functions of an animal's blood is that of providing body cells with oxygen and food nutrients. If enough parasites are living off the blood of an animal, the blood supply to the animal may be greatly diminished. When this occurs, the host animal will become ill because the body cells are not getting enough oxygen and food nutrients. The animals are said to be anemic. The animals are sluggish, feel poorly, and do not grow or perform as they should, Figure 19–1.

Animals that are hosts to parasites are in a weakened condition. This makes the animals more susceptible to disease; disease organisms can more effectively attack weak animals. As will be explained in Chapter 20, an animal's body has an immune system to fight disease organisms that invade its body. In order for the immune system to function properly, the animal

Figure 19-1. Animals infected with parasites do not feel well and do not perform as they should. *North Dakota Department of Education.*

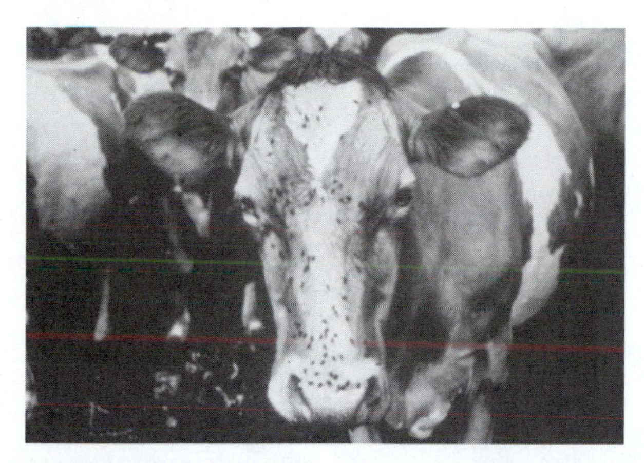

Figure 19-2. Parasites may pass diseases from a sick animal to a healthy one. *North Dakota Department of Education.*

must be strong and in good health. If the animal is in a weakened condition, the immune system will not function as it should, and the animal will get sick more easily.

Parasites often carry disease organisms from one animal to another. An insect may feed on an infected animal and then feed on a healthy animal and transmit disease organisms, Figure 19–2. For example, a very dreaded disease of horses is sleeping sickness. This disease is passed on by an insect that bites and sucks blood from the animal. An insect may bite an infected horse and draw blood from the sick animal. The insect may then fly to another area where it bites a healthy animal, and in doing so, passes disease germs to the healthy animal. Humans can also get diseases this way. A common example is malaria, which is transmitted through the bite of a mosquito.

Animals that are infected with parasites are almost always uncomfortable. Parasites cause irritation of the skin, intestinal tract, or other parts of the body. Animals that are irritated and uncomfortable do not grow as well and are not as efficient. In order for animals to grow, breed, or feed their young, they must feel healthy. If parasites are causing the animal pain or the host animal has to spend most of its time trying to alleviate an itch caused by parasites, the animal will perform poorly.

Animals that are infected with parasites consume more feed per pound of gain. In other words, the feed efficiency of the animal is lowered. This means that the cost to maintain the weight of the animal is increased. If parasites are feeding on an agricultural animal, they are feeding either directly or indirectly on the feed supplied by the producers.

Parasites can generally be broadly divided into two categories: **internal parasites** and **external parasites**. Although some parasites live their entire lives on or in the host animal, most live only a portion of their life on or in the host. In these instances, the host animal supports the parasite through only a phase of its life cycle. For example, insects go through four complete stages from the time they hatch until they are mature adults capable of reproducing. This process of change is called **metamorphosis**. At each one of these stages the insect looks completely different from its appearance during the other three stages, Figure 19–3. It is during one of these four stages in a parasite's life that the host animal is infested.

When the young insect hatches, it is usually in the larval stage. This means that the young insect looks very much like a worm. A larva usually is a voracious eater and that can do a lot of damage to plants or to a host animal.

When the larva matures, it passes into the pupa stage, which is usually a relatively dormant stage. A pupa is the intermediate stage between the larva and the adult. During this stage, the body tissues of the young insect convert from a larva to an adult.

The last stage is the adult. In this stage the insect lays eggs and the cycle begins again.

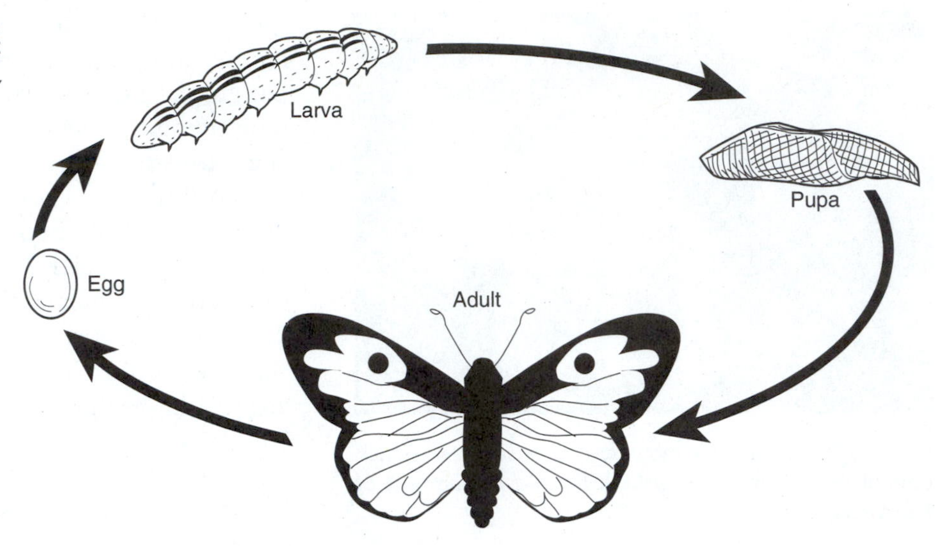

Figure 19-3. Most insects go through four stages of development. *The University of Florida.*

INTERNAL PARASITES

Internal parasites actually live within the animal's body and may feed on the animal's blood or on feed that passes through the animal. Internal parasites are divided into three major groups: roundworms, tapeworms and flukes.

Roundworms

Roundworms cause more damage to agricultural animals than any other group of internal parasites. They infect almost all types of livestock and exist by living in the digestive tracts of their hosts, Figure 19–4. **Stomach worms** infect all classes of livestock and cause damage by the adults' burrowing into the lining of the host's stomach and sucking the animal's blood.

Also, by digging into the stomach lining, the worm damages the tissue of the stomach, enzymes are not produced, and the host animal cannot digest food as well. Poisons are released by the worms as they digest their food and excrete the waste into the host animal. These poisons can cause the host animal to become ill. Generally, an ill animal will not eat or perform nearly as well as a healthy animal.

The worms lay eggs in the stomach of the host and pass out of the animal in the feces. While in the feces, the eggs hatch into larvae and the larvae crawl out onto a blade of grass. A grazing animal then eats the grass and swal-

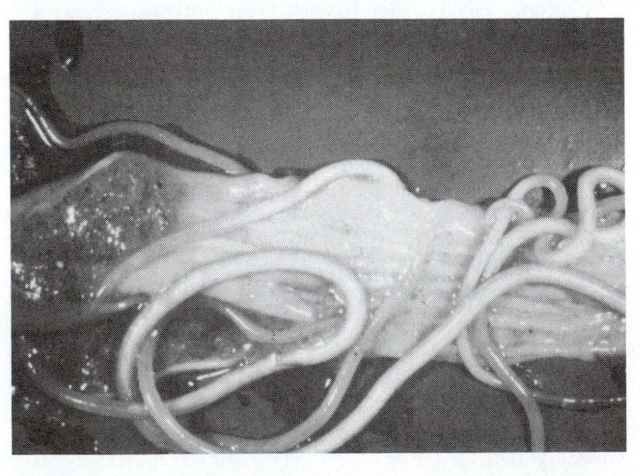

Figure 19-4. Roundworms infect the digestive tracts of their hosts. *North Dakota Department of Education.*

lows the larvae. Once the larvae are swallowed, they settle in the stomach of the host animal and begin to penetrate the stomach lining. The parasites remain there, feeding off the animal's blood and laying eggs. The eggs are passed out of the host animal and the whole process starts again, Figure 19–5.

Another type of roundworm is the **strongyle**. Strongyles are similar to stomach worms in their life cycle, except that instead of living in the stomach lining, they live in the intestines of the host animal. Strongyles cause damage by causing scar tissue in the small intestine and by sucking blood from the host animal. Since the small intestine absorbs food nutrients into the bloodstream, a damaged small intestine reduces the efficiency of the digestive system of the infested animal. These parasites are particularly damaging to horses and can cause a digestive disorder called **colic**.

The largest of the roundworms are the **ascarids**. Ascarids most often attack young animals. Like the stomach worms and strongyles, the larvae of ascarids are ingested by animals grazing on blades of grass to which the larvae have attached themselves. The larvae

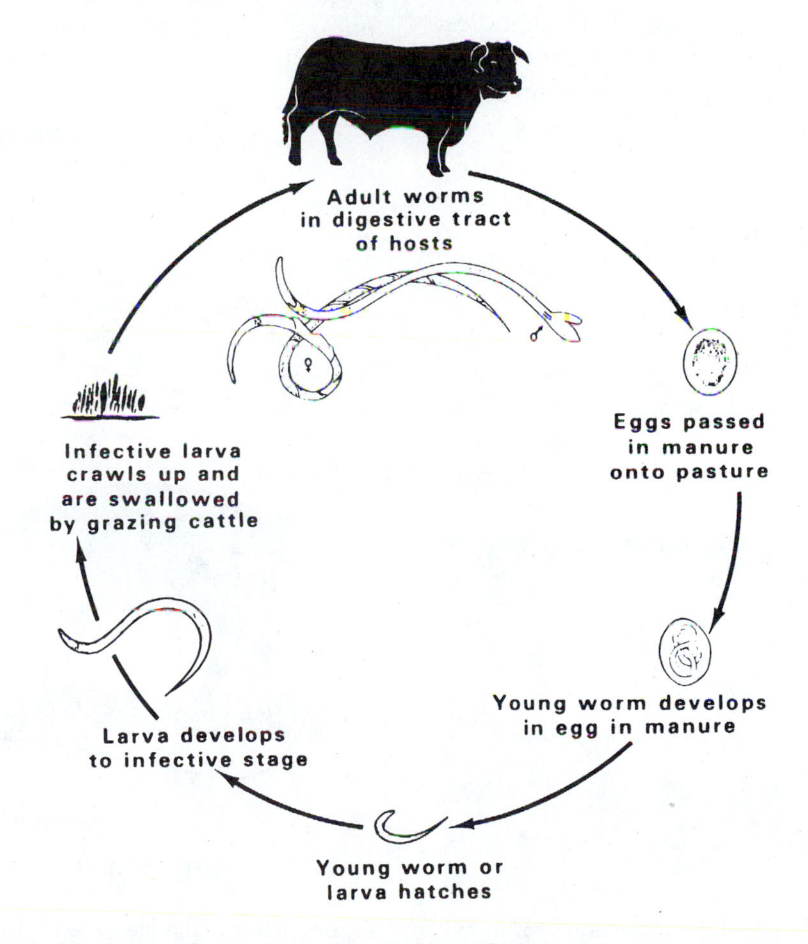

Figure 19-5. Roundworms live only part of their life cycles in host animals. *Cooperative Extension Service, The University of Georgia.*

burrow into the walls of the intestines and from there work their way through the host's heart, liver, and lungs. When they reach the lungs, the worms are coughed up by the host animal and swallowed. The larvae are passed into the small intestine where they develop into adults. The adults lay eggs that are passed out onto the grass in the host animal's feces, the eggs hatch, the larvae attach to blades of grass; the process is renewed.

Tapeworms

Tapeworms belong to a class of worms that are segmented. This means that the bodies of the worms are made up of distinct segments. Each of these segments contain both male and female reproductive organs, and each segment is capable of producing fertilized eggs. These segments break off the body of the worms and reproduce.

Tapeworms cause less damage than roundworms because they do not feed on the animal's blood or cause scarring of the digestive tract. They do, however, cause losses because of the manner in which they feed. The adults of the tapeworm live in the small intestine of the host animal, Figure 19–6. These parasites grow to be quite large, with some species reaching lengths of 25 feet. The tapeworm lives off feed that is passed into the host animal's intestine. It causes the animal harm by devouring the food the animal has eaten. The life cycle begins when the segments of the adult tapeworms break off and pass out in the feces. Each segment contains eggs that hatch in the feces. The eggs are eaten by a small mite called an *oribatid mite* that lives in the grasses found in pastures, Figure 19–7.

The mite serves as an **intermediate host**. An intermediate host is an animal that a parasite uses to support part of its life cycle. An intermediate host is not harmed by the parasite.

Figure 19-6. The adult of the tapeworm lives in the small intestine of the host animal. *North Dakota Department of Education.*

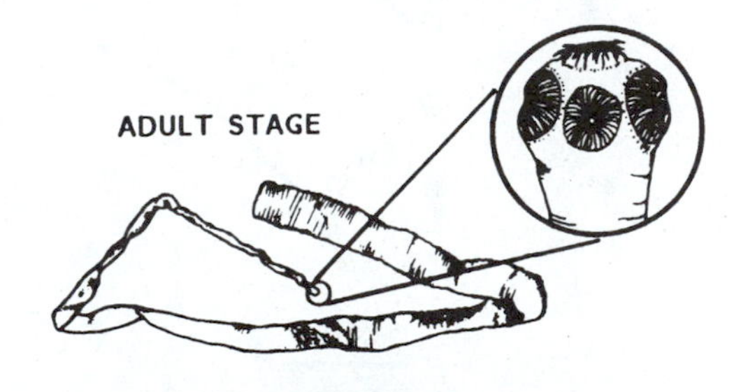

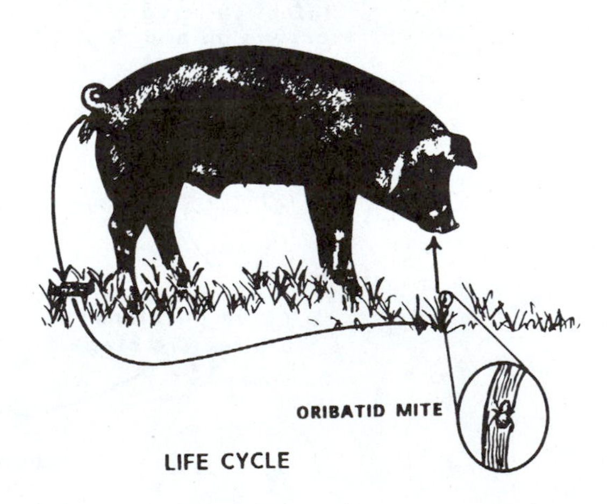

Figure 19-7. The life cycle of the tapeworm involves two hosts. *Instructional Materials Service, Texas A & M University.*

Since the mite lives on grasses, they are swallowed by grazing animals. The eggs are then passed through the animal to the small intestine, where they hatch and live until maturity.

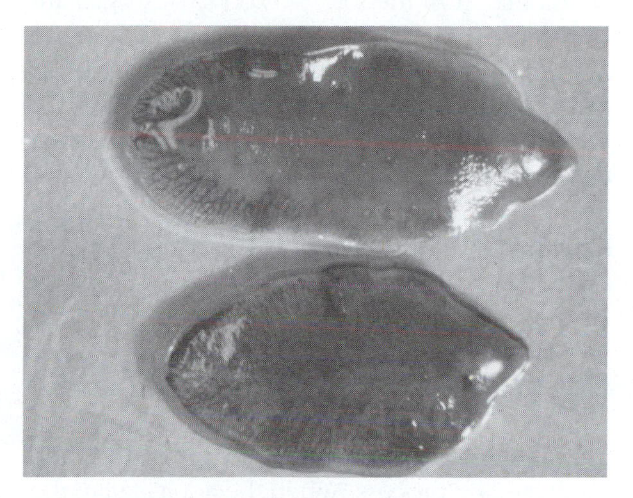

Figure 19-8. Liver flukes cause damage to the liver of the host animal. *North Dakota Department of Education*

Flukes

Flukes are small, seed-shaped flatworms that live in various parts of the host animal. By far the most damaging flukes are those that live in the liver. The adults of the liver fluke live in the bile ducts of the liver where they cause scarring of the liver and bile ducts and general irritation of the liver, Figure 19–8. The adults lay eggs in the bile duct. The eggs are passed through to the intestines and out in the feces. These parasites need an intermediate host. In order for the eggs to hatch, they must land in water. After the larvae hatch, they swim in search of snails that serve as intermediate hosts. Once the snails are located, the larvae enter the snails to develop and reproduce, Figure 19–9. This stage is unusual because the larvae divide and multiply asexually. The larvae divide by themselves to create new organisms without mating and laying eggs. The new larvae emerge

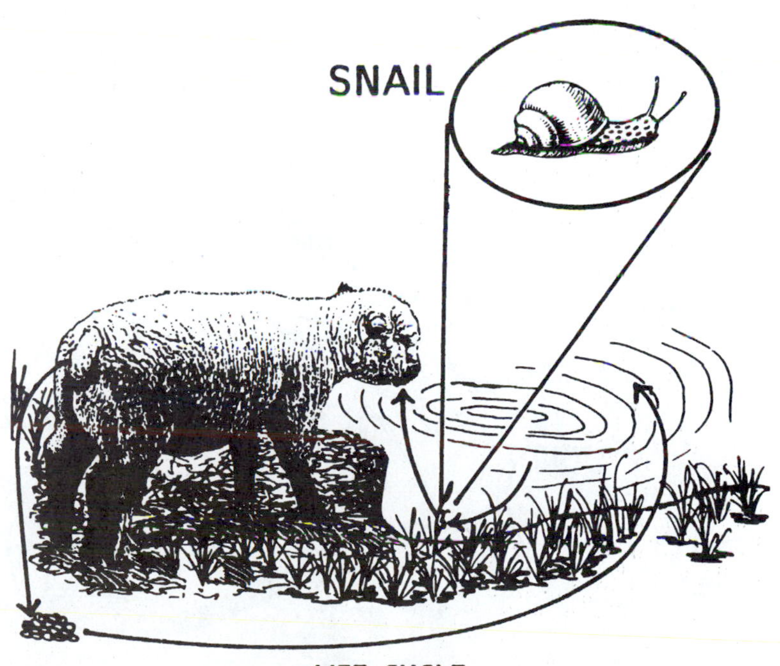

Figure 19-9. Liver flukes use the snail as an intermediate host. *Instructional Materials Service, Texas A & M University.*

and attach to plants in or near the water. Livestock eat the plants and become infected with the flukes. The flukes eat their way through the walls of the digestive tract and migrate to the liver where they feed on the host animal's blood. They begin to lay eggs in about three months. Fluke infestation causes damage to the host animal's liver and causes the bile ducts to thicken and cease normal function. Livers from livestock infested by flukes are unfit for human consumption, and thus a valuable human food source is wasted.

External Parasites

External parasites generally do not cause as much damage to animals as internal parasites. They can, however, cause losses in terms of animal discomfort, the loss of hide quality, and blood loss. External parasites include ticks, lice, and flies.

Ticks

Ticks generally will attach themselves to most **warm-blooded** agricultural animals. They cause damage by penetrating the skin and sucking blood from the host animal. This not only leaves a sore that can be an avenue for disease organisms but also can cause the host animal to be anemic from the loss of blood. Tick eggs are laid in the grass and over winter to hatch in the spring. When the larvae hatch, they climb up onto the grass or into bushes, Figure 19–10. When an animal passes by, the tick larvae attach themselves to the animal and gorge on the animal's blood. They then fall to the ground where they remain until the following spring. In the spring, they undergo metamorphosis. The tick larvae change into **nymphs**. The nymphs climb into bushes, attach themselves to passing animals, and fill themselves with blood. Then the nymphs drop to the ground, overwinter, change into an adult in the spring, and go through the same feeding process. When the adults fall to the ground, eggs are laid, and the life cycle begins again. Three years are required for ticks to complete their life cycles.

When ticks feed, they insert their mouth parts into the host animal's skin and inject saliva into the wound. The saliva contains an anticoag-

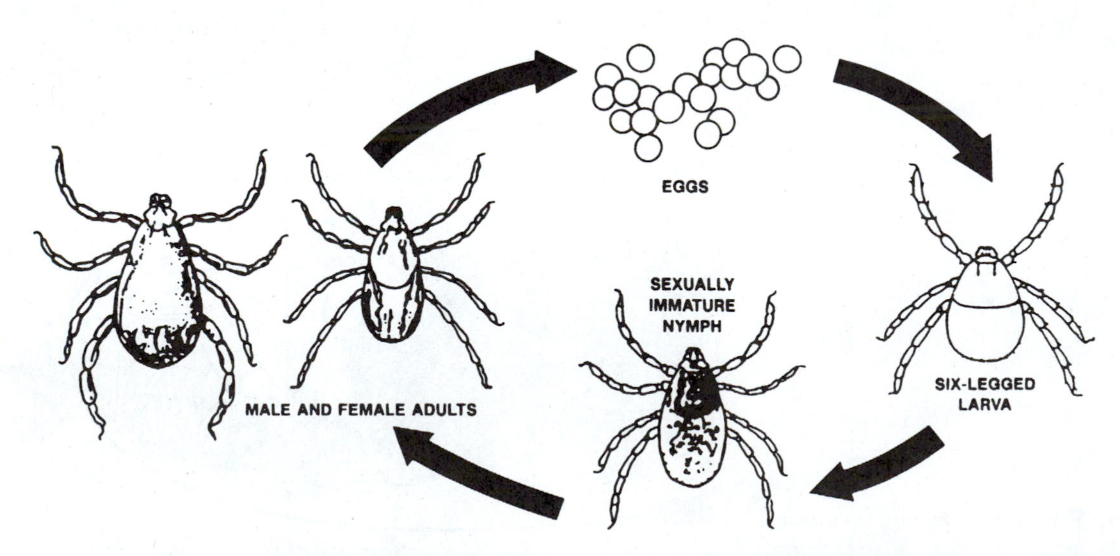

Figure 19-10. The typical life cycle of the tick. *Cooperative Extension Service, The University of Georgia.*

ulant, which is a substance that prevents the blood from clotting and allows the blood to flow freely into the ticks. These pests attack almost all warm-blooded animals, including people. Since the same tick may have as many as three hosts during its life cycle, diseases may be spread from one animal to another. One such disease that is spread to humans is Rocky Mountain spotted fever.

Lice

Lice (*singular,* louse) are tiny, wingless insects that are external parasites of most warm-blooded agricultural animals, Figure 19–11. There are two types; blood-sucking lice that feed by drawing blood through the animal's skin, and biting lice that feed on the hair or skin particles and excretions of the animals. The life cycle of lice is simple compared to the life cycle of some of the other parasites. Their whole lives are lived on the host animal, Figure 19–12. The adult females attach eggs to the hair follicles of host animals. The eggs hatch one to two weeks later, and the newly hatched nymphs live to maturity on the host animal. By biting or piercing the skin of the host animal, lice cause the animal to become very uncomfortable. This results in the animal's rubbing and scratching against posts, trees, or other objects in an attempt to obtain relief from the itching. As a result, time is lost from grazing or eating and a weight loss may occur. In addition, animals that are heavily infested with lice may become so uncomfortable that normal processes like breeding may be interrupted.

Heel Flies

Heel flies, also known as cattle grubs, are a serious parasitic pest of cattle. The adult flies lay eggs on the lower part of the legs of cattle. When the eggs hatch, the larvae penetrate the skin through the hair follicles and begin a journey through the animal's body that may take several months. The larvae burrow through the

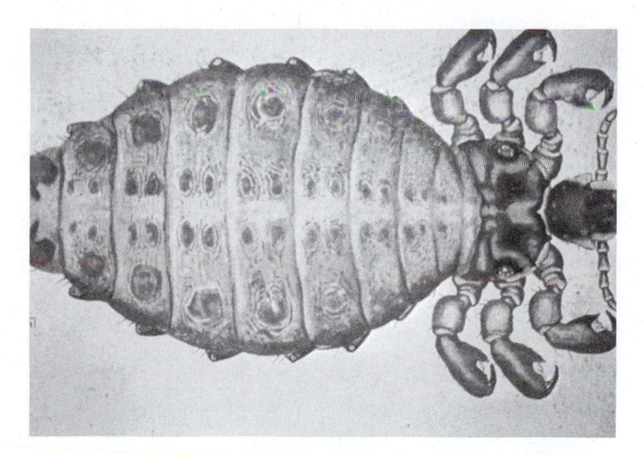

Figure 19-11. Lice are small, wingless insects that cause irritation to their hosts. *North Dakota Department of Education.*

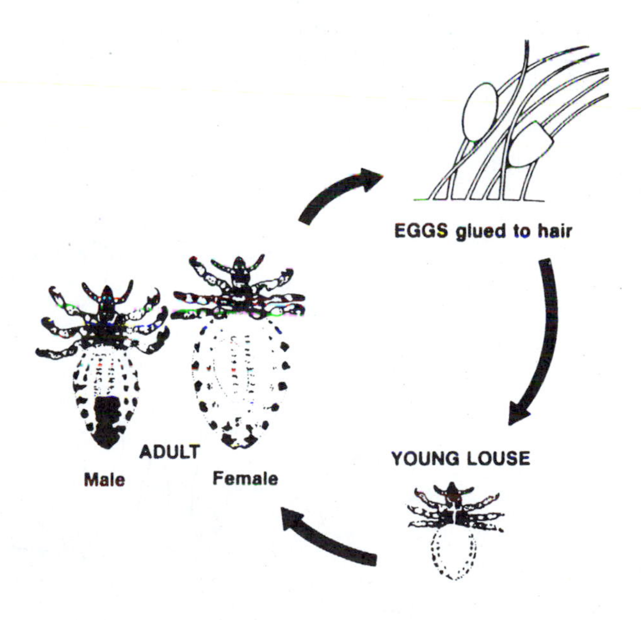

Figure 19-12. The entire life cycle of the louse is lived out on the host animal. *Cooperative Extension Service, The University of Georgia.*

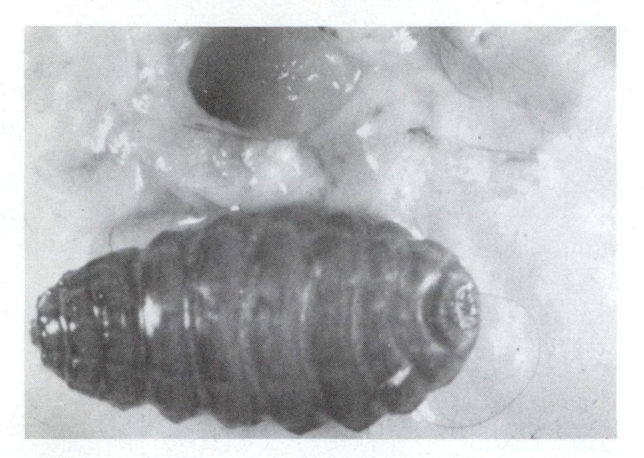

Figure 19-13. Cattle grubs eat a hole through the skin on the animal's back. *North Dakota Department of Education.*

soft tissue of the animal all the way from the leg to the back, where they eat a hole in the skin of the animal's back for a breathing hole, Figure 19–13. The larvae, or grubs, feed on the host animal's flesh until they mature. At this time, they eat their way through the hide and fall to the ground where they live on debris. Here the larvae turn into pupae and mature to adults. This process takes about one to two months. The adult flies emerge, attach themselves to the heel of an animal, lay eggs, and the process starts all over again, Figure 19–14.

Cattle grubs cause damage due to the discomfort suffered by the animal. In addition, meat damaged by the grubs must be trimmed away and is lost. Hides from cattle infected by grubs have holes in them from the larvae's opening holes for breathing and emerging from the animal. This greatly reduces the value of the hide.

Parasite Control

In order to make the animals comfortable so they grow and produce efficiently, both internal and external parasites must be controlled. The most widely used method of control is medicating the host animals. Since the parasites feed on some part of the animal's

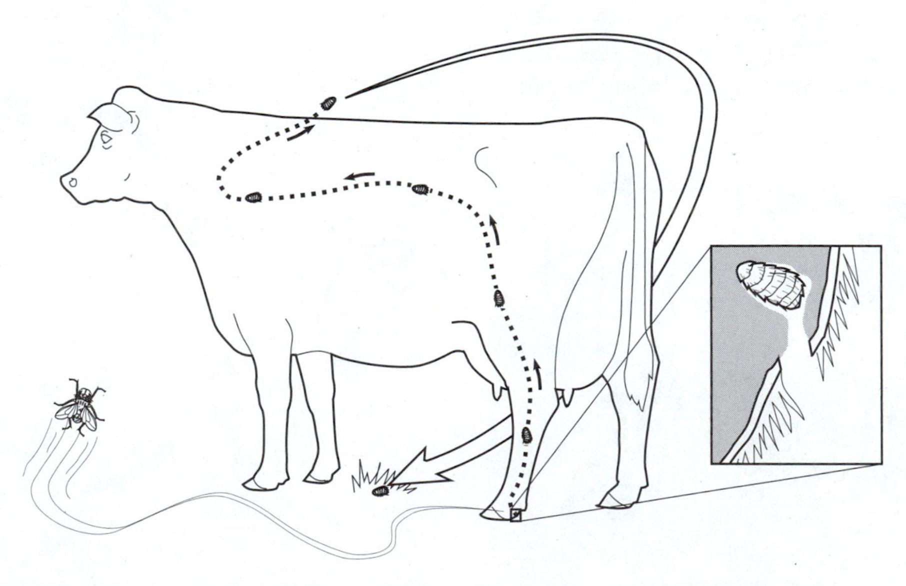

Figure 19-14. The adult heel fly lays eggs on the hair of the animal's heel. The larvae travel from the foot to the back through the body of the animal. *Cooperative Extension Service, The University of Georgia.*

Figure 19-15. External parasites can be controlled by applying insecticides to the animal's skin. *North Dakota Department of Education.*

Figure 19-16. Internal parasites can be controlled by giving medication to the host animal. *Calvin Alford, Cooperative Extension Service, The University of Georgia.*

body or ingested food, medication must be applied to the host to get rid of the parasites. With external parasites, the chemical or medication is applied to the animal's skin through spraying, pouring on, or running the animals through a dipping vat, Figure 19–15.

Recent research has developed a new generation of medications called **systemic pesticides** that are injected into the animal's body. Although the medication has little effect on the host, the parasites are killed or are repelled from the host animal. Internal parasites are controlled by giving the host animal an injection or a large pill (called a **bolus**), or by putting the medication in the animal's feed or water, Figure 19–16. The medications have undergone vigorous tests and regulations by the United States Department of Agriculture and the Food and Drug Administration in order to make sure the drugs are safe for the animal and that the milk, meat, eggs, or other products from the animals are safe for human consumption.

An alternative approach to controlling parasites is the use of **biological control**. At different stages of the parasite's life cycle, it is more vulnerable than at other stages. It is on these vulnerable stages that scientists have concentrated their efforts at controlling or eradicating the pests.

An interesting example is that of the screwworm. At one time the screwworm was a serious pest of agricultural animals. The adult female screwworm flies generally lay eggs in open wounds on animals; however, some infestations have been known to have occurred without the animal's having an open wound. The eggs hatch and the larvae, known as maggots, feed on the flesh of the host animal. The maggots tear out pockets of healthy flesh next to the wound and inject a toxin into the wound to prevent it from healing. The larvae grow and feed in the wound for about a week before dropping to the ground and changing to the pupa stage. The pupae go into the ground where they mature and become adults. Once the adults emerge from the ground, they mate and lay eggs to begin the life cycle again.

The screwworm fly mates only once during its lifetime, and scientists saw this as a weak point in its life cycle. In the late 1950s and early 1960s, a government program was begun to eradicate the screwworm. This was done by treating adult screwworm flies with gamma rays from cobalt-60. This treatment rendered both the male and female flies sterile. Massive numbers of the sterile screwworm flies were released in the areas of the South where the screwworm infestations occurred. When a sterile female mated with either a sterile male or a fertile male, no eggs were produced. When a sterile male mated with a normal female, no viable eggs were produced. After several years of releasing huge numbers of sterile adults, the screwworm was—for all practical purposes—eradicated. This effort was duplicated in the 1970s in a cooperative effort with the Mexican government. This effort was also quite successful. Through scientific research, efforts and control measures like this, pests can be controlled without causing harm to the host animal or the environment.

SUMMARY

Parasites have always been a problem for animal producers. Each year they cost millions of dollars in damage and control measures. In addition, they can cause disease problems to both animals and humans. By studying the life cycles of parasites, new and better control measures have been developed. A combination of methods are now used to control these pests. Through research and development some serious pests have been eradicated.

◆ Review Exercises

TRUE OR FALSE

1. Mutualism is the relationship of animals when one benefits and the other is not harmed. Commensalism is a relationship which is beneficial to both species of animals. ______.
2. Parasites very rarely carry disease organisms from one animal to another. ______.
3. Animals that are infected with parasites consume less feed per pound of gain. ______.
4. When a young insect first hatches, it is usually in the larval stage, looking very much like a worm. ______.
5. Roundworms cause more damage to agricultural animals than any other group of internal parasites. ______.
6. Ascarids most often attack young animals and work their way to the animal's heart, liver, and lungs. ______.
7. An intermediate host is an animal that the parasite uses to support part of its life cycle and which is killed in the process. ______.
8. When flukes use snails as their intermediate hosts, the larvae actually divide and multiply asexually inside the snails. ______.
9. Ticks cause damage by penetrating the skin and sucking the blood from the host animal. This leaves a sore that can be an avenue for disease organisms and can leave the animal anemic. ______.
10. Only half of the life cycle of the louse—that of the larva stage—is lived on the host. ______.
11. Heel flies live out their entire life cycle on the legs of the host, causing lameness that interferes with grazing. ______.
12. In order for animals to be comfortable and grow and produce efficiently, both internal and external parasites must be controlled. ______.
13. Biological control has proven to be completely ineffective in controlling parasites in agricultural animals. ______.
14. The adult female screwworm fly must have an open wound or injury in which to lay her eggs. There have been no known infestations occurring without an open wound. ______.
15. The governments of two separate countries cannot work successfully together to eradicate pests due to duplication of efforts and communication problems. ______.

FILL IN THE BLANKS

1. Parasitism is a ______________ that is ______________ to one animal and ______________ to another and accounts for the vast majority of the incidences of ______________.
2. Anemic conditions in agricultural animals result in a ______________ condition that makes the animals more susceptible to ______________ ______________ due to the ______________ ______________ functioning improperly.
3. Parasites cause irritation of the ______________, ______________ ______________, or other ______________ of the ______________.

4. Internal parasites actually live ________ the animal's ____________ and may feed on the animal's ____________ or on ________ that passes through the ________.
5. Strongyles are a type of ____________ that live in the ____________ of the ________ animal and cause damage by causing ____________ tissue in the ____________ intestine and by sucking ____________ from the animal.
6. Tapeworms are ____________ worms. Each segment contains both ____________ and ____________, reproductive ________, and each is capable of producing ____________ ________.
7. The most damaging flukes are those that live in the ____________ ____________ of the liver, where the adult causes ____________, ____________ of the liver and ________ ________, and general ________ of the ________.
8. External parasites cause losses in terms of animal ____________, the loss of ____________ ____________, and blood ____________.
9. Ticks can have up to ________ hosts during their life ____________, so diseases, such as ____________ ____________ ____________ ________ can be spread from one animal to ____________.
10. There are two types of lice: ________ sucking lice that feed by drawing ________ through the animal's ____________, and biting lice that feed on the ____________ or ____________ particles and ____________ of the animals.
11. Internal parasites are controlled by giving the host animal an ____________, a large pill (called a ________), or by putting the ________ in the animal's ________ or ________.
12. The medications have undergone vigorous tests and regulations by the ____________ ____________ ____________ of ________ and the ____________ and Drug ____________ in order to make sure the drugs are safe for the ________ and that the ________, ________, ____________, or other products from the animals are safe for ____________ ____________.
13. At different stages of the parasite's life, or life ____________, they are more vulnerable than at other ________.
14. Adult screwworm flies were rendered ____________ by treatment with ____________ ____________ from ____________.
15. Through scientific research efforts and control measures, some pests can be ____________ without causing ____________ to the ________ ________ or the ________.

DISCUSSION QUESTIONS

1. What is meant by a symbiotic relationship?
2. Name and give examples for three types of symbiotic relationships.
3. What are three ways that parasites can cause harm to their hosts?

4. What are three major groups of internal parasites that infest agricultural animals?
5. Explain what is meant by a life cycle.
6. What are the four stages in the life of an insect?
7. Explain how the following parasites cause damage to their hosts: roundworms, tapeworms, flukes, ticks, lice, heel flies.
8. List three ways of controlling parasites.

STUDENT LEARNING ACTIVITIES

1. Go to the library and research the life cycle of an insect pest. Develop a plan to control the insect by using the pest's life cycle.
2. Visit with a livestock producer and determine the measures he or she uses to control parasites. Ask the producer to explain how the methods are different from what they were a few years ago.
3. Visit with a local veterinarian and ask the vet to explain what the most serious parasites in the local area are and the measures used to control them.

Chapter 20

Animal Diseases

Student Objectives in Basic Science

As a result of studying this chapter, you should be able to

1. List the types of disease-causing organisms.
2. Describe three types of bacteria.
3. List the characteristics of viruses.
4. List the characteristics of protozoa.
5. Describe how an animal's immune system works.
6. Explain the function of red and white blood cells.
7. Describe how vaccines work.
8. Distinguish between infectious and noninfectious diseases.
9. Describe how diseases are spread.
10. Explain how antigens enter the body.
11. Explain the difference between passive and active immunity.
12. Distinguish between naturally acquired immunity versus artificially acquired immunity.

Student Objectives in Agricultural Science

As a result of studying this chapter, you should be able to

1. Describe the indications that an animal is sick.
2. List examples of diseases of agricultural animals caused by micro-organisms.
3. Explain how livestock diseases are spread.
4. List examples of diseases caused by genetic disorders.
5. Give examples of diseases caused by improper nutrition.
6. Give examples of plants that are poisonous to agricultural animals.
7. Cite examples of government disease-eradication programs.

Key Terms

infectious diseases
contagious diseases
cocci
bacillus
spirilla bacteria
virus
protozoa
phagocytes
lymphocytes
antigens
active immunity
naturally acquired active immunity
artificial active immunity
noninfectious diseases
aflatoxins
ergot

The term *disease* is broadly defined as not being at ease, or uncomfortable. Animals, just like humans, get diseases and have health problems. Producers of agricultural animals have a vested interest in keeping their animals healthy. Healthy animals grow faster and produce more profit for their owners.

Diseases come in a variety of types and have a variety of causes, Table 20–1. Some are mild and only cause minor discomfort to the animal; others are very severe and cause death quickly. Although animals that are sick may *not* show any outward signs or symptoms of being ill, they usually display symptoms that indicate

Table 20-1.

Disease	Cause	Symptoms	Preventive and Control Measures
NUTRITIONAL DEFECTS			
Anemia	All farm animals are susceptible	Characterized by general weakness and a lack of vigor; iron deficiency prevents the formation of hemoglobin, a red iron-containing pigment in the red blood cells responsible for carrying oxygen to the cells	A balanced ration will ordinarily prevent anemia. Baby pigs raised on concrete need iron supplement.
Bloat	Typically occurs when animals are grazing on highly productive pastures during the wetter part of late spring and summer	Swollen abdomen on the left side, labored breathing, profuse salivation, groaning, lack of appetite, and stiffness	Maintain pastures composed of 50 percent or more grass
Colic	Improper feeding	Pain, sweating, and constipation; kicking and groaning	Careful feeding
Enterotoxemia	Bacteria and overeating	Constipation is an early symptom and is sometimes followed by diarrhea	Bacterin or antitoxin vaccine should be used at the beginning of the feeding period
Founder	Overeating of grain or lush, highly improved pasture grasses	Affected animals experience pain and may have fever as high as 106°F	Good management and feeding practices will prevent the disease

Table 20-1. (Continued)

Disease	Cause	Symptoms	Preventive and Control Measures
VIRAL DISEASES			
Cholera	Hog cholera (now eradicated from the United States) is caused by a filterable virus	Loss of appetite, high fever, reddish-purplish patchwork of coloration on the affected stomach, breathing difficulty, and a wobbly gait	A preventive vaccine is available; no effective treatment; producers should use good management
Equine Encephalomyelitis	Viruses classified as group A or B cause the disease; transmitted by bloodsucking insects such as mosquito	Fever, impaired vision, irregular gait, muscle spasms, a pendulous lower lip, walking aimlessly	Control of carrier; use of a vaccine
Hemorrhagic Septicemia	Caused by a bacterium that seems to multiply rapidly when animals are subject to stress conditions	Fever, difficult breathing, cough, discharge from the eyes and nose	Vaccination prior to shipping or other periods of stress
Newcastle	A poultry disease caused by a virus that is spread by contaminated equipment or mechanical means	Chicks will make circular movements, walk backwards, fall, twist their necks so that their heads are lying on their backs, cough, sneeze, high fever, and diarrhea	Several types of Newcastle vaccines are available; antibiotics are used in treating early stages of the disease to prevent secondary infections
Warts	Believed to be caused by a virus	Protruding growths on the skin	No known preventive measures; most effective means is with a vaccine
BACTERIAL DISEASES			
Pneumonia	Bacteria, fungi, dust, or other foreign matter; the bacterium, *Pasteurella multiocida*, is often responsible for the disease	A general dullness, failing appetite, fever, and difficult breathing	Proper housing, ventilation, sanitation, antibiotics
Tetanus	A spore-forming anaerobic bacterium; the spores may be found in the soil and feces of animals	Difficulty swallowing, stiff muscles, and muscle spasms	Immunizing animals with a tetanus toxoid

Table 20-1. (Continued)

Disease	Cause	Symptoms	Preventive and Control Measures
Anthrax	A spore-forming bacterium	Fever, swelling in the lower body, a bloody discharge, staggering, trembling, difficult breathing, convulsive movements	An annual vaccination; manure and contaminated materials should be burned and area disinfected; insects should be controlled
Blackleg	A disease of cattle and sheep caused by a spore-forming bacterium, which remains permanently in an area; the germ has an incubation period of one to five days and is taken into the body from contaminated soils and water	Lameness, followed by depression and fever; the muscles in the hip, shoulder, chest, back, and neck swell; sudden death within three days of onset of symptoms	A preventive vaccine
Brucellosis	Caused by the bacterium, *Brucella abortus*	The abortion of the immature fetus is the only sign in some animals	Vaccinating calves with *Brucella abortus* will prevent cattle from contracting this disease; infected cattle must be slaughtered
Distemper	A disease of horses; exposure to cold, wet weather, fatigue, and an infection of the respiratory tract aid in spreading the disease	Increased respiratory rate, depression, loss of appetite, and discharge of pus from the nose are visible symptoms; infected animals will have a fever and swollen lymph glands, located under the jaw	Animals with disease should be isolated, provided with rest, protected from the weather, and treated with antibiotics
Erysipelas	A resistant bacterium capable of living several months in barnyard litter	Three forms: acute, subacute, and diamond skin form; acute symptoms are a high fever, constipation, diarrhea, and reddish patches on the skin; subacute is usually localized in an organ such as the heart, bladder, and joints; sloughing off of the skin is common	An antiswine erysipelasserum is available

Table 20-1. (Continued)

Disease	Cause	Symptoms	Preventive and Control Measures
Leptospirosis	A bacterium found in the blood, urine, and milk of infected animals	Abortion and sterility; symptoms are blood-tinged milk and urine	Susceptible animals should be vaccinated
Tuberculosis	The three types of tubercle bacilli causing the disease are human, bovine, and avian; the human type rarely produces tuberculosis in lower animals, but the bovine type is capable of producing the disease in most warmblooded vertebrates.	Lungs are affected; however, other organs may be affected; some animals will show no symptoms; others will appear unhealthy and have a cough	Maintaining a sanitary environment and comfortable quarters will help in preventing the disease
Pullorum	A poultry disease caused by a bacterium which is capable of living for months in a dormant state in damp, sheltered places; the germs infect the ovary and are transmitted to the chicks through the eggs	Infected chicks huddle together with their eyes closed, wings drooped, feathers ruffled, and have foamy, white droppings	Blood test is required for positive identification of the disease; disposal of infected hens will aid in preventing the disease; chicks should be purchased from a certified pullorum-free hatchery
Foot Rot	A fungi common to filth is responsible for foot rot; animals are most apt to contact foot rot when they are forced to live in wet, muddy, unsanitary lots for long periods of time	Skin near the hoofline is red, swollen, and often has small lesions	Maintaining clean, well-drained lots is an easy method of preventing foot rot
Calf Diphtheria	A fungal disease that lives in soil, litter, and unclean stables; it enters the body through small scratches or wounds	Difficulty breathing, eating, drinking; patches of yellowish, dead tissue appear on the edges of the tongue, gums, and throat; there is often a nasal discharge	The diseased tissue is removed to expose healthy tissue, which is treated by swabbing it with tincture of iodine

Table 20-1. (Continued)

Disease	Cause	Symptoms	Preventive and Control Measures
PROTOZOA			
Coccidiosis (pertaining to poultry)	Several species of protozoa are responsible	Occurs in two forms; cecae and intestinal: cecae is the acute form that develops rapidly and causes a high mortality rate; bloody droppings and sudden death are symptoms; intestinal coccidiosis is chronic in nature, and its symptoms are loss of appetite, weakness, pale comb, and low production; few deaths occur from the latter form	Since the disease is transmitted by the droppings of infested birds, maintaining sanitary conditions and the feeding of a coccidiostat will prevent the disease
UNKNOWN CAUSES			
Atrophic Rhinitis	Causes have not been determined; several different-bacteria are involved; it is contagious, especially in young pigs, and is spread by direct contact	Affects the bone structure of the nasal passages; the snout will become twisted and wrinkled	Sanitation is important in preventing the disease; there is not a specific treatment; the use of sulfamethazine may help

Instructional Materials Service, Texas A & M University

they are not feeling well. The animal may be droopy, go off feed and water, be restless, or have a dull coat, Figure 20–1. In some cases, the animal may have a fever. This means that the body temperature of the animal is higher than normal.

Figure 20-1. Sick animals may appear droopy. This turkey suffers from fowl pox. *Dr. Jean Sander, Department of Avian Medicine, The University of Georgia.*

INFECTIOUS DISEASES

Infectious diseases are those caused by microorganisms that invade the animal's body. These are usually **contagious diseases**, which means that the infected animal can pass the disease on to a healthy animal. There are many types of microorganisms that cause diseases in animals.

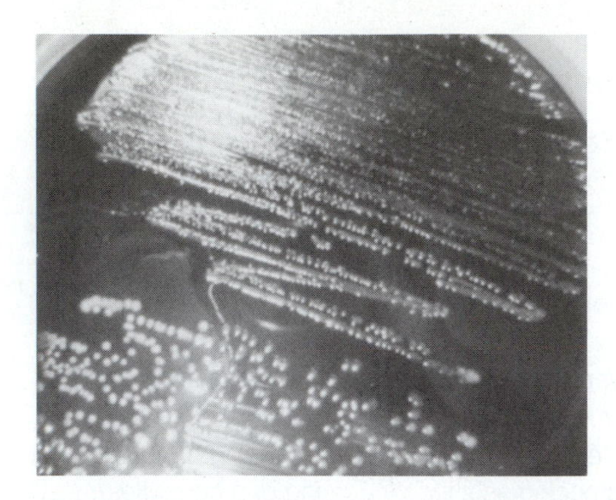

Figure 20-2. Bacteria are all around us. These bacteria are growing on media in a petri dish. *Dr. Jean Sander, Department of Avian Medicine, The University of Georgia.*

Bacteria

One of the most common types of disease-causing organisms is bacteria. Bacteria are all around us, Figure 20–2. They can be found in the hottest of deserts and buried deep in polar ice. They live on and in the bodies of all animals and are probably more numerous than the cells of the animal's body. Many of the bacteria around us are beneficial. Those living in the stomachs of ruminants help the animals digest food. Bacteria are useful in the production of cheese and such foods as sauerkraut.

There are, however, many types of bacteria that cause harm. Harmful bacteria invade the cells of an animal's body. Parasitic bacteria may harm the animal by feeding off the cells of the body or by secreting a material known as a toxin. A toxin is a substance that causes harm to an organism. In other words, it is a poison. When large numbers of harmful bacteria invade an animal's body, the animal becomes ill. The type and form of the illness depends on the type of bacteria that invades the animal.

Cocci are round, spherical-shaped bacteria. Diseases such as some forms of pneumonia and strep infections are caused by this bacteria.

Bacillus bacteria are rod-shaped organisms that may be single, in pairs, or may be arranged in chains. They cause many serious diseases in agricultural animals; a few are anthrax, tetanus, blackleg, intestinal coliform, salmonella, and tuberculosis.

Spirilla bacteria are shaped like spirals or corkscrews. These bacteria are very motile, which means they can move about very easily. They also require a moist atmosphere to survive; consequently, they live very well in the reproductive tracts of animals. Some of the diseases they cause are leptospirosis, vibriosis, spirochetosis, and many others.

Most bacteria can be controlled by the use of antibiotics. The first of these medicines was penicillin, which was produced from extracts of molds. Many different forms of penicillin are now produced artificially and are very effective against bacterial infections; however, some bacteria have developed resistance to antibiotics.

Viruses

A **virus** is a very tiny particle of matter composed of a core of nucleic acid and a covering of protein that protects the virus. Viruses have characteristics of both living and nonliving material. It could be said that viruses are on the borderline between living and nonliving. They are made up of some of the material found in cells, but they are not cells because they do not have nuclei or other cell parts.

Viruses do not grow and cannot reproduce outside a living cell. Once inside a living cell, the virus reproduces using the energy and materials of the invaded cell. Viruses harm cells by causing them to burst during the reproduction process of the virus and by using material in the cell that the cell needs to function properly, Figure 20–3. Therefore, viral diseases cause the animal to be sick by preventing certain cells in the animal's body from functioning properly.

There are many different types of viruses that cause a variety of serious diseases in agricultural animals. Viral diseases are more difficult to treat than diseases caused by bacteria. The antibiotic drugs that have proven effective against bacteria are of no use against viruses. Some of the more serious livestock diseases caused by viruses are foot-and-mouth disease, influenza, hog cholera, and pseudorabies. Many viral diseases are incurable. The best means of dealing with them is prevention.

Protozoa

Another type of microorganism that causes diseases in agricultural animals is the protozoan. **Protozoa** are single-celled organisms that often are parasitic. They cause harm to animals by feeding on cells or by producing toxins. Examples of diseases caused by protozoa are African sleeping sickness and anaplasmosis. Coccidiosis is one of the most costly diseases in the poultry industry, Figure 20–4. This disease is caused by several different species of protozoa and causes diarrhea and weight loss in chickens. Most protozoa can be controlled by drugs.

Figure 20-3. This chicken suffers from fowl pox. Note how the viruses have destroyed cells in the comb. *Dr. Jean Sander, Department of Avian Medicine, The University of Georgia.*

The Immune System

Disease-causing viruses, bacteria, and protozoa are all around the environments of animals and people. They are so prevalent that they are ingested into the body almost constantly. These organisms can enter the body through regular body openings such as the mouth, nose, eyes, the reproductive system, or any other natural opening. They may enter through the skin or through a wound in the body as well. If the animal's body did not have a means for defending itself against these disease agents, the animal would live a short, miserable life.

Fortunately, animals (including people) have several lines of defense in fighting disease. The first line is physical barriers that keep the disease-causing agents out. For instance, the nostrils are lined with hairs that attract particles that harbor germs before they can enter the body. When an animal sneezes, the particles are expelled. Almost all body openings and many internal organs are lined

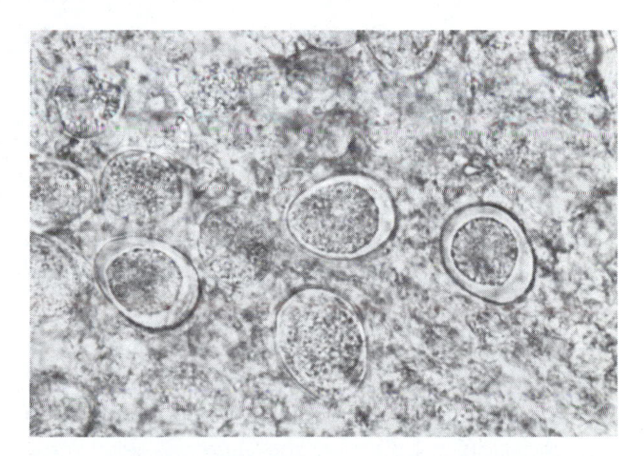

Figure 20-4. These protozoa cause coccidiosis in poultry. *Dr. Jean Sander, Department of Avian Medicine, The University of Georgia.*

Figure 20-5. Animals come in contact with the ground. Disease-causing germs can live in the soil for many years. *Rick Jones, Cooperative Extension Service, The University of Georgia.*

with mucous membrane. These are tissues that secrete a viscous, watery substance that traps and destroys bacteria and viruses.

The greatest avenue for organisms to enter the body is through the digestive and respiratory systems. Countless billions of microorganisms live in the soil, and some of them are disease-causing germs. Certain disease germs can live in the soil for many years.

Most animals come in contact with the ground as they eat, Figure 20–5. Cattle, sheep, and horses graze and pull grass from the ground, and pigs root in the ground for food. Even processed feed is far from being sterile. Every minute animals breathe in large amounts of air laden with all sorts of particles and organisms. Fortunately, the body has ways to destroy harmful organisms. For instance, if the germs swallowed with feed are not trapped by the mucous membrane of the digestive tract, they are most often killed by the digestive enzymes. Germs ingested by breathing are trapped in the mucous membranes of the respiratory tract.

Disease agents that get through the first line of defense are usually destroyed by the second line of defense. This line is composed of cells and chemicals in the bloodstream. Blood is basically composed of two types of cells—white blood cells and red blood cells. The red blood cells carry oxygen and nourishment to the other body cells. White blood cells are produced in the bone marrow and circulate throughout the body to get rid of worn-out body cells. Certain of these white blood cells, called **phagocytes**, intercept and destroy disease-causing agents, Figure 20–6. These cells also migrate to certain organs, such as the liver, lymph nodes, and the spleen, and remain there to intercept disease-causing agents. White blood cells also circulate through other body fluids and the mucous membranes.

When phagocytes encounter foreign organisms, they release chemicals that induce the production of more white blood cells to help fight the disease organism. In fact, one important way a veterinarian can tell if an ani-

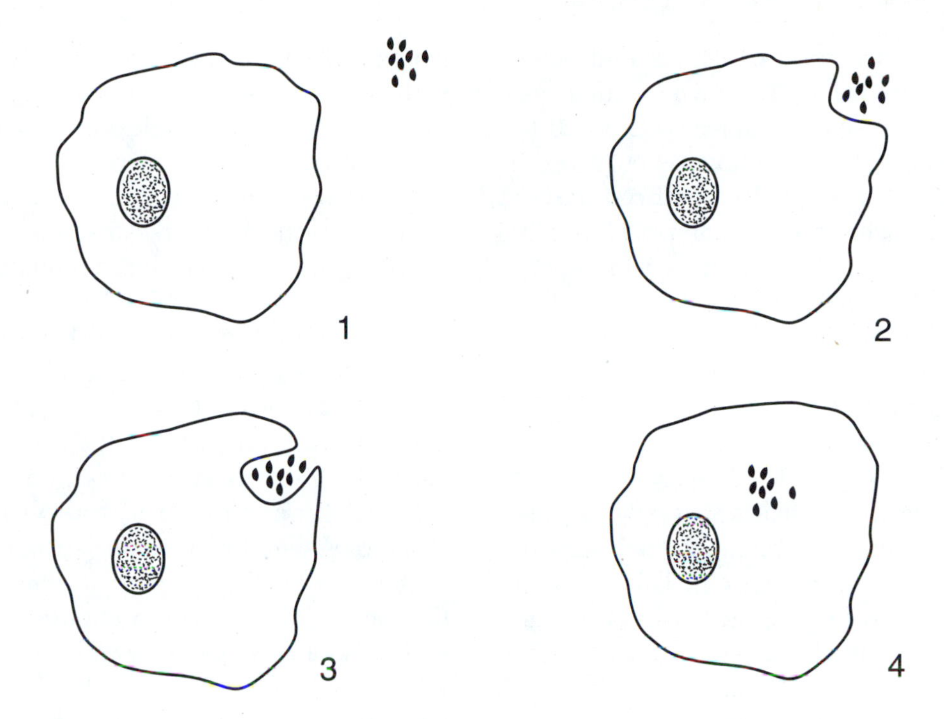

Figure 20-6. Phagocytes intercept, engulf, and destroy disease-causing agents.

mal is sick is by counting the number of white blood cells in the animal's bloodstream. A count larger than normal indicates that there are disease organisms present in the animal's body, and a large number of phagocytes have been produced to combat them.

Certain white blood cells are produced by the lymph glands and are called **lymphocytes**. These cells react to foreign substances by releasing chemicals that kill the disease-causing organisms or inactivate the foreign substance. The substances that cause the release of the chemicals are called **antigens**. Antigens may be viruses, bacteria, toxins, or other substances. The chemicals released by the lymphocytes are known as antibodies. The lymphocyte can also become a "memory cell" that is ready to release an antibody if the same type of antigen enters the body at a later time. When this happens, it is known as a secondary immune response. This response occurs much more rapidly and lasts longer than the primary response.

Immunity

Immunity means that an animal is protected from catching a certain disease. This is because the animal's body is capable of producing sufficient antibodies in sufficient time to neutralize the disease-causing agent before the animal becomes sick. Immunity can be either active or passive. **Active immunity** means that the animal is more or less permanently immune to the disease. Passive immunity means the animal is only temporarily immune.

Animals are born with some immunity to diseases. In mammals, the first milk, called *colostrum*, given to the newborn animal is rich in antibodies from the mother. These antibodies serve the new animal until its own immune system can take over. As the animal is exposed to more and more antigens, antibodies build up in the animal's body. **Naturally acquired active immunity** is obtained by the animal's actually having a disease and recovering. The memory phagocytes react quickly when the antigen that causes that specific disease enters the animal's body and the antigens are overwhelmed.

Artificial active immunity can be induced in the animal by injecting antigens into the animal that causes the phagocytes to react without making the animal seriously ill, Figure 20–7. This process was begun by an Englishman named Edward Jenner in the late 1700s. At that time, smallpox epidemics swept through many parts of the world, killing over half of the people that contracted the disease. Those who survived were permanently immune to the disease, which meant they would never be sick with it again. Jenner knew that those who contracted cowpox never came down with smallpox. Cowpox was, as the name implies, a disease of cattle. Humans could also get the disease, but it was usually mild. Jenner collected material from sores that developed on people with cowpox and injected healthy people with the material. The people injected became ill with a mild case of cowpox, but were then immune to smallpox. The Latin word for cow is *vacca,* and the word *vaccination* was coined based on the fact that the immunity originated from cows.

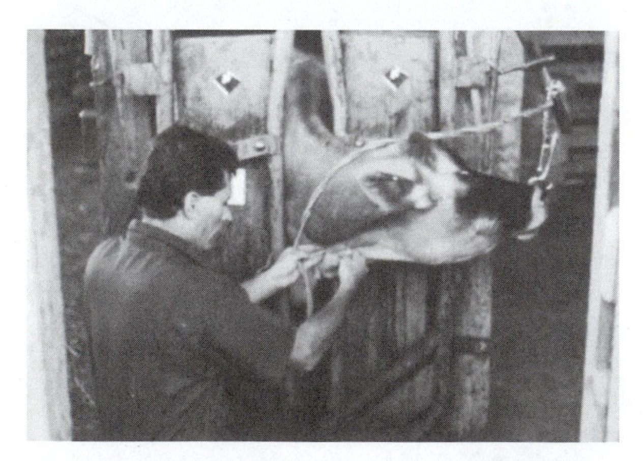

Figure 20-7. Artificial immunity can be induced by injecting antigens into an animal. *Calvin Alford, Cooperative Extension Service, The University of Georgia.*

This concept was used later by Louis Pasteur to develop several vaccines. All of the modern vaccines, whether given to humans or animals, work on basically the same principles. When he used materials from cowpox sores, Edward Jenner injected live viruses. This worked well in this particular case, but as more vaccines were developed, it was discovered that often the vaccination could cause the disease. Also, many of the viruses can live for long periods of time in the soil. If bottles of live vaccine are dropped and broken, the soil can become contaminated. Research has proven that weakened or killed viruses can be effective vaccines against many diseases. These materials act as an antigen in stimulating the production of antibodies in much the same way as a live virus, but without the dangers incurred with using live viruses.

NONINFECTIOUS DISEASES

Not all diseases of agricultural animals are caused by being infected with microorganisms. Diseases can be caused by means other than contact with infected animals. In other words, these are **noninfectious diseases.** They are not contagious.

Genetic Diseases

Some diseases are caused by defects in the genes that were transferred from the animal's parents. Usually animals with genetic disorders will also pass the problem on to the next generation, so the disease stays in certain breeds or bloodlines of animals. One example is a condition known as white heifer's disease in Shorthorn cattle. Certain heifers that are solid white in color have a genetic defect that causes them to be sterile. Since the cause is purely genetic, other cattle or other animals cannot contract the disease from the heifers with the disease. In certain lines of Holstein cattle, calves are born with a condition known as mule foot in which the hooves are shaped like a horse's or mule's foot rather than having two toes like normal cattle. Again, this disorder is purely genetic and cannot be spread through contact with other cattle.

At the present, the only way to control genetic diseases is by using good selection practices and avoiding breeding animals that are known to have genetic defects in their line. Perhaps in the near future the process of genetic engineering will be developed to the point where these problems can be removed from animals.

Nutritional Diseases

Animals can become sick from faulty nutrition. All animals need certain amounts of a variety of nutrients. If these nutrients are absent from the animal's diet, the animal can become very ill, Figure 20–8. An example of a nutritional disease is milk fever in dairy cattle. Cows with this disease lie down and are unable to stand. The condition is caused by an insufficient amount of calcium in the bloodstream. Since milk is rich in calcium and milk comes from the bloodstream, cows producing heavily sometimes have this problem. The disease is usually cured by the injection of calcium salts directly into the animal's bloodstream. The effects are immediate and dramatic. Animals that have been down for hours suddenly are able to stand and move about.

Other nutrition-related diseases can be caused by overeating. Cattle that are turned in on lush, green grazing can have a problem known as bloat. Bloat is caused when the sudden ingestion of large amounts of green forage (usually legumes) causes foaming in the animal's digestive tract. The foamy bubbles block the openings of the tract and prevent the passage of gas. The gas can build up to such an extent that it can cause death.

Horses, cattle, and sheep can get a condition known as founder if they eat too much grain, especially if the ration is changed too rapidly. This condition causes the feet to become inflamed and the hooves to grow upward and outward. The animals don't eat very well and lose weight rapidly.

POISONING

Like any other animals, agricultural animals can be made sick by ingesting toxic materials. These materials can be picked up in a

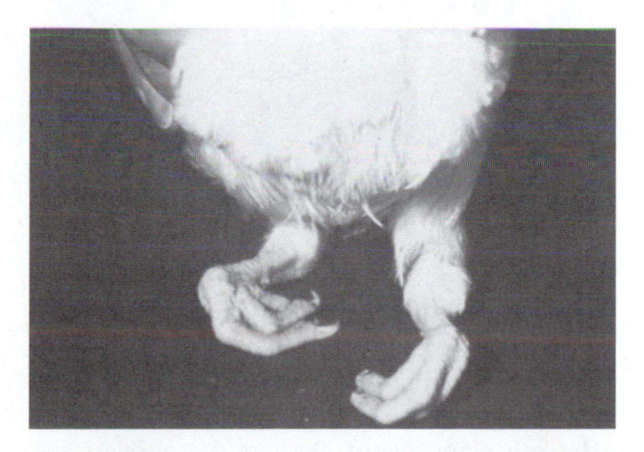

Figure 20-8. Lack of proper nutrition can cause disease. The deformity in this chicken's feet was caused by a deficiency of riboflavin in the diet. *Dr. Jean Sander, Department of Avian Medicine, The University of Georgia.*

Figure 20-9. Common plants such as cocklebur (left) and milkweed (right) are poisonous to livestock. *USDA photo.*

variety of ways. The animal can eat feed that is contaminated. If feed becomes moldy, certain toxins can develop that can make livestock quite sick and may even kill them. Among the most potent of the toxins from moldy feed are **aflatoxins** and **ergot**. Both of these toxins are developed from fungi that grow on grains. Producers are careful that feed fed to agricultural animals is free from mold.

Another type of poisoning that causes problems with animals is that of poison plants. Animals that graze can readily pick up plants that contain toxins. Poisonous plants occur all over the United States. Losses are incurred in all states from plant poisoning. The western region of the country sustains the heaviest loss because of so much grazing on uncultivated range land.

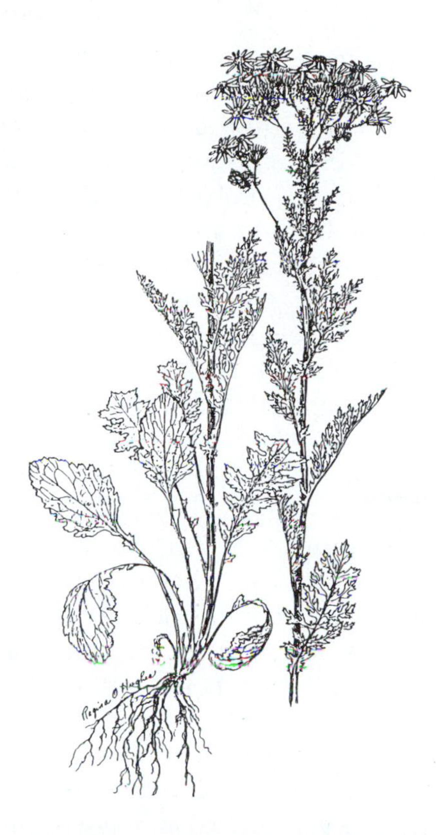

Figure 20-10. Tansy ragwort is deadly to horses and cattle, yet practically harmless to sheep. *USDA photo.*

Some fairly common plants, such as ferns, bitterweed, buttercups, cocklebur, and milkweed, can be poisonous to some species of agricultural animals, Figure 20–9. Certain plants can be very toxic to some animals and almost harmless to others. For example, tanzy ragwort, that grows in the Northwest, is deadly to horses; yet sheep can eat the plant with little or no harm, Figure 20–10.

DISEASE PREVENTION

Producers of agricultural animals take careful measures to protect their animals from diseases. Most have strict schedules of vaccinations for all their animals to prevent them from contracting diseases, Figure 20–11. Even though there may have been no outbreaks of a particular disease, producers still use vaccinations because they are aware that disease organisms are spread in a variety of manners.

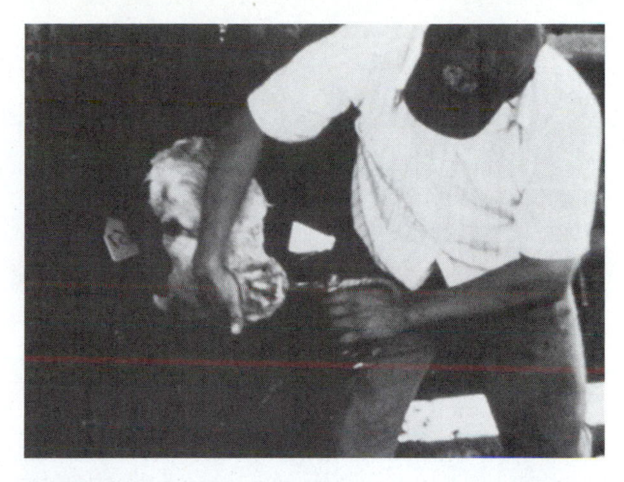

Figure 20-11. Most producers have strict vaccination schedules to protect their animals from diseases. *Calvin Alford, Cooperative Extension Service, The University of Georgia.*

Infectious disease organisms can be transported by wildlife. Deer can transmit certain diseases to cattle and sheep. Wild pigs can spread disease to domesticated herds, and wild horses can infect domesticated animals. Many species of birds can transmit disease to animals, especially to chickens. At one time, birds were held responsible for spreading hog cholera from one farm to the next.

Diseases can also be spread by humans. People moving from one farm to another can carry disease organisms on shoes or clothes. Many modern swine and poultry producers no longer allow visitors in their production houses, Figure 20–12. Those who do allow visitors insist that disposable boots be worn or that shoes be thoroughly disinfected.

Newly purchased animals may also be a source of disease outbreak. Producers usually

Figure 20-12. Many producers do not allow visitors in their production houses. *Dr. Frank Flanders, Agricultural Education, The University of Georgia.*

keep new animals away from the other animals until they are certain that the new animals are disease-free. Animals that come from foreign countries are kept in quarantine until they can be declared disease-free. They are kept in isolation areas outside the country during this period. It is economically feasible for only very valuable animals to be kept in quarantine.

Government regulations such as quarantining help deter the spread of livestock diseases. Livestock that are transported across state lines require a certificate showing that the animal has been examined, tested, and declared disease-free. Also, animals entering fairs and livestock shows are required to have health certificates. The federal government has eradication programs to eliminate diseases. An example is the brucellosis (Bangs) program. Brucellosis is a disease of several agricultural animals, but it is a particularly serious problem in cattle. This disease causes abortion and accounts for large losses in profit among cattle producers. States that have herds of cattle with brucellosis require that all animals being sold to producers be tested for the disease. If an animal is tested positive, it is branded and usually sent to slaughter.

SUMMARY

Since humans began raising livestock, problems with diseases have arisen. Many of the diseases are minor and cause relatively few problems while other diseases can completely devastate large herds of animals. There are thousands of different types of pathogens and substances that can cause disease. Through research and development, humans have learned very effective ways of preventing and treating animal diseases. In the future, as in the past, new diseases will surface and new and better control methods will have to be developed.

◆ Review Exercises

TRUE OR FALSE

1. The health of an animal does not affect the profit made by its owner. ______.
2. Infectious diseases, which are usually contagious, are caused by microorganisms. ______.
3. Protozoa cause harm to animals by feeding on cells or by producing toxins. ______.
4. The first line of defense against disease includes both hairs (such as those found in the nostrils) and mucous membranes (found in many of the internal organs). ______.
5. A very large number of white blood cells in an animal can tell a veterinarian that an animal is healthy and at no risk for disease. ______.
6. Naturally acquired active immunity is obtained by the animal's actually having the disease and recovering. ______.
7. Artificial active immunity can be induced in an animal or person by injecting antigens that cause the phagocytes to react without making the animal or person seriously ill. ______.

8. Genetic diseases are contagious. ______.
9. Diseases such as white heifer's disease and mule foot can be prevented with vaccines. ______.
10. Although sickness due to faulty nutrition, such as bloat, can be serious, it is never fatal. ______.
11. The western region of the country sustains the heaviest loss from poisonous plants because of so much grazing on uncultivated rangeland. ______.
12. Plants that are very toxic to some animals can be almost harmless to others. ______.
13. Infectious disease organisms can be transported by wildlife and humans. ______.
14. Animals that come from foreign countries are kept in quarantine until they can be declared disease-free. ______.
15. All animals entering fairs and livestock shows are required to be kept in quarantine for at least 48 hours. ______.

FILL IN THE BLANKS

1. Bacteria may harm animals by feeding off the ______ of the ______ or by secreting a material known as a ______, which causes ______ to an organism.
2. One of the first antibiotics developed to fight bacteria was ______, which was produced from ______ of ______.
3. Viruses harm cells by causing them to ______ during the ______ process of the virus and by using ______ in the cell that the cell needs to ______ properly.
4. Disease-causing viruses, ______, and protozoa can enter the body through the ______ body openings such as the ______, ______, ______, the ______ system, or any other ______ opening as well as through the skin or through a ______ in the body.
5. Some white blood cells, called phagocytes, intercept and ______ disease-causing ______ as well as migrating to certain organs such as the ______, ______ nodes, and the ______.
6. Lymphocytes are cells that react to ______ substances by releasing ______ that kill the ______-causing organisms or ______ the foreign substance.
7. Active immunity means that the ______ is more or less ______ immune to the disease, while passive immunity means the animal is only ______ immune.
8. In mammals, the ______ milk given to the ______ ______ animal is rich in ______ from the mother, which serves the new animal until its own ______ ______ can take over.
9. Research has proven that weakened or ______ viruses can be effective ______ against many ______.
10. The only way known to prevent genetic diseases is by using good ______ practices and avoiding ______ animals that are known to have genetic ______ in their line.
11. Diseases such as ______ ______ in dairy cattle (caused by lack of calcium), ______ (caused by cattle being turned out on lush, green grazing), and ______ (caused by animals eating too much grain) are all caused by incorrect nutrition.

12. Among the most potent of the toxins from moldy ________________ are ________________ and ________________, which develop from ________________ that grow on ________________.
13. Modern swine producers who allow visitors usually insist that ________________ boots be worn or that ________________ be thoroughly ________________.
14. Livestock that are transported across state lines require a ________________ showing that the animal has been ________________, ________________, and declared ________________-________________.
15. If an animal is tested positive for brucellosis (Bangs), it is ________________ and usually sent to ________________.

DISCUSSION QUESTIONS

1. What are some of the indications that an animal is not well?
2. What is meant by an infectious disease?
3. List three types of bacteria according to shape.
4. How do viruses cause an animal to be sick?
5. What is an animal's first line of defense in fighting infectious disease?
6. What role do white blood cells play in fighting disease?
7. What are antigens?
8. What is the difference between active and passive immunity?
9. List three types of noninfectious diseases.
10. Explain at least two ways diseases are spread.

STUDENT LEARNING ACTIVITIES

1. Visit with a local veterinarian who treats large animals. Determine what livestock diseases he/she has encountered in your area. Find out the procedures used in diagnosing diseases.
2. Visit with a producer and find out what measures he/she takes to prevent diseases. Ask him/her which diseases he/she fears the most.
3. Prepare a list of the plants in your area that are poisonous to livestock. Your local county Extension Office should have helpful information.

Chapter 21

The Issue of Animal Welfare

Student Objectives in Basic Science

As a result of studying this chapter, you should be able to

1. List potential problems brought about by animals being raised in confinement.
2. Determine why animals raised in an agricultural setting are healthy and efficiently grown.
3. Explain a potential problem associated with the continuous ingestion of antibiotics.
4. Cite examples of how the use of animals in research has helped humans.
5. List the laws that govern the use of laboratory animals for research.

Student Objectives in Agricultural Science

As a result of studying this chapter, you should be able to

1. List the reasons why some people object to the raising of farm animals.
2. Defend the use of confinement operations.
3. Defend the use of management practices associated with the raising of agricultural animals.
4. Explain how producers benefit when their animals are content and healthy.
5. List the laws governing the use of agricultural animals.

Key Terms

withdraw periods
debeaking
docking
dehorning
castration
pecking order
elastrator
scrotum
freeze branding
hot branding

People have used animals for as long as they have been on the earth. Early humans hunted animals to eat and to use their hides for clothing and shelter. Later, as civilizations began to develop, humans began to raise animals in order to have a ready and abundant source of food and clothing. When this happened, people began to control all aspects of the lives of animals. This meant that the animals depended on the people who raised them for food and protection.

Almost from the very beginning, people have been concerned about the well-being of animals they raised and controlled, Figure 21–1. After all, animals are living creatures with the ability to feel pain and to suffer distress, and people have always been concerned that animals not suffer needlessly. In fact, as far back as the time of the Greek mathematician Phythagoras, arguments have been made that people should not eat animals because animals should not be killed. Later Greek philosophers such as Plutarch argued that animals should be treated with justice. In the seventeenth and eighteenth centuries, protection societies began to develop that sought to prevent animals from being mistreated.

In the United States, the Animal Rights Movement began in the 1970s. Today, various animal rights organizations are very active politically, trying to pass laws governing how animals may be treated. There are at least two main lines of thought associated with this movement. One philosophy is that animals should have the same rights as humans. People who espouse this philosophy believe that animals should be free to live their lives without interference from people, that it is not right to kill animals for food or to obtain their skins for clothing or any other purpose. These people believe that killing animals is just as wrong as killing humans, that animals should have the same rights as humans. This group is known broadly as animal rights activists.

The other line of thinking is that it is moral to raise animals for human use, but that animals should not be abused or mistreated in any way. These people believe that although animals are raised for slaughter, the animals should be made as comfortable and as "happy" as possible while they are alive. This group is generally known as animal welfare activists.

In theory, livestock producers do not argue with the animal welfare activists. The controver-

Figure 21-1. Producers have always been concerned over the welfare of their animals. *Cooperative Extension Service, The University of Georgia.*

sy centers around what constitutes the abuse or mistreatment of animals. The animal welfare activists object to what they refer to as livestock factories in which animals are mass produced in conditions that totally neglect the welfare of the animals. They see the only motivation in producing the animals to be profit. According to the animal welfare activists, there are several areas that cause problems for animals.

Confinement Operations

Animal welfare activists feel that modern livestock operations are nothing more than animal factories where animals are mass produced like nonliving things. They feel that the modern farm where animals are produced is a lot different from farms of several years ago. The activists are of the opinion that all the modern producer cares about is making money, even if it causes animals to suffer. The production of animals in a confined space is viewed as being cruel and causing animals to suffer, Figure 21–2. They object to pigs being raised in crowded pens, where they never leave the pen and have little room for exercise. The placing of sows into farrowing crates where they cannot turn around or take a step is considered cruel.

Animal welfare activists disapprove of layer hens being kept in cages for their entire lives. They feel that the hens are put under stress because they do not have enough room to stretch their wings or to get any exercise. The hens are seen as being similar to a factory production line where all feed and water are brought to the hens and their sole function in life is that of producing eggs. Cattle feedlots are viewed as objectionable when the animals are crowded together and no shade is provided to protect them from the sun and no shelter is provided against the rain and the cold.

Livestock producers contend that almost all livestock produced in the United States come from family-owned farms and ranches. Producers are in the business because they enjoy working with animals and have the animals' best interests at heart.

Producers point out that in order to stay in business, they must make a profit. More profit can be made if the animals are healthy and well

Figure 21-2. Animal welfare activists object to animals being raised in a confinement operation. *James Strawser, Cooperative Extension Service, The University of Georgia.*

Figure 21-3. Healthy animals that are not under stress produce better and make more profit for the producer. *James Strawser, Cooperative Extension Service, The University of Georgia.*

cared for, Figure 21–3. Animals that are under stress cannot grow and produce well. In fact, the more comfortable an animal is, the more profit can be made because the animal is growing more rapidly. Producers point out that sows are put in farrowing crates to protect the piglets that might otherwise be crushed by the mother, Figure 21–4.

Another argument for the use of confinement operations is that over the years animals have been specially bred for confinement operations. These animals are vastly different from their relatives in the wild. Also, animals in confinement are easier to care for because the producer can see each animal every day and often several times a day, Figure 21–5. Housing provides shelter from the elements and from predators.

Many millions of dollars and countless hours of effort have gone into research to design housing and facilities that make animals comfortable. In other words, a hog house is scientifically designed for hogs. The design takes into account the well-being of the animal, Figure 21–6.

An uncomfortable, stressed-out animal simply will not grow as efficiently as an animal that is content. The fact that animals are far

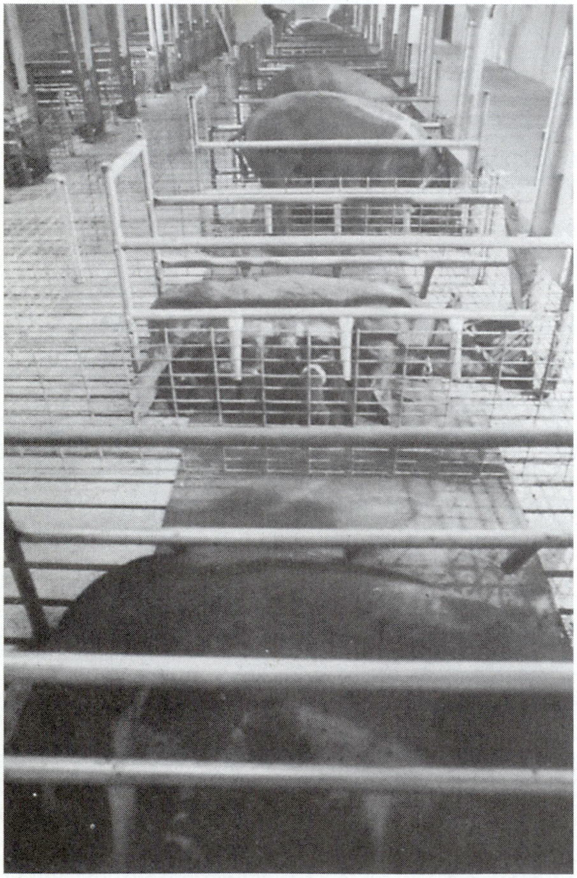

Figure 21-4. Sows are placed in farrowing crates to protect the piglets. *James Strawser, Cooperative Extension Service, The University of Georgia.*

Figure 21-5. Animals raised in confinement operations are easier to care for. *Rick Jones, Cooperative Extension Service, The University of Georgia.*

Figure 21-6. Hog facilities are designed to provide comfort for the animals. *Rick Jones, Cooperative Extension Service, The University of Georgia.*

more efficient in terms of production and growth than they have been in years past adds credence to the producer's arguments.

THE USE OF DRUGS

The animal welfare activists disapprove of feeding drugs such as antibiotics to animals as a preventative measure. They point out that traces of the drugs may possibly show up in the meat that is to be consumed by people. They feel that the drugs will have an adverse effect because bacteria that the drugs are guarding against may become immune to the antibiotics as a result of the prolonged feeding of the drugs. They are also concerned that the bacteria will develop into strains of organisms that will not respond to modern antibiotics. This could cause serious health problems not only for animals but also for humans.

On the other hand, producers counter that the addition of medication to the feed makes the animals healthier than they would be if they were allowed to roam free in nature. Because of the medication, the animals remain not only free from disease but also free from parasites, Figure 21–7. A healthy animal that is free from external and internal parasites suffers less than an animal that does not have the benefit of medication.

Producers point out that in order for a drug to be approved for use in animals, the Food and Drug Administration puts the proposed drug through exhaustive testing to prove

Figure 21-7. Animals that are fed medications are free from parasites and are healthier. *USDA photo.*

that the medication is not only safe and beneficial to the animals, but that it is also safe for humans who consume the meat, eggs, or dairy products from those animals. Laws mandate strict **withdrawal periods** for the drugs. This means that there is a minimum number of days that an animal must be off a particular medication before it can be slaughtered or before dairy products from the animal can be used.

Figure 21-8. Animal welfare activists consider many management practices to be cruel to the animals. *Rick Jones, Cooperative Extension Service, The University of Georgia.*

Management Practices

Animal welfare activists are concerned that animals undergo such management practices as **debeaking**, tail **docking**, **dehorning**, and **castration**. They point out that almost always these practices are done without anesthesia, and this causes the animal to suffer, Figure 21–8. The activists take the position that the animal would be better off if these operations were not performed. They reason that nature had a reason for creating each animal "as is" and that it should be left in its natural condition.

Producers explain that these practices are necessary for the well-being of the animals. For example, an animal with no horns is far less likely to cause injury to another animal or to a human than an animal that has horns. Since nature intended the horns for use in self-defense, animals under the protection and care of humans have no real need or use for the horns; therefore, dehorning is beneficial to the animal.

The debeaking of chickens is done to prevent them from injuring others as they establish a natural **pecking order**. In modern operations, the animals are far less likely to be injured if the chicks are debeaked, Figure 21–9. (Actually only the tip of the beak is removed.)

Tails are docked in sheep because research has shown that removing the tail allows the animal to remain cleaner and healthier. Mud, dirt, and manure cling to the tail and cause stress to the animal.

Castration is done at an early age to prevent problems associated with male animals fighting for dominance and to provide a higher-quality meat when the animal is slaughtered. Although the animals undergo temporary discomfort with these procedures, producers argue that in the long run the practices are in the best interests of the animals.

Figure 21-9. Chicks are debeaked to prevent them from injuring each other. *James Strawser, Cooperative Extension Service, The University of Georgia.*

Producers have always looked for management techniques that cause less stress to their animals. Procedures such as tail docking and castration are done at an early age because the younger animals suffer less trauma than older animals that undergo these operations.

The use of caustic soda or other solutions on the horn buds of young calves is usually thought of as being less stressful than cutting the horns out when the animal is older. Many producers use a technique like the **elastrator** band to castrate animals and dock their tails. This method does not require opening a wound on the animal. It consists of placing a rubber band around the tail or **scrotum** of the animal to prevent the flow of blood. Consequently, the scrotum or tail eventually drops off.

One technique that has lessened stress on animals is the use of **freeze branding** instead of **hot branding**. Producers must have some way of identifying individual animals. The use of the hot branding iron has been used by cattle producers for many generations. Critics of this process say that the procedure causes severe pain for the animals. The hot iron causes a burn that scars over and leaves a permanent mark on the animal.

Figure 21-10. To freeze brand, irons are placed in a container of liquid nitrogen to make them tremendously cold. *James Strawser, Cooperative Extension Service, The University of Georgia.*

A newer technique called *freeze branding* uses a tremendously cold iron to do the marking. In this method, an iron is immersed in a container of liquid nitrogen. Nitrogen in the liquid state is approximately –320°F and it cools the iron to an extremely cold temperature, Figure 21–10. The iron is then applied to a part of the animal's skin that has had the hair closely clipped. This procedure kills the pigment-producing cells at the base of the hair follicle and results in the growth of white hair. A longer application of the iron will result in the killing of the entire follicle and will leave a bald mark where the iron touched. The advantage of this method is that the animals feel very little pain as they are permanently marked, Figure 21–11.

Figure 21-11. Freeze branding kills the hair pigment, thus marking the animal. *James Strawser, Cooperative Extension Service, The University of Georgia.*

Figure 21-12. In some parts of the world, shepherds still live with their sheep.

Livestock producers have always been known for their concern and care for the animals they raise. They have always taken pride in producing strong, healthy animals. Throughout all of recorded history, accounts have been left of how producers carefully tended their animals—often at the risk of peril to themselves. Shepherds have traditionally lived with their sheep and guarded them against wild animals, Figure 21–12. Even though methods have now changed and the raising of animals is much more intensive, most animals are still cared for by families who earn their livelihoods caring for animals.

According to the Animal Industry Foundation, 97 percent of the farms in the United States are family owned and operated. The United States Department of Agriculture (USDA) reports that there are only seven thousand non-family-owned farms in the entire United States. A tour of any of the thousands of livestock shows that are conducted each year in the United States will show the amount of pride these farm families take in the livestock they produce.

At the 1990 National Cattlemen's Association Convention the producers adopted the following statement of principles on animal care, environmental stewardship, and food safety:

I believe in the humane treatment of farm animals and in continued stewardship of all natural resources.

I believe my cattle will be healthier and more productive when good husbandry practices are used.

I believe that my and future generations will benefit from my ability to sustain and conserve natural resources.

I will support research efforts directed toward more efficient production of a wholesome food supply.

I believe it is my responsibility to produce a safe and wholesome product.

I believe it is the purpose of food animals to serve mankind and it is the responsibility of all human beings to care for animals in their care.

Careers in Animal Science

Animal science is a broad and diverse industry that deals with the biological sciences. If you enjoy studying about and working with animals, you might think about a career in animal science. As with any career, preparation for the occupation is essential. There are hundreds of different jobs available in the area of animal science that require many different types and levels of education and training. As with most careers, salaries and working conditions are usually better in jobs that require more education.

You may wish to start work as soon as you graduate from high school; you may wish to attend a two year post-secondary institution; or you may wish to major in animal science or a related area at a university. You may even wish to continue your education after graduating from a university. Many exciting careers dealing with animals can be obtained by attending graduate school and obtaining a Master's Degree or a Ph.D. in animal science. You could even become a research scientist who discovers new and better methods for growing agricultural animals! No matter what level of career you wish to enter, you will need to prepare yourself with an education. Then after you begin your career, you will continue to learn as you improve and advance in your chosen area.

It is never too early to begin thinking about a career in animal science. Your parents, the school guidance counselor, your high school agriculture teacher, and the faculty in an animal science department at a community college or university are all good sources of information on a career in animal science.

Careers with a High School Diploma

If you choose to go to work immediately after graduating from high school, there are several jobs available to you in the area of animal science. You should take all of the courses in agriculture, agriscience, and agribusiness that your school offers. In these courses you will learn the basics of how plants and animals live, grow, and reproduce. In addition, you will learn the essentials of properly caring for animals, as well as the responsibility involved in caring for them. There will be opportunities for you to learn skills in actual job situations through a Supervised Agricultural Experience Program (SAEP). You will also obtain leadership and personal development skills through the FFA Organization. You can participate in such activities as livestock judging contests, dairy products judging, poultry judging, livestock and showmanship shows, and proficiency awards. Most of these activities are available to you whether you live on a farm or in the center of a city.

Some of the jobs that can be secured with a high school diploma follow:

Herdsman
Feed Mill Worker
Milking Machine Operator
Sheep Shearer
Groomer
Chick Grader
Egg Candler
Small Animal Producer
Slaughterhouse Worker
Milk Hauler
Poultry Processing Plant Worker

FIGURE 1. Egg candlers carefully check eggs for defects such as blood spots. *James Strawser, Cooperative Extension Service, The University of Georgia.*

FIGURE 2. Slaughter house workers perform a wide variety of duties such as removing the internal organs of a carcass. *Dr. Estes Reynolds, Cooperative Extension Service, The University of Georgia.*

FIGURE 3. Egg handlers clean, grade, and pack eggs for shipment to distribution centers. *U.S.D.A. photo.*

FIGURE 4. Milkers clean the udders of cows and attach and remove the cups that draw out the milk. They also clean and maintain the equipment. *James Strawser, Cooperative Extension Service, The University of Georgia.*

FIGURE 5. Workers in a poultry processing plant operate machines that dress and prepare the poultry for sale to consumers. *James Strawser, Cooperative Extension Service, The University of Georgia.*

FIGURE 6. Producers of small animals raise animals for use in research laboratories. *Agricultural Communications Department, The University of Georgia.*

Careers with an Associate's Degree

An associate's degree is a two-year degree from a community college or two year institution. There are many programs that teach animal science in community colleges all across the country. If you enroll in one of these programs you will study practical courses in the fundamentals of producing and caring for animals. You will also study the sciences of chemistry, biology, and zoology as well as other courses such as math and English. Many of the courses taken at a community college may be transferred to a university if you later decide to continue your education.

If you enjoyed livestock judging in high school, you may continue your interest with competitive livestock evaluation at the community college level. You may also get involved in student organizations such as the Post-secondary Agriculture Students Organization that participate in many activities involved with animal science.

Listed below are jobs that are available with an associate's degree.

Veterinarian Assistant
Computer Operator
Farrier
Poultry Vaccinator
Producer
Artificial Insemination Technician
Meat Cutter
Embryo Implant Technician
Wool Grader
Animal Buyer

FIGURE 7. Meat cutters break carcasses into wholesale cuts then divide these cuts into retail cuts ready for sale at the grocery store. *U.S.D.A. photo.*

FIGURE 8. Artificial insemination technicians have to store, properly thaw, and use semen in breeding programs. *James Strawser, Cooperative Extension Service, The University of Georgia.*

FIGURE 9. Computer operators may keep track of breed registries or may balance feed formulas. *Courtesy of USDA/ARS #K-5668-11.*

FIGURE 10. Farriers trim and care for horses' hooves as well as fitting shoes. *Kentucky Horse Park.*

FIGURE 11. Fish producers hire many workers to feed, care for, and supervise the harvest of the fish. *George Lewis, Cooperative Extension Service, The University of Georgia.*

FIGURE 12. Embryo transplant technicians must flush embryos from donor animals and place them in recipient animals. *Calvin Alford, Cooperative Extension Service, The University of Georgia.*

Careers with a Bachelor's Degree

A bachelor's degree requires four years of education at a college or university. There are a broad array of choices in majors that work with animals. These include Animal Science, Dairy Science, Poultry Science, and Agricultural Education.

You will study courses in science such as chemistry, biology, and zoology. At a large number of universities the departments that teach these courses are housed in the College of Agriculture. In agricultural science you will study Animal Anatomy, Nutrition, Animal Growth and Development, and other courses dealing with how animals live, grow and reproduce. By taking courses in education, you can qualify to teach agriculture at a high school.

You can also participate in such competitive events as livestock evaluation and meat evaluation. Student organizations include Block and Bridle, Collegiate FFA, and Collegiate 4-H. Careers with a bachelor's degree include:

Farm or Ranch Manager
Producer
High School Agriculture Teacher
Extension Agent
Agricultural Journalist
Meat Grader
Company Representatives for Animal Feed and Health Products
Field Service Technician
Hatchery Manager
Dairy Inspector

FIGURE 13. Hatchery managers are responsible for ensuring that the operation is run efficiently and that a high percentage of the eggs hatch. *Cooperative Extension Service, The University of Georgia.*

FIGURE 14. Inspectors must monitor the quality of milk and dairy products to ensure that the products are safe and wholesome. *James Strawser, Cooperative Extension Service, The University of Georgia.*

FIGURE 15. Meat graders determine the quality grade of beef carcasses and assign a grade to the carcass. *U.S.D.A. photo.*

FIGURE 16. County agents advise youths as well as adults on matters relating to livestock. *Cooperative Extension Service, The University of Georgia.*

FIGURE 17. Farm or ranch managers oversee and keep records on all the operations involved with producing animals. *James Strawser, Cooperative Extension Service, The University of Georgia.*

FIGURE 18. High school agriculture teachers instruct students in the basics of animal science. *U.S.D.A. photo.*

Careers with a Graduate Degree

Once you complete your bachelor's degree you may want to continue your education and get a master's degree and then a Ph.D. degree. With these graduate degrees you will be able to conduct scientific research or continue with a degree in Veterinarian Medicine. You will choose a specific area of animal science in which to concentrate your studies and most of your course work will be in that area. For example, you might want to study in the area of nutrition or animal reproduction. Your course work will include study in statistics and research methodology so you will be able to understand, design, and conduct scientific research. These degrees require good grades in college and a determination to study hard to reach your career objective.

Examples of jobs requiring an advanced degree such as a Ph.D. or Doctorate of Veterinary Medicine are:

Veterinarian
Meat Inspector
Animal Geneticist
Animal Nutritionist
Reproductive Physiologist
Microbiologist
Research Scientist
College Professor

FIGURE 19. Meat inspectors carefully examine carcasses for signs of disease or other unwholesomeness. *Dr. Estes Reynolds, Cooperative Extension Service, The University of Georgia.*

FIGURE 20. Reproductive physiologists research ways to improve the reproductive efficiency of animals. *Courtesy of USDA/ARS #K-1968-13.*

FIGURE 21. A poultry research scientist might investigate ways of improving the hatchability of eggs. *James Strawser, Cooperative Extension Service, The University of Georgia.*

FIGURE 22. One project of a microbiologist might be to analyze blood samples from cattle. *Agricultural Research Service, U.S.D.A.*

FIGURE 23. A research scientist must analyze and interpret data gathered from scientific experimentation. *Agricultural Research Service, U.S.D.A.*

FIGURE 24. An animal nutritionist may develop feeds that are more efficient in developing growth or maintaining an animal. *James Stawser, Cooperative Extension Service, The University of Georgia.*

Similarly, the National Pork Producers Council has adopted the following Pork Producers Creed:

I believe in the kind and humane treatment of farm animals and that the most efficient production practices are those that are designed to provide comfort.

I believe my livestock operation will be more efficient and profitable if managed in a manner consistent with good husbandry practices as known and recommended by the animal husbandry community.

I believe in an open-door policy to visitors to my farm, to all those who are sincerely interested in production methods and the welfare of animals, so long as they do not endanger the health and welfare of my animals and do not interrupt my production routine or impair the production process.

I believe in and will support research efforts designed to measure stress of farm animals and directed toward more efficient production of food and enhancement of the welfare of animals and man.

I believe it is the animal's purpose to serve man; it is man's responsibility to care for the animals in his charge. I will vigorously oppose any legislation or regulatory activity that states or implies interference with that responsibility.

Research

Another issue of great concern for animal welfare activists is that of the use of animals for research. As far back as the second century A.D., scientists have used animals in research. Most of the medical advances made by humans during the past 100 years have come about through the use of animals to test treatments and medication. Before any type of medical treatment can be tested on humans, it must first be tested using animals.

Most people realize that research is carried out using mice, rabbits, and guinea pigs. However, during recent years much controversy has come about over the use of cats, dogs, and primates. Those who oppose the use of animals in research contend that the knowledge gained through the research cannot justify the suffering the animals must undergo to test a treatment or to experiment with a new theory. These people are particularly emphatic over the use of animals to test such products as cosmetics. They cite examples of how rabbits have chemicals placed in their eyes to determine how irritating an ingredient in an eye shadow for humans might be. The activists insist that the suffering of animals should not be allowed merely to produce new products that are used only to make people appear more attractive.

People and groups opposed to the use of animals in research argue that animals really do not have enough in common with humans to be used in research. They also contend that much of the research using animals could be done through the use of computer models and through the use of cell cultures.

Scientists who use animals in research counter these arguments with the point that animals cannot be elevated to the same level as humans. The scientists argue that the use of the animals is well justified by the advances in medicine and health care that have come about through the use of experiments with laboratory animals. They cite the examples of diseases such as polio that have been almost eradicated through the use of research using animals. They do agree that advances have been made in the use of cell cultures and the use of computer simulations and that the use of these techniques can reduce the necessity of using animals in some instances. However, they also point out that these techniques have limited use. The use of live animals is unavoidable because no method has been developed that adequately substitutes for a live animal.

SUMMARY

The controversy over animal welfare will likely continue. As the number of people who are actively involved in the production of animals becomes fewer and the farms and ranches become larger, there will be less understanding on the part of the public concerning the production of animals for food. The producers will have to take on increasing responsibility to educate the public about production methods. They will also be compelled to ensure that the practices employed are truly in the best interests of the animals. Efforts must always be put forth to keep facilities clean and comfortable for the animals. Doing so will help maintain the image of the livestock producer as someone who truly cares for the animals' welfare.

◆ Review Exercises

TRUE OR FALSE

1. Concern for the welfare of agricultural animals is relatively recent. ______.
2. The animal welfare activists believe that it is moral to raise animals for human use, but that animals should not be abused or mistreated in any way. ______.
3. Animal welfare activists feel that modern livestock operations are just the same as they were several years ago. ______.
4. According to the animal welfare activists, pigs, layer hens, and cattle in some feedlots are treated cruelly and are caused to suffer. ______.
5. The fact that animals are far more efficient in terms of production and growth than they have been in years past adds credence to the producers' arguments that the animals are content. ______.
6. Animal welfare activists point out that debeaking, tail docking, dehorning, and castration are done without anesthesia and therefore cause the animals to suffer. ______.
7. A new technique for identifying individual animals is freeze branding, in which a severe freeze burn is caused that scars over and leaves a permanent mark on the animal. ______.
8. Livestock producers have always been known for their neglect and uncaring attitude towards the animals they raise. ______.
9. Both the National Cattlemen's Association and the National Pork Producers Council have adopted creeds outlining their beliefs about the fair treatment of animals and the environment. ______.
10. Animal welfare activists have concentrated solely on farm animals, completely ignoring animals used for research. ______.
11. People opposed to the use of animals in research contend that computer models as well as cell cultures can be used. ______.
12. Although no specific examples can be shown, scientists argue that research must be continued, even if no good comes out of it. ______.
13. As animal production changes, producers will have to take an increasing responsibility to educate the public about protection measures. ______.

14. The controversy over animal welfare is just about ended, with both sides compromising. ______.

15. Both the producers and the animal welfare activists have good points, and neither group is 100 percent right. ______.

FILL IN THE BLANKS

1. Early humans hunted animals for ________ and to use their ____________ for ____________ and ______________.

2. The animal rights activists believe that killing __________ is just as wrong as __________ humans and that ___________ should have the same ____________ as humans.

3. The animal welfare activists object to what they refer to as livestock _________ in which animals are ___________ ____________ in conditions that totally ___________ the welfare of the _______________.

4. Producers are in the business because they enjoy _____________ with _____________ and have the animals' best _____________ at ___________________.

5. Many millions of dollars and countless ________________ of effort have gone into ________________ to design housing and _________________ that make animals _________________.

6. Producers point out that in order for a drug to be approved for use in animals, the Food and Drug Administration puts the ____________ drug through ____________ testing in order to prove that the __________________ is not only safe and _________________ for the animals, but it is also safe for humans who _____________ meat, eggs, or _________ ___________ from those animals.

7. Castration is done at an _____________ age to prevent problems associated with ______________ animals fighting for ______________ and to provide a higher quality ____________ when the animal is ________________.

8. In the United States, ____________ of the farms are ___________ owned and operated, and there are only _______________ non-family owned _____________ in the entire country.

9. Those who oppose the use of animals in research contend that the _______________ gained through the ______________ cannot justify the _____________ the animals must undergo to test a ____________ or to experiment with a new _______________.

10. The activists cite examples of how rabbits have ________________ placed in their _________________ to determine how __________________ an ingredient in an ________ ____________ for humans might be.

11. Scientists cite examples of diseases such as ______________ that have been almost ______________ through the use of research using ______________ in experiments.

12. Scientists also agree that advances have been made in the use of cell ______________ and the use of ____________ simulations and that the use of these ______________ can reduce the necessity of using _______________ in some instances.

13. As the number of people who are actively involved in the production of ___________ becomes ___________ and the ___________ and ranches become ____________, there will be less ____________ on the part of the public concerning the production of animals for ______________.

14. Producers will be compelled to ensure that the ___________ used are truly in the best ______________ of the _________________.

DISCUSSION QUESTIONS

1. What is the difference in the positions of people who advocate animal rights and those who advocate animal welfare?
2. Why do animal welfare activists oppose raising animals in confinement?
3. What arguments do producers offer in defense of confinement operations?
4. How do livestock producers justify the use of antibiotics in feed?
5. Why is the practice of freeze branding considered more humane than hot branding?
6. What percentage are of the farms in the United States are owned and operated by families?
7. What two livestock associations have adopted statements that deal with the treatment of animals?
8. Why is the controversy over animal welfare likely to continue?
9. Why are some individuals and groups opposed to using animals in research?
10. What justifications do scientists offer in defense of the use of animals in laboratory experiments?
11. What are two new technologies that lessened the need for the use of animals in experimental research?

STUDENT LEARNING ACTIVITIES

1. Locate and read an article in a magazine, newspaper, or other publication that advocates animal rights. Make a list of the points made by the author that you feel are correct and factual. Also make a list of the author's points that you feel are not based on fact. Compare your lists with the lists of other students in your class.
2. Make a list of management practices that might be criticized by animal welfare activists. Think of and write down methods or alternatives that the producer might employ that will lessen the criticism.
3. Write a brief report on at least one scientific advancement that used animals in the research. Be sure to include how the animals were used and how people have benefited from the advancement.
4. Take a stance either for or against the use of animals to test cosmetics. Present your arguments to the class. Compare your arguments to the arguments of those in the class who took the opposing view.

Chapter 22

Consumer Concerns

Student Objectives In Basic Science

As a result of studying this chapter, you should be able to

1. Explain the rationale for consumer concern over food safety.
2. Define and explain the role of *cholesterol.*
3. Define *genetic engineering.*
4. Explain the rationale for concern over genetic engineering.
5. Discuss the concept of the greenhouse effect.
6. Explain how the balance of oxygen and carbon dioxide is maintained in the atmosphere.
7. Explain how bacteria can be beneficial to the environment.

Student Objectives in Agricultural Science

As a result of studying this chapter, you should be able to

1. Explain why agriculturalists must be more sensitive to the concerns of consumers.
2. Tell the difference between meat grading and meat inspection.
3. Summarize how meat is inspected.
4. Tell how research has shown that the meat supplied to consumers is safe and nutritious.
5. Give examples of how genetic engineering has benefited the producer.
6. Describe how producers of agricultural animals are good caretakers of the environment.

Key Terms

curing
antemortem
rendering
antibiotics
Food and Drug Administration (FDA)
cholesterol
marbling
genetic engineering
genetics
RST
BST
coliforms
colon
public lands
Bureau of Land Management
greenhouse effect

At one time in the history of our country, almost everyone knew how food was produced and processed. This was because the growing and processing of food was done right at home. Before the end of the nineteenth century, the vast majority of people in the United States lived in rural areas and produced most of the food they ate. Crops were gathered and dried, pickled, or canned for the family to use during the winter months. When the weather turned cool, livestock was slaughtered and the meat preserved by drying, canning, or **curing**. Since the people processed their own food, they knew how the food was processed and what went into the food. Today, relatively few people process their own food. The only form of food that most people ever see is the finished product in the grocery store, Figure 22–1. Trends in food processing follow what the consumer demands. Consumers are those people who buy and use products. Since both the husband and wife are working outside the home in most families, consumers want food that is more processed than in the past. This means that food products have to be closer to being ready to eat than at any time in our history. As more and more steps are added in the processing of food, consumers become more concerned about how the processing was done and what ingredients went into the product, Figure 22–2.

At the same time, consumers do not understand how crops and animals are grown. Since the consumers want food products that are relatively inexpensive, producers have to use means that can get as much efficiency as possible from the crops and animals they grow. Chemicals and other substances are used in the production and processing of food that consumers do not generally understand. As modern technology increases and new discoveries help growers produce more efficiently, concerns are raised among consumers about the safety and wholesomeness of the food they buy and consume. Much of this concern centers around the animal industry and the products such as meat, milk, and eggs produced by the industry. Animal products are very susceptible to spoilage. They can easily pick up microorganisms from the processing that can cause spoilage.

Figure 22-1. Most people only see the finished product in the grocery store. *National Livestock and Meat Board.*

Figure 22-2. As more steps are added in the processing of foods, the more concerned consumers become. *James Strawser, Cooperative Extension Service, The University of Georgia.*

Recently, public concern over food poisoning has been raised as a result of contaminated meats. In these cases, the cause of the illnesses was hamburger containing the *E. coli* bacteria. This bacteria inhabits the colon of animals and humans and certain strains of the bacteria can cause illness and even death. Problems arise during the slaughter process when the internal organs are removed from the animal. Sometimes the carcass is contaminated when it comes in contact with *E. coli* bacteria from the viscera. Hamburger is particularly susceptible because in the grinding process the bacteria on the surface of meat are ground and mixed with all of the meat in the batch. Given a warm, moist environment, the bacteria grow and reproduce rapidly, and eventually enough bacteria are produced to cause illness if the bacteria are not destroyed. Poultry has also come under careful scrutiny because of the possibility of contamination with salmonella bacteria. Most often all harmful microorganisms can be destroyed by cooking the meat thoroughly.

As a result of sickness due to contaminated meats, new regulations are in effect for the inspection and handling of meat products. The first step in prevention of food-borne bacteria on meat is sanitation and precautions at the meat-processing plants. Slaughterhouses are under new and more rigorous requirements for sanitizing the plant and preventing meat from contacting fecal material. In addition, new scientific tests are conducted on the meat to determine the presence of bacteria. Although we have had serious outbreaks of sickness due to contaminated meat, our supply of meat is still considered to be safe when handled and cooked properly. Only a very tiny percent of the meat on the market has been found to contain large numbers of harmful bacteria.

The Meat Inspection Act was passed in 1906. It ensures that all meat products processed and sold in this country will pass inspection by the United States Department of Agriculture. Since that time, the law has been revised and updated several times to ensure that only the very best, most wholesome meat reaches the consumer. Meat inspection is not the same thing as meat grading. Meat inspection simply guarantees that the meat will be safe, wholesome, and accurately labeled. All meat that is sold must, by law, be inspected.

Figure 22-3. Before they are slaughtered, animals are inspected in a well-lighted area. *James Strawser, Cooperative Extension Service, The University of Georgia.*

Grading refers to the eating quality and degree of yield expected from a carcass. Grading is optional. Meat inspection undergoes several phases. First, the animals that are to be slaughtered must be inspected while they are alive. This is called **antemortem** inspection (*ante* means before and *mortem* means death). As the animals are brought in prior to slaughter, a government inspector examines them. Animals that are down, disabled, diseased, or dead are condemned as unsafe for human consumption, Figure 22–3. Animals that the inspector thinks may have a problem are sent aside for further examination. The plant where the animals are to be processed must provide the inspector a well-lighted, clean area to examine those animals the inspector suspects are not in the best of health. These animals are examined

Figure 22-4. All meat sold commercially must, by law, be inspected. *USDA photo.*

thoroughly and, if found to be ill, are tagged as condemned and are not allowed to be slaughtered for human consumption.

After the animals are slaughtered, they must again undergo inspection. Slaughter plants are required to provide adequate lighting (50 footcandles) for the inspector to see in order to thoroughly inspect the carcasses. The head, lungs, heart, spleen, and liver are inspected for signs of disease, parasites, or other problems that might render the meat of cattle, sheep and hogs less than wholesome. The internal and external cavity of slaughtered poultry must be examined as well as the air sacs, kidneys, sex organs, heart, liver, and spleen, Figure 22–4.

Carcasses that do not pass inspection are declared to be condemned and are not allowed to be used for human consumption. These condemned carcasses undergo a process called **rendering** during which they are placed under heat severe enough to kill any organism that could cause problems. The rendered meat is then used as a by-product not intended for human consumption. By-products can be used as pet food or for other products such as fertilizers. Condemned carcasses usually represent less than one percent of the carcasses inspected. Producers, buyers, and packers all try to avoid sending animals to slaughter that they know will not pass inspection. Those that do get by are discovered by the federal inspectors at the processing plant.

Animal Medications

Animals grown in large numbers are fed medication to prevent and cure disease and to ward off parasites. This can result in a better, safer product for the consumer. Animals that are kept healthy all their lives reach the processing plant healthier and yield a more healthy product. Consumers are sometimes worried about the residue of these medications that are fed to livestock. For instance, concern is raised over the amount of **antibiotics** fed to cattle, hogs, and poultry, Figure 22–5. The worry is that as humans consume meat from animals that were fed antibiotics, these medications might enter the human body and build up a residue that will

Figure 22-5. Consumers are sometimes worried over the medication given to livestock in their feed. *National Livestock and Meat Board.*

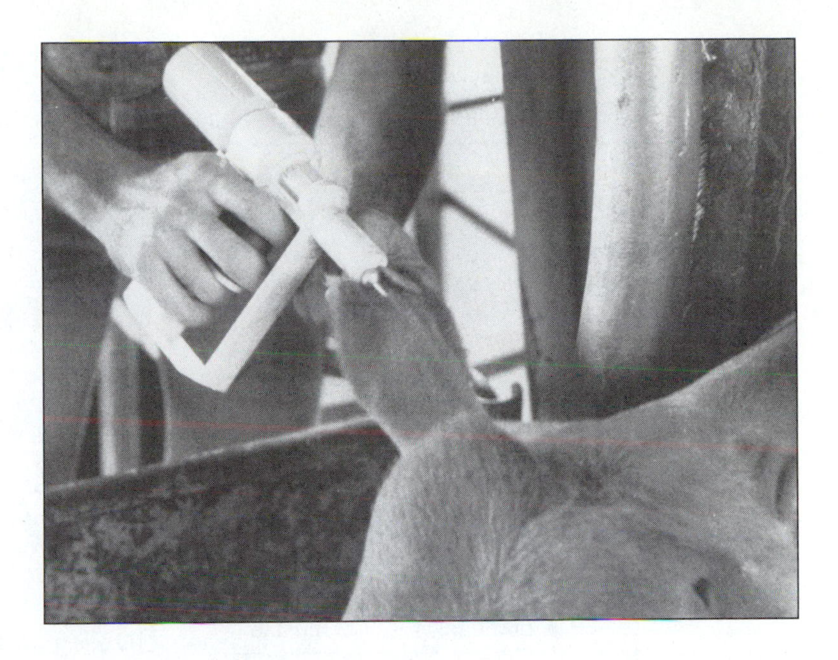

Figure 22-6. Ear implants release hormones very slowly. The ear is discarded after slaughter. *Syntex Animal Health Inc.*

create an intolerance to medication. Since the types of antibiotics given to agricultural animals are the same type as given to humans, some believe that bacteria may develop strains that are resistant to the medications, which would make the medications ineffective when needed by humans. Research by the United States Department of Agriculture (USDA) and the **Food and Drug Administration (FDA)** has clearly shown that adding antibiotics to animal feed in the proper amount does not result in antibiotic residue in meat.

HORMONES

Animals are also given growth hormones to aid in feed efficiency and growth rates. Many consumers are concerned over the long-range effects of the hormones fed to animals that might be residual in meat, eggs, milk, or other products. According to the National Cattlemen's Association, there is no evidence of any human health problem from the use of any natural or synthetic hormones fed to livestock. The FDA closely regulates the amount of hormone residue allowed in meat. When cattle are given growth hormones, the dosage is administered by means of implanting a small capsule underneath the skin that releases the hormone very slowly. The capsule is implanted under the skin of the ear because at slaughter the ears are put into the waste bin and are not processed for consumption, Figure 22–6.

No more than one percent of the human body's daily production of an implanted hormone is allowed to be in the daily intake of meat. Since the hormone level in the meat is greatly reduced (by 90 percent) through the digestive process, the amount of hormones absorbed by the body through meat is extremely small. The charts in Figure 22–7 show the relative amounts of estrogen produced by people and the estrogen in a serving of beef from a steer that has been implanted with hormones and one that has not. In fact, a person will obtain thousands of times the amount of estrogen from a gram of soybean oil than from a gram of beef from an implanted steer.

Table 1 Daily Human Estrogen Production (Nanograms)

Human	Amount
Female child, before puberty	54,000
Male child, before puberty	41,000
Nonpregnant woman	480,000
Pregnant woman	20,000,000
Adult male	136,000

Table 2 shows the estrogen level in beef from nonimplanted and implanted steers.

Table 2 Estrogen Levels in Beef

Beef Source	Nanograms per gram of muscle	Nanograms per 3 oz. of muscle
Steer, implanted	0.022	1.9
Steer, nonimplanted	0.015	1.3

Table 3 shows estrogen levels in various food products.

Table 3 Estrogen Levels in Foods

Food	Nanograms per gram of food	Nanograms per 3 oz. of food
Beef from implanted steer	0.022	1.9
Wheat germ	40	3,400
Soybean oil	20,000	1,680,000
Milk	0.13	11

Figure 22-7. Beef contains relatively small amounts of estrogen. *National Cattlemen's Association.*

In order for a product to be released and sold for use by animal producers, the product must be researched and tested through rigorous standards before it is declared safe by the government. The product must be so thoroughly studied and tested that there is no significant chance that the product will cause any health problems for the consumer. The USDA closely monitors animal products for traces of residues resulting from feed additives or medications given to animals. The allowable amounts of these residues in foods would have to be hundreds and, in some cases, thousands of times greater before any effect at all could be detected in humans who consume the animal product, Figure 22–8.

Figure 22-8. The USDA closely monitors animal products for impurities and residues. *James Strawser, Cooperative Extension Service, The University of Georgia.*

CHOLESTEROL

Several years ago, some medical research indicated that a high intake of a substance known as **cholesterol** was a contributing cause of heart disease. Cholesterol is a fatlike substance found in animal tissue. It is an essential part of nerve tissue and cell membranes of all animals, including humans. Cholesterol plays an important role in the manufacture of hormones by the body and in the production of the bile that is used in digestion. Although cholesterol is essential to life, studies have shown a correlation between high cholesterol levels and clogged heart arteries. Since meat contains cholesterol, consumers have been concerned about eating meat. Recent studies have been somewhat contradictory about how important a role cholesterol actually plays in the formation of deposits in the coronary system. Some studies indicate that these deposits are more related to the amount of exercise a person gets and the person's heredity than the amount of cholesterol intake. Modern meats contain less fat than the meat produced in the past. Both the pork producer and the beef producer organizations are promoting pork and beef as healthful foods that are leaner and trimmer than meats of a few years ago. For example, a choice cut of beef today has a smaller fat content (less **marbling**) than a choice cut of beef ten years ago, Figure 22–9. Although fat is what gives meat its flavor, research is constantly being conducted to produce a lean meat product that has a low-fat content.

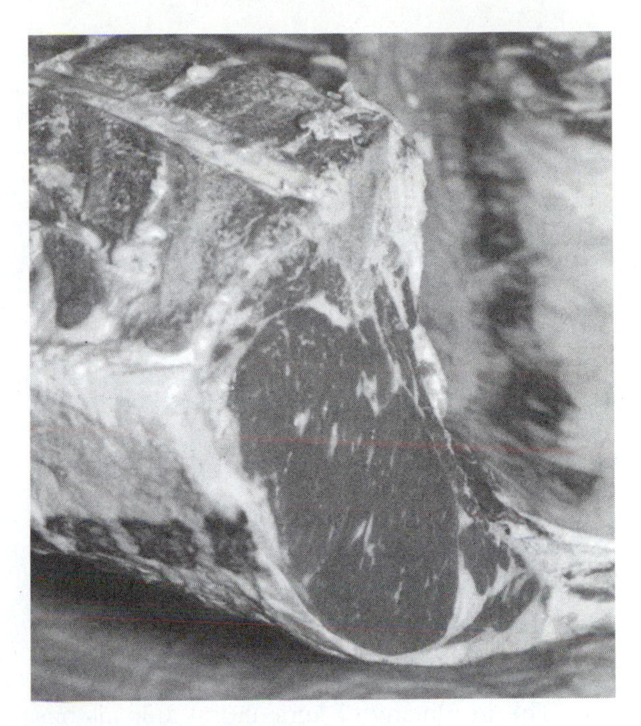

Figure 22-9. Choice cuts of beef have less fat content than those of a few years ago. *North American Limousin Association.*

GENETIC ENGINEERING

Genetic engineering is the alteration of the genetic makeup of an organism to produce a desired effect. The development of modern technology has given scientists the ability to enter an organism's genetic makeup and insert, remove, and alter genes that are responsible for the organism's characteristics. This has the promise of tremendous benefits to the producers of agricultural animals. If we are able to improve the **genetics** of animals by simply changing the genes, we can boost the efficiency of producing them and, in turn, produce better and cheaper products for the consumer. At the same time, consumers are concerned over the use of genetic engineering. Some people see this effort as interfering with nature. They fear that by altering the genetic makeup of an animal, the potential exists to create animals that might not be in the best interest of humans. Many science fiction movies have been made that depict monstrous animals created by a scientist who disturbs the natural order of the animal's makeup. In addition, there is a fear that products from genetically engineered animals will contain substances that will prove harmful to the people who consume them. One such example is the use of a substance called *recombinant bovine somatotropin (***RST** or **BST***)*. This substance is produced by genetic engineering and is a naturally occurring hormone. Scientists have known for many years that cows receiving additional amounts of BST significantly increase their milk production. Until recently, this substance was scarce and very expensive. However, due to genetic engineering, the hormone can now be produced quickly and cheaply. The use of BST has been and is now being debated in regard to its safety. Some say that no studies have been conducted to determine the long-term effects of drinking the milk from cows that have been given BST to increase their milk production. They point to the fact that large doses can cause inflammation of the cow's udders and that this may be a sign that the hormone is not safe. Proponents point out that the cows are only given very small doses and that no ill effects have been discovered. Also, the National Institute of Health and the Food and Drug Administration have declared BST to be safe both for the cows and for the humans who consume the milk.

Environmental Concerns

The producing of agricultural animals has been criticized by some as being harmful to the environment. They point out that animals, particularly in a confinement operation, generate a lot of waste. They fear that this waste will get into the water supply and contaminate water designated for human consumption. This concern is well-founded. Animal wastes contain bacteria called **coliforms**, which means "bacteria from the **colon**." These bacteria can carry disease to humans and other animals if allowed to escape into the water supply. Animals in confinement do create a lot of waste that could possibly contaminate the water supply; however, there are stringent laws that prevent this from happening. It is illegal for any producer to discharge animal waste into a stream. In fact, the Environmental Protection Agency has regulations that help prevent this from happening even accidentally. The use of lagoons and holding ponds have helped protect the environment. A lagoon is a body of water made especially for holding animal wastes from confinement operations, Figure 22–10.

Modern lagoons are designed to allow the breakdown of harmful substances through the use of beneficial bacteria that work to decompose the waste material into less complex substances. The decay of any organic material comes about through the action of microorganisms such as bacteria. To aid in this process, the construction of lagoons is regulated in order to provide for the correct depth of water that will

Figure 22-10. The use of lagoons and holding ponds helps protect the environment. *Soil Conservation Service.*

Figure 22-11. Manure is spread out on land for use as fertilizer. *Case International.*

allow for the most efficient growth of beneficial bacteria. The lagoon is periodically pumped out and spread on fields. Manure is a natural, high-quality fertilizer that is valuable in the production of many crops, and once the decomposition process has taken place, there is no harm to the environment. The use of manure actually can cut down on pollution by reducing the amount of commercially produced fertilizers required to grow crops, Figure 22–11. Almost all manufacturing plants create a degree of pollution.

The Overgrazing of Public Lands

Some people are also concerned that our **public lands** are being overgrazed and ruined by producers who do not take care of the land. In the western part of the United States, government lands are leased to animal producers,

Figure 22-12. Although leased for grazing, public rangelands are in better shape than at any time this century. *Soil Conservation Service.*

Figure 22–12. A recent report of the **Bureau of Land Management** pointed out that our public rangelands are in better shape than at any time during this century. Cattle and sheep producers know that they will be using the land for a long time and that it is in their best interest to care for the land. Most producers want to pass their operations on to subsequent generations. The only way for this to happen is to carefully manage the land. Land that is well cared for will produce animals more efficiently. According to the USDA, a sound management system for grazing builds the soil and enhances wildlife habitats. Wildlife numbers on government rangeland have improved dramatically over the past 30 years. When properly managed, the grazing of livestock on public lands can actually be good for wildlife. Through the control of undesirable species of brush and plants, better grazing is provided for the wildlife as well as for the cattle. Also water has to be provided for the agricultural animals and, in doing so, water is provided for wildlife as well.

Global Warming

In recent years, concern has been raised over a phenomenon known as the **greenhouse effect**. The theory behind this effect is that the earth is gradually warming due to problems encountered in the destruction of parts of our environment. (The effect was named for the extra heat inside a house made from glass or greenhouse.) Proponents of the theory say that a higher concentration of carbon dioxide in the atmosphere causes the heat from the sun to become more intense and the temperature of the earth to be raised. The proper balance of carbon dioxide and oxygen in the air is brought about by plants and animals. Animals breathe in air, absorb oxygen, and give off carbon dioxide. Plants take in air, absorb carbon dioxide, and give off oxygen. Vast acres of forest in the tropical areas of the world have always helped keep this balance in order. As trees are destroyed, some of the world's potential for absorbing carbon dioxide and releasing oxygen is lost. Environmentalists point out that forest areas are being cleared in order to raise livestock and this, along with the methane gas produced by the animals, is contributing to global warming. In South America especially, tropical rain forests are being destroyed in order to make room for the production of cattle, Figure 22–13.

Figure 22-13. Tropical rain forests are being destroyed to make room for agriculture. *Carol Hoffman, Environmental Studies, The University of Georgia.*

However, the majority of the land being cleared is for crops and other uses. This country imports very little meat from South and Central America. Land cleared for grazing in the United States is very small. Land statistics show that the amount of forest land today is only slightly less than it was in 1850.

Confinement operations produce a lot of manure. Due to natural biological processes, the manure produced in these operations gives off methane gas. Some environmentalists have suggested that this may be a contributing factor to the greenhouse effect. Research indicates that the amount of methane given off by these animals does not contribute to global warming. As

Figure 22-14. This generator is used to produce methane gas from animal waste in Kenya.

a matter of fact, in some parts of the world this methane gas has been collected and used as a fuel source, Figure 22–14.

SUMMARY

Agriculture is an essential industry that will have to continue to expand as the population grows. Our modern society is not as knowledgeable about this industry as past generations when the vast majority of people grew up on a farm. In addition, science has equipped us with new insights into the food we eat and the environment surrounding us. Concern arises as people read reports outlining problems with the food supply and the environment. The agricultural industry has two very important responsibilities. First, we must make sure that we supply a plentiful, wholesome, and safe food supply for the population. Secondly, we must ensure that the practices we use are environmentally sound and that the world we live in can be preserved for generations to come.

◆ Review Exercises

TRUE OR FALSE

1. Trends in food processing follow what is easiest for the producers to deliver. ______.
2. Almost all of the problems of salmonella poisoning from poultry can be traced to the producers. ______.
3. Meat inspection and meat grading are essentially the same thing. ______.
4. Animals that are down, disabled, diseased, or dead are condemned as unsafe for human consumption. ______.
5. Medication helps animals fight diseases and ward off parasites. ______.
6. A person will obtain thousands of times the amount of estrogen from a gram of soybean oil than from a gram of beef from an implanted steer. ______.
7. Recent studies have been somewhat contradictory as to how important a role cholesterol actually plays in the formation of deposits in the coronary system. ______.
8. Cows that get additional recombinant bovine somatotropin give less milk, sometimes drying up completely. ______.
9. Cows are given huge amounts of BST and both the National Institute of Health and the Food and Drug Administration have declared BST to be harmful to lactating cows. ______.
10. The producing of agricultural animals has been criticized by some as being harmful to the environment. ______.
11. Bacteria found in animal wastes cannot carry diseases to humans and other animals if allowed to escape into the water supply. ______.

12. The depth of the water in a lagoon has no effect on the growth of beneficial bacteria. ______.
13. A recent report of the Bureau of Land Management pointed out that our public rangelands are in worse shape than at any time this century. ______.
14. As trees are destroyed, some of the world's potential for absorbing carbon dioxide and releasing oxygen is lost. ______.
15. Because of their natural biological processes, agricultural animals give off methane gas from the large amounts of manure. ______.

FILL IN THE BLANKS

1. Before the end of the nineteenth century, the vast ______________ of people in the United States lived in _____________ areas and produced most of the _____________ they ___________.
2. As modern technology increases and new ____________ help growers to produce more __________________, concerns are raised among _________________ about the safety and _______________ of the food they buy and consume.
3. The 1906 Meat Inspection Act ensures that all _____________ products processed and _____________ in this country will pass ________________ by the United States Department of ______________________.
4. Condemned carcasses undergo a process called _________________ where they are placed under enough ______________ that any _______________ that can cause ____________________ are ______________.
5. Research by the USDA and the Food and Drug Administration has clearly shown that adding ___________________ to animal _____________ in the proper ____________ does not result in antibiotic ____________ in the meat.
6. When cattle are given growth hormones, the _________________ is given by means of ______________________________ a small _______________ underneath the skin that releases the _________________ very slowly.
7. Genetic engineering is the _______________ of the ________________ makeup of an __________________ to produce a desired ________________.
8. Some consumers fear that the products from ________________ engineered animals will contain __________________ that will prove harmful to the ___________________ who _________________ them.
9. Animal wastes contain bacteria called ________________________, which means _______________ from the ______________.
10. A lagoon is a body of _________ made especially for holding animal _______________ from _______________ _________________.
11. The decay of any __________________ material comes about through the action of _____________________________ such as ________________.
12. Most producers want their _______________ to pass on to the next __________________ and know that the only way for this to happen is to carefully ______________ the land.
13. The theory behind the greenhouse effect is that the ________________ is gradually ____________ due to problems encountered in the _________________ of parts of our ___________________.
14. Environmentalists point out that __________ areas are being _______________ in order to raise ________________ and this, along with the ________________ gas produced by the animals, is contributing to _____________ ________________.
15. Research indicates that the amount of methane gas given off by agricultural animals does _______________ contribute to ____________ _______________.

DISCUSSION QUESTIONS

1. Explain why consumers are more concerned with the quality and safety of food than they were 50 years ago.
2. What are consumer concerns over meat safety?
3. What is the Meat Inspection Act?
4. What is the difference between antemortem and postmortem inspection?
5. Why are animals given growth hormones?
6. What hormone level is allowed by the Food and Drug Administration in a person's daily intake of meat?
7. What role does cholesterol play in our bodies?
8. Why are consumers concerned about the intake of cholesterol?
9. What is BST and what does it do?
10. List three areas of consumer concern over the environment.

STUDENT LEARNING ACTIVITIES

1. Talk to at least ten people who buy meat. Make a list of the concerns these people have about the meat they buy. Share your list with the class.
2. Visit a slaughterhouse when carcasses are being inspected. Ask the inspector to explain what he or she looks for in a healthy carcass.
3. Formulate some ideas as to what could be changed about agricultural animals through genetic engineering. List the benefits of the changes and also list possible problems.
4. Visit a confinement operation and observe the manure-disposal system. List possible environmental problems that you observe. Also list the efforts made to protect the environment.

Glossary

abomasum — the fourth, or true, stomach division of a ruminant animal

absorption — the passage of food from the digestive system to the bloodstream

active immunity — the type of immunity in an animal that is permanent

adipose — the technical term for fat tissue

aerobic organisms — grow only in the presence of oxygen

aflatoxin — the highly toxic substance produced by some strains of the fungus *Aspergillus flavus*. It is found in feed grains.

aging — the process by which meats are hung in a cool environment for a specific period to improve the flavor and tenderness; also, the process of maturing and getting older

agricultural animals — animals raised for the purpose of making a profit

agriculture — the broad industry engaged in the production of plants and animals for food and fiber; the provision of agricultural supplies and services; and the processing, marketing, and distribution of agricultural products

albumen — the white of the egg

alimentary canal — tract extending from the mouth to the anus, through which food passes and where it is exposed to the various digestive processes

allele — an alternative form of a gene. For example one allele may control red coat color and another may control black coat color.

alternative animal agriculture — production of animals other than the traditional agricultural animals

alveoli — small grape-like structures in the udder of a cow that produces milk

amino acid — the basic building block of protein

amniotic fluid — the fluid that surrounds a fetus before birth

anabolism — the growth process by which tissues are built up

anaerobic organisms — grow without the presence of oxygen

anemia — a disease caused by a deficiency of hemoglobin, iron, or red blood cells

animalia — the highest level of scientific classification (kingdom) to which all animals belong

animal rights — a line of thinking that proposes that animals have the same rights as people

animal welfare — a line of thinking that proposes that animals should be treated well and that their comfort and well-being should be considered in their production

antemortem — preceding death

antibiotics — a group of drugs used to fight bacterial infections

antibodies — substances produced by an animal's body that fight disease or foreign materials in the bloodstream or other places in an animal's body

antigen — any substance that stimulates the production of antibodies in an animal's body

aquaculture — the production of animals that live predominantly in the water

artificial active immunity — immunity that comes about as a result of a vaccination

artificial hormone — a manufactured substance that is used in place of a naturally produced hormone

artificial insemination — the placing of sperm in the reproductive tract of the female by means other than that of the natural breeding process

artificial vagina — a tubelike device used to collect semen from a male animal

ascarid — largest of the parasitic roundworms, most often attacking young animals

asexual reproduction — the production of young by only one parent

assistant dogs — dogs that are used to help humans perform any task they cannot do or need assistance in doing.

bacillus — rod-shaped bacterium

backfat — fat tissue that is deposited under the skin of an animal

balance — general proportions in the physical structure of an animal

balanced ration — a diet designed to provide an animal with all the necessary nutrients

barrow — a male pig that has been castrated before reaching sexual maturity

beef — the meat from cattle over a year old

bee space — the space (about 3/8 inch) in a beehive that allows bees to work back-to-back

binomial nomenclature — a system of scientific classification of living organisms that uses two names, the genus and the species. The names usually come from Latin derivatives.

biological control — the use of natural means rather than chemicals to control pests

blast freezing — most common method of commercial freezing, utilizing high-velocity air and temperatures of -10°C to -40°C

blastula — a mass of cells with a cavity that occurs from the dividing of a fertilized egg. From this stage the cells begin to differentiate.

blind nipples — nonfunctional nipples on the mammary system of a female pig

bloat — a condition in cattle caused by gas being trapped in the digestive system. Left untreated, the condition can be fatal.

blood typing — analyzing an animal's blood to determine the animal's ancestry

boar — a male pig that has not been castrated

bolus — a large pill; also, a soft mass of chewed food

breed association — an organization that promotes a certain breed of animal. They control the registration process of purebred animals of that breed.

breeding true — offspring almost always looking like the parents

brisket — the breast or lower chest of a four-legged animal

broiler — a chicken approximately eight weeks old that weighs 2 1/2 pounds or more

broiler industry — raising chickens for their meat

brood cells — cells in the hive where the queen bee lays her eggs

brood chamber — that portion of a bee hive where the queen lays eggs and the young bees are hatched and raised

BST — bovine somatotropin, a naturally occurring hormone that aids in stimulating the production of milk in cows

bull — a male bovine that has not been castrated

Bureau of Land Management — the federal agency that oversees the management of government lands

by-product — a product that is created as the result of producing another product

cage operation — an operation in which hens are kept in cages all their lives as they produce eggs

candling — the use of light shined through an egg to determine defects in the egg

cannibalism — the habit of some birds in a poultry flock of repeatedly pecking and clawing other birds in the flock, often causing injury and death

cannon bone — a bone in hoofed mammals that extends from the knee or the hock to the fetlock or pastern

carcass — that part of a meat animal that is left after the hide and hair, feet, head, and entrails have been removed

carcass merit — quality and yield of a carcass

carding — one of the first steps in the processing of wool. The fibers are separated from other fibers in the locks or bunches of wool.

carnivore — an animal whose diet consists mainly of other animals

carotene — an orange or red pigment found in green leafy plants, especially carrots. It can be converted to vitamin A by an animal's body.

cartilage — firm but pliant tissue in an animal's body that may turn to bone as the animal ages

castings — manure from worms

castrated — condition in which an animal's testicles have been removed

catabolism — the process of breaking down tissues from the complex to the simple as in the digestive process

catheter — a tube that is inserted into an animal's body to inject or withdraw fluid

cecum — the enlargement on the digestive tract of animals such as the horse that allows them to digest large amounts of roughages

cell — the basic building block of living tissue. It generally consists of a membrane wall, a nucleus, and a cytoplasm.

cellulose — an inert complex carbohydrate that makes up the bulk of the cell walls of plants

cervix — the organ that serves as an opening to the uterus

chalazae — ropelike structures inside an egg that hold the yolk in the center of the egg

chine — the backbone of an animal

cholesterol — a fat-soluble substance found in the fat, liver, nervous system, and other areas of an animal's body. It plays an important role in the synthesis of bile, sex hormones, and vitamin D.

chromosomes — a linear arrangement of genes that determines the characteristics of an organism

chromotid — one strand of a double chromosome

chronological age — the actual age of an animal in days, weeks, months, or years

classes — further divisions within phyla or subphyla

cleavage — the splitting of one cell into two parts

climate-controlled houses — livestock (or poultry) houses that are kept at the proper temperature, lighting, and humidity for optimal growth and comfort of the animals

clitoris — small, sensitive organ within the vulva that provides stimulation during the mating process

cloaca — the opening in a hen's body through which the egg is expelled

clone — an organism, produced by asexual means, with the exact same genetic makeup as another

cocci — round, spherical-shaped bacteria

codominant genes — genes that are neither dominant nor recessive

cold-blooded — *See* **ectothermic**.

cold-water fish — a fish that will not thrive in water temperatures above 70°F

colic — a condition in horses caused by a blockage in the digestive system. The intestine becomes distended and causes pain to the animal.

coliforms — a group of bacteria that inhabit the colons of people and animals

collagen — a protein that forms the main component of connective tissues in animals

colon — the large intestine

colony — a group of bees consisting of workers, queen, and drones that live together as a unit

colostrum — the first milk that a mammal gives to the young following birth. It is rich in nutrients and imparts immunity from the mother to the offspring.

combination curing — a combination of dry curing and injection curing

commensalism — a symbiotic relationship in which one species benefits and the other is not harmed

companion animals — animals whose main purpose is serving as pets or friends to humans

concentrate — a feed that is high in carbohydrates and low in fiber

conception — uniting of sperm and egg

conditioning — process of learning by associating a certain response with a certain stimulus

confinement operation — a system of raising animals in a relatively small space

conformation — the shape or proportional dimensions of an animal

consumers — those who buy or use food, manufactured goods, or other products

contagious disease — a disease that may be passed from one organism to another

control group — a group of animals or plants (in a scientific experiment) that does not receive the treatment under study

copulation — the act of sexual union between two mating animals

corpus luteum — a swelling of tissue that develops on the follicle at the site where an ovum has been shed

cortex — the outer layer or region of any organ; also, in wool fibers the tissue immediately external to the xylem

cow — a female bovine that has had a calf

cow-calf operation — a system of raising cattle, the main purpose of which is the production on calves that are sold at weaning

cow-hocked — a condition in which an animal's back feet are splayed out and the hocks are turned in

cowper's gland — a gland in the male reproductive tract that produces a fluid that is added to the ejaculate

crimp — the amount of waves in wool fiber

crossbred — an animal that is the result of the mating of parents of different breeds

crude protein content — total amount of crude protein in a feed, calculated by analyzing the nitrogen content and multiplying that percentage by 6.25

crustaceans — aquatic animals with a rigid outer covering, jointed appendages, and gills

cryogenics — method of freezing, utilizing condensed gases in direct contact with the product being frozen

cud — a small wad of regurgitated feed in the mouth of a ruminant that is rechewed and swallowed

curd — the coagulated part of milk that results when the milk is clotted by adding rennet, by natural souring, or by adding a starter

curing — treating meat to retard spoilage

cutability — percent of lean cuts a carcass will produce

cuticle — outer lager of cells of wool fibers

cytoplasm — the living material within a plant or animal cell excluding the nucleus

dam — the mother of an animal

dam breeds — those breeds of agricultural animals that are used as dams in a cross-breeding program

debeaking — removing the tip of a chicken's beak to aid in the prevention of cannibalism

dehorning — permanently removing an animal's horns

dental pad — a hard pad in the upper mouth of cattle and other animals that serves in the place of upper teeth

deoxyribonucleic acid (DNA) — a genetic acid that controls inheritance

differentiation — the development of different tissues from the division of cells

diffusion — in the process of absorption, the passing of particles through a semipermeable membrane

digestion — the changes that food undergoes within the digestive tract to prepare it for absorption and use in the body

disaccharides — the more complex sugars

discriminate breeder — an animal that will only breed with a certain mate

dissolved oxygen — oxygen in water that is available for the use of animals with gills (such as fish)

docile — having a quiet, gentle nature

docking — the removal of an animal's tail

domesticated — raised under the care of humans

dominant gene — a gene that expresses its characteristics over the characteristics of the gene with which it is paired

donor cow — a cow of superior genetics from which an embryo is taken to implant in a cow of inferior genetics

double muscling — a condition in beef animals that is characterized by large, bulging, round muscles

draft horse — a horse that is used mainly for pulling loads or for working

drone — a male honeybee

dry curing — the process of curing meat by rubbing the cure ingredients onto the surface of the product and allowing it to move into the product by osmosis

dual-purpose animal — an animal that is raised for more than one purpose, e.g. sheep for wool and mutton

duodenum — the first portion of the small intestine

ecological balance — the balance nature has regarding the living things in a given area

ectoderm — the outer of the three basic layers of the embryo, which gives rise to the skin, hair, and nervous system

ectothermic — animals whose body temperature adjusts to the air and water around them. Also known as cold blooded.

efficiency — ability of the animal to gain on the least amount of feed and other necessities

ejaculation — release of semen (the ejaculate)

elastin — a protein substance found in tendons, connective tissue, and bone

elastrator — a device that is used to stretch a rubber band over the scrotum or tail of an animal. The blood circulation is cut off and the testicles or tail drops off.

embryo — an organism in the earliest stage of development

embryo transplant — removing an embryo from a female of superior genetics and placing the embryo in the reproductive tract of a female of inferior genetics

endocrine system — the system of glands in an animal's body that secrete substances that control certain bodily processes

endoderm — the innermost layer of cells of an embryo, which develops into internal organs

endothermic — an animal whose body temperature is warmer than its surroundings

energy — the capacity to do work

environment — the total of all the external conditions that may act upon an organism or community to influence its development or existence

enzyme — a protein that is produced by an animal's body that stimulates or speeds up various chemical reactions

epididymis — a small tube, leading from the testicles, where sperm mature and are stored

epinephrine — a hormone that is released when an animal gets frightened or upset. In cows it inhibits the milk letdown process.

epistasis — interaction of genes that are not matched pairs to cause an expression different from the coding of the genes

ergot — a fungus disease of grains that produces a toxin

esophagus — the tube leading from the mouth to the stomach

essential amino acid — any of the amino acids that cannot be synthesized by an animal's body and must be supplied from the animal's diet

estimated breeding value — in beef cattle, an estimate of the value of an animal as a parent

estrogen — a hormone that stimulates the female sex drive and controls the development of female characteristics

estrus — the period of sexual excitement (heat) when the female will accept the male

estrus cycle — the reproductive cycle of female animals measured from the beginning of one heat period until the beginning of the next

estrus synchronization — using synthetic hormones to make a group of females come into heat (estrus) at the same time

ethology — the science of animal behavior

ewe — a female sheep

expected progeny difference — an estimate of the expected performance of an animal's offspring

exotics — animals that are out of the ordinary, such as an unusual breed

experiment — an operation carried out under controlled conditions to discover an unknown entity, to test a hypothesis, or demonstrate something known

experimental group — the group used to test a hypothesis; the group subject to experimentation

exsanguination — the removal of an animal's blood during the slaughter process

extender — a substance added to semen to increase the volume

external parasite — a parasite that lives in the hair or on the skin of an animal

facultative — microbes that can grow with or without free oxygen

fallopian tubes — the tubes leading from the ovaries to the uterus

families — smaller divisions within classes

farrow — to give birth to a litter of pigs

farrowing crate — a crate or cage in which a sow is placed at the time of farrowing to protect the newborn pigs

feed conversion ratio — the rate at which an animal converts feed to meat

feeder pig — a young pig weighing less than 120 pounds that is of sufficient quality for finishing as a market hog

feedlot — a pen in which cattle are placed for fattening prior to slaughter

feedlot operation — an agricultural enterprise where beef animals are placed in pens and fed grain to fatten them

feedstuff — a basic ingredient of a feed that would not ordinarily be fed as a feed by itself

felting — the property of wool fibers to interlock when rubbed together under conditions of heat, moisture, and pressure

fermentation — the processing of food by the use of yeasts, molds, or bacteria

fertile — capable of producing viable offspring

fertilization — the union of the sperm and egg

fertilization membrane — a membrane surrounding an egg that is formed after the egg is fertilized. This prevents another sperm from entering.

fertilized egg — an egg that has united with a sperm

fingerling — a small fish that is of sufficient size to use for stocking

finish — the amount of fat on an animal that is ready for slaughter

finishing — fattening of animals prior to slaughter

fluke — small, seed-shaped parasitic flatworm, the most damaging of which live in the host's liver

flushing — the process of removing embryos from the donor cow by injecting a fluid by means of a catheter passed through the cervix and into the uterine horn

follicle — a small blister-like structure that develops on the ovary that contains the developing ovum

follicle-stimulating hormone (FSH) — the naturally occurring hormone that stimulates the development of the follicle on the ovary

Food and Drug Administration (FDA) — a federal agency that regulates the production, manufacture, and distribution of food and drugs

forage — livestock feed that consists mainly of the leaves and stalks of plants

foundation comb — a sheet of honeycomb placed onto frames on which the bees complete the comb to fill with honey

frame size — a score that depicts the size and weight of an animal at maturity. The measure is taken at the shoulder or at the hip.

free choice — feeding an animal with an unlimited supply of feed. The animal is free to eat whenever it wants.

freeze branding — a method of marking cattle by using a super cold metal that kills the pigment-producing ability of the hair contacted

fructose — the sugar found in fruit

fry — small, newly hatched fish

fumigate — to kill pathogens, insects, etc. by the use of certain poisonous liquids or solids that form a vapor

fungi — the kingdom to which multi-celled organisms such as fungi belong

galactose — the sugar in milk

gamete — the sex cell, either an egg or sperm

gastrointestinal tract — the digestive system, made up of the stomach and intestines

gelding — a male horse that has been castrated

genes — units of inheritance, composed of DNA

genetic base — the breeding animals available for a producer to use

genetic defect — an impairment of an animal that was passed by the parents to the offspring

genetic engineering — the alteration of the genetic components of organisms by human intervention

genetics — the science that deals with the processes of inheritance in plants and animals; also, the genetic makeup of an organism

genetic variation — the difference between animals due to their genetic makeup

genotype — the genetic makeup of an organism

genus — a class or group marked by common characteristics and comprised of structurally related organisms; the first name in the binomial nomenclature identifying an organism

germinal disk — a spot in the yolk portion of the egg that contains the genetic material from the female

gestation — the length of time from conception to birth

gilt — a female pig that has not given birth

gland cistern — where the milk is stored in the cow

glucose — a common sugar that serves as the building blocks for many complex carbohydrates

grease wool — wool as it comes from the sheep

greenhouse effect — an effect supposedly caused by an increase of carbon dioxide and pollutants in the air. The effect is supposed to cause the climate of the earth to warm.

growing operation — in swine production, the phase between the time they are weaned and the time they are finished for market

growthability — the ability of an animal to make efficient rapid growth

grub — the larva stage of some insects, particularly beetles

hand breeding — a system of breeding horses where the mares and stallions are kept separate until they are bred

heifer — a female bovine that has not produced a calf

helix — strands consisting of molecules of DNA that are shaped like a corkscrew

herbivore — an animal that eats plants as the main part of its diet

heritability — the portion of the differences in animals that is transmitted from parent to offspring

heterosis — the amount of superiority in a crossbred animal compared with the average of their purebred parents; also called **hybrid vigor**

heterozygous — two parental genes calling for a specific characteristic (e.g., hair color) that are not identical (e.g., one calls for black hair, the other for white hair). The dominant gene will override the effect of the other gene.

hip height — a measurement taken on the highest point of the hip of cattle at a given age. This is an indication of the frame size and the weight of an animal at maturity.

hippotherapy — a type of physical therapy using animals with physically challenged humans to help improve mobility

hive — a structure used to house bees

homogenization — the process of forcing the large cream globules through a screen at high pressure, reducing them to the size of the milk globules

homogenized milk — milk that has been blended to dissolve the fat molecules so that the fat (cream) will not become separated from the rest of the milk

homozygous — two parental genes calling for a specific characteristic (e.g., hair color) that are identical (e.g., both call for black hair)

honeycomb — six-sided cells joined together, used to store nectar

hormones — chemical substances, secreted by various glands in an animal's body, that produce a certain effect

host — an animal on which another organism depends for its existence

hot branding — using a hot iron to burn a permanent identifying mark onto an animal

hutches — cubicles used to house rabbits

hybrid — an animal produced from the mating of parents of different breeds

hybrid vigor — See **heterosis**.

hyperplasia — an increase in the number of cells in the tissues of organisms

hypertrophy — growth due to an increase in the size of cells

hypothesis — a theory by a scientist as to the cause or effect of a phenomena. This is tested by experimentation or other types of research.

ileum — last division of the small intestine

immobilization — the process of rendering an animal oblivious to pain during the slaughter process.

immunity — resistance to catching a disease

imprinting — a kind of behavior common to some newly hatched birds or newly born animals that causes them to adopt the first person, animal, or object they see as their parent

incubation — the process of the development of a fertilized poultry egg into a newly hatched bird. The eggs must have the proper heat, humidity, and length of time.

index — a system of comparing animals within a group with the group average. A score of 100 is used for the average.

indiscriminate breeder — an animal that will breed with any animal of the same type and the opposite sex

infantile vulva — a condition in gilts in which the vulva is very small and underdeveloped. Gilts with this condition are generally infertile.

infectious disease — a disease that is contagious

infundibulum — the enlarged funnel-shaped structure on the end of the fallopian tube that functions in collecting the ova during ovulation

ingestive behavior — the mannerisms or habits that an animal uses during the intake of food

injection curing — pumping a curing solution into a meat product

inorganic — not containing carbon and usually derived from nonliving sources

insect — an animal of the class Insecta. They have three body parts and six legs.

instinct — the ability of an animal, based on its genetic makeup, to respond to an environmental stimulus

intelligence — the ability to learn

intermediate host — an animal, other than the primary host, that a parasite uses to support part of its life cycle

internal parasite — a parasite that lives inside the body of the host animal

inverted nipples — a condition in female pigs in which the opening of the nipples on the mammary system appears to be inverted or to have a crater in the center. These are usually nonfunctional.

irradiation — a food preservation process that uses low levels of radiation to kill pathogens in food products.

isthmus — the part of the fallopian tubes between the ampulla and the uterus

jejunum — part of the small intestine

killer bees — honeybees of African origin that are reputed to have a very aggressive nature

kingdoms — the five common divisions into which natural objects are classified

kosher — designates any food produced, killed, or prepared according to Jewish dietary laws

laboratory animal — an animal that is raised for the purpose of being used for laboratory experimentation

lactation — the process of an animal's giving milk

lactose — a sugar obtained from milk

lagoon — a body of water used for the decomposition of animal wastes

lamb — referring to meat, that which comes from a sheep that is less than one year old

lanolin — the fatty substance removed from grease wool when it is scoured and cleaned

lard — the processed fat from swine

larva — the immature stage of an insect from hatching to the pupal stage

layer — a chicken raised primarily for egg production

lean-to-fat ratio — the amount of lean meat in a carcass compared to the amount of fat

letdown process — relaxation process (initiated by the release of oxytocin) allowing milk to pass out of the cow through the teat

libido — the sexual drive of an animal

life cycle — the changes in the form of life an organism goes through in its lifetime

ligaments — the tough, dense fibrous bands of tissue that connect bones or support viscera

light horse — a horse that weighs between 900 and 1,400 pounds at maturity

linear evaluation — a method of evaluating the degree of a trait in an animal. Certain traits are given a score based on the ideal.

lipid — a fat or fatty tissue

lobule — cluster of alveoli

loin eye — a cross section of the Longissimus (the muscle running the length of the backbone) of an animal's carcass

lumen — hollow cavity in an organ (*pl* **lumens** or **lumina**)

luteinizing hormone — the hormone that stimulates ovulation

lymphocyte — a kind of white blood cell produced by the lymph glands and certain other tissues. It is associated with the production of antibodies.

macromineral — minerals that are required in relatively large amounts in an animal's diet

magnum — the part of the oviduct of a bird located between the infundibulum and the isthmus. This is where the albumin of the egg is produced.

maiden flight — the new queen bee's flight during which she mates with the drones

maintenance ration — the feed mixed in the proper proportions and amounts for an animal to maintain its weight and other bodily functions

manure — excrement from animals

marbling — the desired distribution of fat in the muscular tissue of meat that gives it a spotted appearance. Marbling is used in the quality grading of a carcass.

mare — a female horse that has produced a foal

mastication — the act of chewing food

mastitis — a disease involving the inflammation of the udder of milk-producing females

maturity — the point in an animal's life when it is old enough to reproduce, also refers to the age of an animal or carcass

meat animals — animals that are raised primarily for the meat in their carcass

medium wool type — a breed of sheep raised primarily for meat

meiosis — cell division that results in the production of eggs and sperm

mesoderm — the central layer of cells in a developing embryo, which gives rise to the circulatory system and certain other organs

mesophiles — microbes that grow at medium temperatures (20°–45°C)

metabolism — the chemical changes in cells, organs, and the entire body that provide energy for the animal

metamorphosis — the process by which organisms, especially insects, change in form and structure in their lives

micromineral — minerals that are required in relatively small amounts in an animal's diet

micronutrients — nutrients that are required in relatively small amounts in an animal's diet

microbe — minute plant or animal life. Some cause disease; others are beneficial.

micromanipulator — a very small instrument that is used to dissect cells and embryos in the cloning process

milking parlor — milking area

mitosis — cell division involving the formation of chromosomes

mohair — the long lustrous hair from the Angora goat

molting — the process of poultry casting off old feathers before a new growth occurs

monera — the kingdom to which singular celled organisms such as bacteria belong

monogastric — refers to an animal that has only one stomach compartment, such as swine

monosaccharides — the simplest sugars, e.g., glucose, fructose, and galactose

morphogenesis — process of cell development into different tissues and organs

morula — a spherical mass of cells that develops into an embryo

most probable producing ability — an estimate of a cow's future productivity for a trait

mother breeds — those breeds of animals that make the best mothers, such as the Yorkshire and Landrace breeds of swine

motile — able to move about

mucin — substance (secreted by cells in the magnum) that develops into the white or albumen of the egg

mucous membrane — a form of tissue in the body openings and digestive tract that secrete a viscous, watery substance called mucus

mule — a cross between a horse and a donkey. The mother is a mare and the father is a jack.

muscling — the degree and thickness of muscle on an animal's body

mutation — an accident of heredity in which an offspring has different characteristics than the genetic code intended

mutton — the flesh of a sheep older than one year of age

mutualism — a symbiotic relationship that is beneficial to both species

naturally acquired active immunity — immunity to a disease that is acquired by the animal's having had a disease

natural selection — the natural process that results in the survival of those individuals or groups best adjusted to the conditions under which they live; commonly called *survival of the fittest*

nonessential amino acid — amino acid that can be synthesized by the animal's body

noninfectious disease — a disease that cannot be transmitted from one animal to another

notochord — in animals of the phylum Chordata a stringy rodlike structure that is made of tough elastic tissue which is present in the embryo

nucleotide — a basic structural component of DNA and RNA

nucleus — the central portion of the cell that contains the genetic material

nukes — small hives in which queen bees are commercially produced

nursery — a facility for caring for pigs after they are weaned

nutrients — substances that aid in the support of life

nutritional disease — a disease that is caused by not enough or too much of a certain nutrient in an animal's diet

nymph — a stage in the development of some insects that immediately precedes the adult stage

offspring — the young produced by animals

omasum — the third compartment of the ruminant stomach, where a lot of the grinding of the feed occurs

omnivorous — describing an animal that eats both plants and other animals

oogenesis — the process of egg production in the female

orders — smaller divisions within classes

organic — containing carbon or being of living origin

organism — any living being—plant or animal

ossification — the process of forming bone

ovary — the female organ that produces the egg and certain hormones

ovulation — the process of releasing eggs from the ovarian follicles

ovum — an egg

oxidation — any chemical change that involves the addition of oxygen

oxytocin — the hormone that stimulates constriction. It activates the egg-laying process in hens. It also causes the alveoli to release milk in cows.

palatability — the degree to which a feed or food is liked or accepted by an animal or human

papillae — any small nipplelike projections

paraffin — a waxy substance

parasitism — a symbiotic relationship in which one organism lives on or in another organism at that organism's expense

passive immunity — immunity that is temporary

pasterns — the part of an animal's leg that connects the cannon with the foot or hoof

pasteurization — the process of heat treating milk to kill microbes

pasture breeding — a system of breeding horses in which the stallion runs free in the pasture with the mares

pecking order — the order in which some poultry in a flock may peck others without being pecked in return

pedigree record — the record of an animal's ancestry

pelt — the natural whole skin covering including the hair, wool, or fur of an animal

pelvic capacity — the dimensions of a female's pelvic area that is an indication of its ability to give birth easily

penis — the male organ of copulation

pepsin — a digestive enzyme secreted by the stomach

per capita consumption — the amount of a product that is consumed by a person over the period of a year

performance data — the record of an individual animal for reproduction, growth, and production

perissodactyl — an animal with only one toe on its foot such as the horse, donkey, zebra, etc.

pH — a measure of the acidity or alkalinity of a substance

phagocyte — an animal cell capable of ingesting microorganisms or other foreign bodies

pharmaceuticals — medicines or drugs used in human or animal health care

phenotype — the observed characteristic of an animal without regard to its genetic makeup

photosynthesis — the process by which green plants, using chlorophyll and the energy of sunlight, produce carbohydrates from water and carbon dioxide

phulon — the greek word for race or kind

phyla — the primary divisions of the kingdom Animalia

physiological age — the age of an animal as determined by an examination of the carcass

pigeon-toed — condition in which the front feet are turned in

pigmentation — the naturally occurring color in the hair and skin of an animal

pin nipples — small, underdeveloped nipples on the teats of a pig. They are usually nonfunctional.

pituitary gland — a small gland at the base of the brain that secretes hormones that stimulate growth and other functions

placenta — the membranous tissue that envelops a fetus in the uterus

plantae — the highest level of scientific classification (kingdom) to which all plants belong

polar bodies — produced during oogenesis as the result of the cytoplasm going to the cell that becomes the egg. Polar bodies function to provide sustenance for the egg until conception.

polled — an animal that is naturally hornless

pony — a horse that weighs 500–900 pounds at maturity

Porcine Stress Syndrome — a condition in swine characterized by extreme muscling, nervousness, tail twitching, skin blotching, and sudden death

post-legged — a condition in animals in which the rear legs are too straight

postmortem — after death

postnatal — after birth

poultry — any domesticated fowl, such as chickens, ducks, geese or turkeys, that are raised for their meat, eggs, or feathers

predators — animals that kill and eat other animals

prenatal — before birth

prepuce — *See* **sheath**.

primal cuts — the most valuable cuts on a carcass, usually the leg (or hindquarter), loin, and rib

progeny — the offspring of animals

progeny testing — determining the breeding value of animals by testing their offspring

progesterone — a hormone produced by the ovaries that functions in preparing the uterus for pregnancy and maintaining it if it occurs

prolactin — a hormone that stimulates the production of milk

propolis — a glue or resin collected from trees and plants by bees. It is used to close holes in the hive.

prostaglandin — a group of fatty acids that perform various physiological effects in an animal's body. Artificial prostaglandin is used in heat synchronization of cattle.

prostate gland — the male reproductive gland that ejects the semen from the male reproductive tract

protectant — a substance added to semen to protect it during freezing and storage

protista — the kingdom to which singular celled organisms such as protozoa belong

protoplasm — the material of plant and animal tissues in which all life activities occur

protozoa — single-celled organisms that are often parasitic

PSE pork — pale, soft, and exudative pork; the meat is a very light pink in color and soft and dry in texture when cooked

psychrophiles — microbes that grow well in cooler temperatures (0°–20°C)

public lands — lands that are owned by the government

pullet — a young hen

pupa — the stage in an insect's life between the larva stage and the adult stage

purebred — an animal that belongs to one of the recognized breeds and has only that breed in its ancestry

purebred operation — a cattle operation that raises purebred animals to be used in breeding programs

quality grade — the grade given to a beef carcass that indicates the eating quality of the meat

quarantine — the isolation of an animal to prevent the spread of an infectious disease

queen cells — special large cells in the hive in which new queen bees are developed

queen excluder — a device placed in a beehive to prevent the queen from leaving the brood chamber

ram — a male sheep that has not been castrated

rancid — the putrefied state of foods

ration — the feed allowed for an animal in a twenty-four hour period

recessive gene — a gene that is masked by another gene that is dominant

recipient cow — a genetically inferior cow in which an embryo from a genetically superior cow is placed

refrigerated trucks — trucks that contain their own refrigeration unit used for transporting meat or other perishable products

rendering process — process during which condemned carcasses are placed under heat severe enough to kill any organisms that could cause problems

rennet (rennin) — an enzyme extracted from the stomach of cattle used in the cheese-making process

reproductive efficiency — the capability of producing offspring in a timely and efficient manner

retail cuts — cuts of meat that are ready for purchase and use by the consumer

reticulum — the second compartment of a ruminant's stomach

rib eye — the exposed muscle surface that results when a side of beef is cut between the 12th and 13th rib

ribonucleic acid (RNA) — a nucleic acid associated with the control of cellular chemical activities

rigor mortis — a physiological process following the death of an animal in which the muscles stiffen and lock into place

roughage — a feed low in carbohydrates and high in fiber content

royal jelly — secreted from bees, this food causes larvae to develop into queen bees

RST — See **BST**.

rumen — the largest compartment of the stomach system of a ruminant. This is where a large amount of bacterial fermentation of feed occurs.

ruminant — any of a class of animals having multicompartmented stomachs that are capable of digesting large amounts of roughages

salmonella — a large group of bacteria, some of which cause food poisoning

scientific selection — the selection of breeding or market animals based on the results of scientific research

scoured wool — wool after the fibers have been cleaned in the scouring process

scouring — cleaning of grease wool by gently washing it in detergent

scout bees — bees that locate nectar sources and report to the colony

scrotal circumference — a measurement taken around the scrotum of a bull. It is an indication of the fertility of the bull.

scrotum — the pouch that contains the testicles

seed stock cattle — the cattle to be used as the dams and sires of calves that will be grown for market

seines — large nets used to harvest fish from ponds

selective breeding — choosing the best and desired animals and using those animals for breeding purposes

semen — a fluid substance produced by the male reproductive system that contains the sperm and secretions of the accessory glands

seminal vesicles — a gland attached to the urethra that produces fluids to carry and nourish the sperm

semipermeable membrane — a membrane that permits the diffusion of some components and not others. Usually water is allowed to pass but solids are not.

serum — the clear portion of any animal fluid

service animals — animals who aid humans with disabilities. They may serve as aids in hearing, seeing, sensing danger, or other functions.

sex character — the physical characteristics that distinguish males from females

sexual reproduction — reproduction that requires the uniting of an egg and a sperm

sheath — the covering of the male penis

shell gland — another name for the uterus of a hen

shroud — a cloth used to wrap a carcass during the aging process

siblings — brothers and sisters

sickle-hocked — a condition in animals in which the back legs have too much curve

silage — a crop, such as corn, that has been preserved in its succulent condition by partial fermentation

sire — the father of an animal

sire breeds — those breeds of agricultural animals that are used as sires in a cross breeding program

smoothness — lack of awkward bone structure and a smooth, even finish along the top and sides of an animal

social behavior — how animals act when they interact with each other

soundness — structural strength and stability

sow — a female pig that has had a litter of pigs

species — a category of individuals having common attributes as a logical division of a genus; the second name in the binomial nomenclature identifying an organism

specific gravity — the density of a substance compared to the density of water

sperm — the male reproductive cell that unites with an egg

spermatogenesis — the development of the sperm cell

spermatogomia — primitive male germ cells

spermatozoa — mature male gametes

sperm nests — pockets for storing sperm inside the oviducts

sphincter muscle — a ring-shaped muscle that closes an orifice

spirilla — spiral-shaped bacteria

splayfooted — condition in a animal when its front feet are turned out

stallion — a mature male horse that has not been castrated

stanchion — a loose-fitting device that goes around a cow's neck and limits the animal's mobility in the stall or milking area

starter culture — culture that starts the process of fermentation for making cheese

steer — a male bovine that has been castrated before sexual maturity

sterile — not capable of producing offspring

stimuli — any agent that causes a response

stocker — a calf that has been weaned and is being conditioned prior to entering the feed lot

stocker operation — an operation that conditions calves after weaning and before they enter the feed lot

stomach worm — parasite roundworm that burrows into the stomach lining and sucks the host animal's blood

straw — a tube in which semen is frozen and stored

strongyle — parasitic roundworm that lives in the intestine of the host animal

style — the way an animal carries itself

sucrose — common table sugar (disaccharide composed of fructose and glucose)

suint — solid deposits from sheep perspiration found in wool

super — a boxlike structure that makes up part of a beehive. It is removed during honey harvest.

superovulation — the stimulation of more than the usual number of ovulations during a single estrus cycle due to the injection of certain hormones

swarm — a group of bees that have left the hive because of overcrowding

symbiosis — the close association of two dissimilar organisms

synapsis — the process by which chromotids come together and are matched up in pairs

synthetic lines — boars or sows used for breeding that resulted from the blending of several different breeds

systemic pesticides — pesticides that are taken into the body of an animal and are part of the animal's system

tankage — dried animal residues usually freed from fats and gelatin

teat — the portion of the mammary gland that expels milk

tertiary ducts — part of the duct system that carries milk from the alveoli to the gland cistern where the milk is stored

testicles — the male organs that produce sperm and certain hormones

testosterone — the male hormone responsible for the male sex drive and the development of the male sex characteristics

thermophiles — microbes that grow at higher temperatures (45°–65°C)

thurl — the thigh of an animal

toxic — poisonous

trace mineral — a mineral that is needed in relatively minute amounts in an animal's diet

treatment group — in a scientific experiment, the group of animals and plants that receives the treatment that is being researched

type — the total of all the characteristics of an animal that make it and others like it unique

underline — the belly of an animal. In swine it refers to the mammary system of the female.

urethra — the tube that carries urine from the bladder and serves as a duct for the passage of the male's semen

USDA — United States Department of Agriculture

uterus — the female reproductive organ in which the fetus develops before birth

vaccinating — the process of injecting an animal with certain microorganisms in an effort to make the animal immune to specific diseases

vaccine — a substance that contains live, modified, or dead organisms that is injected into an animal to make it immune to a specific disease

vacuum packaging — a means of packaging meat by wrapping it in plastic wrap and drawing the air from it

vagina — the canal in the female reaching from the uterus to the vulva

vas deferens — the tube connecting the epididymis of the testicles to the urethra

veal — the meat from calves slaughtered before they are three months of age

vertebra — the backbone of an animal

vertebrate — an animal having a backbone

viable — capable of living

villi — microscopic, fingerlike projections of the inner lining of the digestive tract

virus — disease-causing agent

vulva — the external reproductive organ of the female

warm-blooded — *See* endothermic

warm-water fish — a fish that does not thrive in water colder than 60°F

weaned — a young animal no longer dependent on its mother's milk

weaning weight — the weight of a calf at weaning, usually considered to be about 500 pounds

wether — a male sheep that has been castrated

whey — watery part of milk that is separated from the curd in the cheese-making process

wholesale cuts — the major parts of a carcass that are boxed and sold to wholesale distributors

withdrawal period — the length of time that must transpire between the time an animal is given a certain drug and the time the animal's milk can be used or the animal is slaughtered

wool type — a breed of sheep raised primarily for wool

work animal — an animal that is raised primarily for the work it can perform

yearling — an animal that is a year old

yearling weight — the weight of a beef animal at 365 days of age

yield grade — a grade in meat animals that refers to the amount of lean meat produced in a carcass

yogurt — a semisolid, fermented milk product

yolk — the yellowish part of a fowl's egg that contains the germinal disk; also the substances such as wool grease in a fleece

zoonoses — diseases and infections that can be transmitted from animal to humans

zygote — a fertilized egg

References

Bastien, R.W. (1991) *Ostrich Management. Guide.* Athens, GA: Cooperative Extension Service.

Gillespie, J.R. (1997) *Modem livestock and poultry production.* (5th ed.). Albany: Delmar Publishers.

Herren, R.V. and Donahue, R.L. (1991) *The agriculture dictionary.* Albany, New York: Delmar Publishers.

Landau, M. (1992) *Introduction to aquaculture.* New York: John Wiley and Sons, Inc.

Legates, J.E. & Warwich, E.J. (1990) *Breeding and improvement of farm animals.* (8th ed.). New York: McGraw Hill Publishing Company.

National FFA Foundation. (1995) *Animal welfare instructional materials.* Alexandria, VA: author

National FFA Foundation. (1996) *Equine science.* Alexandria, VA: author

Oklahoma State University. (1992) *Meat and poultry processing.* Department of Vocational and Technical Education Curriculum and Instructional Materials Center. Stillwater, OK: author

Parker, R. (1995) *Aquaculture science.* Albany, NY: Delmar Publishers.

Parker, R. (1998) *Equine Science.* Albany NY: Delmar Publishers

Taylor, R.E. & Field, T.G. (1998) *Scientific farm animal production.* (6th ed.). Upper Saddle River, New Jersey: Prentice Hall.

Warren, D.M. (1995) *Small animal care and management.* Boston: Delmar Publishers.

Index

Note: Page numbers in **bold type** reference non-text material.

L

M

N